Urban Mining and Sustainable Waste Management

Sadhan Kumar Ghosh
Editor

Urban Mining and Sustainable Waste Management

Springer

Editor
Sadhan Kumar Ghosh
Department of Mechanical Engineering
Faculty of Engineering and Technology
Jadavpur University
Kolkata, West Bengal, India

ISBN 978-981-15-0531-7 ISBN 978-981-15-0532-4 (eBook)
https://doi.org/10.1007/978-981-15-0532-4

This Springer imprint is published by the registered company Springer Nature Singapore Pte Ltd.
The registered company address is: 152 Beach Road, #21-01/04 Gateway East, Singapore 189721, Singapore

Preface

The reality of increasing resource constraints is pushing industries to modify the business patterns and the researchers to think of alternatives and recycling. Concerns have been raised for the future supply of critical metals, from rare earth elements to indium, cobalt, gold, silver, rhodium and zinc, some of which have been flagged, at current production rates, as becoming unavailable to mine in the ensuing decades, unless major new discoveries emerge. One of the key resource issues impacting the electronics industry is how to reduce its reliance on earth metals such as yttrium and scandium, which are minerals key to the production of computers and mobile phones, but difficult to extract and hard to recycle. The potential recovery of rare earths from scrap electronics could become a substantial source of supply, with real business opportunities for companies that look to reuse these metals, rather than pay the high prices demanded for virgin materials.

In the period from 1970 to 2017, the global extraction of biotic and abiotic materials increased more than 240%, by 65 billion tons since 1970 reaching an amount of 92 billion tons in 2017 as per the IPCC data. The wide range of materials consumed can be aggregated into the four main categories, namely biomass, minerals, fossil fuels and metals. Nearly, one-third of the global extraction of raw materials takes place in China and Asia as a whole is responsible for even two thirds. Domestic material consumption (DMC) is currently the most widely used and accepted consumption indicator. In 2015, around 84.4 Gt of raw materials such as minerals, biomass or fossil fuels were extracted worldwide. In comparison, only 8.4 Gt (about 10%) of recycled materials reentered the economic system. This relationship underlines the considerable potential of circularity including urban mining to shift from extraction to reuse and recycling of raw materials in all the spheres of materials used. Material flows and resource productivity indicators are among the most important indicators available for monitoring changes in the patterns and rates of resource use as economies grow. They are used for reporting progress toward SDG targets 8.4 "Resource Productivity" and 12.2 "Sustainable Use of Natural Resources."

Given the high volatility of resource prices, heavy pollution of primary production and above all the resource productivity and sustainable use of natural resources, recycling becomes mandatory. A new approach toward recycling is "urban mining," extends landfill mining to the process of reclaiming compounds and elements from any kind of anthropogenic stocks, including buildings, infrastructure, industries, products (in and out of use), environmental media receiving anthropogenic emissions. The stocked materials may represent a significant source of resources, with concentrations of elements often comparable to or exceeding natural stocks. As for natural ores, extraction and processing of anthropogenic stocks is necessary and the generation of an economic benefit is essential. For these reasons, urban mining originally focused on electrical and electronic wastes (WEEE) which contain relatively high concentrations of expensive metals and rare earth elements. Urban mining largely considers technospheric material stocks as potential assets; landfill mining originates from the waste sector in which such accumulations often have been considered a problem for various reasons. The fact that such technospheric stocks of elements—copper, for example, are now globally comparable in size to the amount remaining in known geological ores illustrates the magnitude of this ongoing relocation process of natural capital. Perhaps even more notable is that when it comes to base metals such as iron and copper, approximately half of the amounts extracted to date are no longer in use.

The overall goals of urban mining are resource conservation and environmental protection, as well as generating economic benefits. Cities hold large stocks of materials, contained in buildings, infrastructure, landfills and also in each household. Material resources remain a major pillar in economic activities. Raw materials are used in building and processing in the factories, transforming them into new products and consumer goods. The International Solid Waste Association (ISWA) and the International Society of Waste Management, Air and Water (ISWMAW) recognize that informal and microenterprise recycling, reuse and repair systems achieve significant recycling rates, with 20–30% in low-income countries. This activity saves local authorities around 20% or more of what they would otherwise spend additionally on waste management, representing many millions of dollars every year in large cities. All the types of waste are being considered for recycling to compensate the loss in resource reserves. In Santa Cruz, Bolivia, the informal waste collectors service 37% of the population. In Delhi, India, only 34% of the city's refuse is recycled, 27% of which is performed by informal collection services. Urban mining not only attends to materials but represents a more comprehensive approach that includes energy. UN's Global E-waste Monitor 2017, published in December, the 44.7 million tons of e-waste (electronic waste) produced worldwide in 2016 contained metals and other materials worth €55 bn. To facilitate effective urban mining, we require comprehensive information about materials and substances. For the exploitation of primary ores, intrinsic properties, such as element concentration, abundance, availability, speciation and partner minerals, determine whether a particular substance can be economically mined. The information necessary for decisions about urban mining comprises all relevant flows and stocks of a particular substance, from production to utilization and disposal at the end of the lifetime.

The 8th IconSWM 2018 has received 380 abstracts and 320 full papers from 30 countries. Three hundred accepted full papers have been presented oral and poster presentations in November at ANU, Guntur, AP, India. The chapters finally selected by the board have been thorough reviewed by experts for the book, "Urban Mining and Sustainable Waste Management" dealing with the biomedical waste management, recycling of construction and demolition waste and industrial inorganic wastes, co-processing, hazardous and other wastes as alternate fuel and raw materials, extraction and recycling of metals from waste PCB and WEEE, slate mine waste water, waste coal utilization, agro-industrial waste, paper mill lime sludge valorisation, use of granite industry's waste, fine fraction recovery in landfill mining, remediation, utilization of plastic waste, pyrolysis, ELV recycling, hydrothermal carbonization, etc.

The IconSWM movement was initiated for better waste management, resource circulation and environmental protection since the year 2009 through generating awareness and bringing all the stakeholders together from all over the world under the aegis of the International Society of Waste Management, Air and Water (ISWMAW). It established a few research projects across the country; those include the CST at Indian Institute of Science, Jadavpur University, KIIT, Calcutta University, etc. Consortium of researchers in international collaboration (CRIC) and many other organizations across the world help in the IconSWM-CE movement. IconSWM has become significantly one of the biggest platforms in India for knowledge sharing, awareness generation and encouraging the urban local bodies, government departments, researchers, industries, NGOs, communities and other stakeholders in the area of waste management. The primary agenda of this conference is to reduce the waste generation encouraging the implementation of Reduce, Reuse and Recycle (3R) and circular economy concept and management of the generated waste ensuring resource circulation. The conference will show a paradigm and provide holistic pathways to waste management and resource circulation conforming to urban mining and circular economy.

The success of the 8th IconSWM is the result of significant contribution of many organizations and individuals, especially the government of Andhra Pradesh, several industry associations and chamber of commerce and industries and the AP Higher Education Council, Swachh Andhra Mission and various organization in India and in different countries as our partners including the UNEP, UNIDO and UNCRD. The 8th IconSWM 2018 was attended by nearly 823 delegates from 22 countries. The 9th IconSWM-CE 2019 was held at KIIT, Bhubaneswar, Odisha, during November 27–30, 2019, participated by 21 countries. The Shri Venkateswara University (SVU) has expressed their willingness to organize the 10th IconSWM-CE at SVU in the temple city Tirupati, Andhra Pradesh, tentatively during December 02–05, 2020. This book will be helpful for the researchers, educational and research institutes, policymakers, government, implementers, ULBs and NGOs. Hope to see you all in 10th IconSWM-CE 2020 at Tirupati.

Kolkata, India
February 2020

Sadhan Kumar Ghosh

Acknowledgements

Shri. Chandra Babu Naidu, Hon'ble Chief Minister and Dr. P. Narayana hon'ble Minister of MA&UD for taking personal interest in this conference.

I am indebted to Shri. R. Valavan Karikal, IAS; Dr. C. L. Venkata Rao; Shri. B. S. S. Prasad, IFS (Retd.); Prof. S. Vijaya Raju; and Prof. A. Rajendra Prasad, VC, ANU for their unconditional support and guidance for preparing the platform for successful 8th IconSWM at Guntur, Vijayawada, AP.

I must express my gratitude to Mr. Vinod Kumar Jindal, ICoAS; Shri. D. Muralidhar Reddy, IAS; Shri. K. Kanna Babu, IAS; Mr. Vivek Jadav, IAS; Mr. Anjum Parwez, IAS; Prof. S. Varadarajan; Mr. Bala Kishore and Mr. K. Vinayakam; Prof. Shinichi Sakai, Kyoto University, JSMCWM; Prof. Y. C. Seo and Prof. S. W. Rhee of KSWM; Shri. CRC Mohanty of UNCRD, members of Industry Associations in Andhra Pradesh; Prof. P. Agamuthu, WM&R; Prof. M. Nelles, Rostock University; Dr. Rene Van Berkel of UNIDO; Ms. Kakuko Nagatani-Yoshida; and Mr. Atul Bagai of UNEP and UN Delegation to India for their active support.

IconSWM-ISWMAW Committee acknowledges the contribution and interest of all the sponsors, industry partners, industries, co-organizers, organizing partners around the world, the Government of Andhra Pradesh, Swachh Andhra Corporation as the Principal Collaborator, the Vice Chancellor and all the professors and academic community at Acharya Nagarjuna University (ANU), the Chairman, Vice Chairman, Secretary and other officers of AP State Council of Higher Education for involving all the universities in the state, the Chairman, Member Secretary and the officers of the AP Pollution Control Board, the Director of Factories, the Director of Boilers, Director of Mines and officers of different ports in Andhra Pradesh and the delegates and service providers for making a successful 8th IconSWM.

I must specially mention the support and guidance by each of the members of the international scientific committee, CRIC members, the core group members and the local organizing committee members of 8th IconSWM who are the pillars for the success of the program. The editorial board members including the reviewers, authors and speakers and Mr. Aninda Bose and the team of M/s. Springer India Pvt. Ltd. deserve thanks who were very enthusiastic in giving me inputs to bring this book.

I must mention the active participation of all the team members in IconSWM Secretariat across the country with special mention of Prof. H. N. Chanakya and his team in IISc Bangalore, Ms. Sheetal Singh and Dr. Sandhya Jaykumar and their her team in CMAK and BBMP, Mr. Saikesh Paruchuri, Mr. Anjaneyulu, Ms. Senophiah Mary, Mr. Rahul Baidya, Dr. Asit Aich, Ms. Ipsita Saha, Mr. Suresh Mondal, Mr. Bisweswar Ghosh, Mr. Gobinda Debnath and the research team members in mechanical engineering department and ISWMAW, Kolkata HQ for various activities for the success of the 8th IconSWM 2018.

I express my special thanks to Sannidhya Kumar Ghosh, being the governing body member of ISWMAW supported the activities from the USA. I am indebted to Mrs. Pranati Ghosh who gave me guidance and moral support in achieving the success of the event. Once again the IconSWM and ISWMAW express gratitude to all the stakeholders and delegates, speakers who are the part of the success of 8th IconSWM 2018.

Contents

About the Editor

Dr. Sadhan Kumar Ghosh is a Professor & Former Head of the Mechanical Engineering Department, as well as the Founder Coordinator of the Centre for QMS at Jadavpur University, India. A prominent figure in the fields of Waste Management, Circular Economy, SME Sustainability, Green Manufacturing, Green Factories and TQM, he has served as the Director, CBWE, Ministry of Labour and Employment, Government of India and L&T Ltd. Prof Ghosh is the Founder and Chairman of the IconSWM and the President of the International Society of Waste Management, Air and Water, as well as the Chairman of the Indian Congress on Quality, Environment, Energy and Safety Management Systems (ICQESMS). He was awarded a Distinguished Visiting Fellowship by the Royal Academy of Engineering, UK, to work on "Energy Recovery from Municipal Solid Waste" in 2012. He received the Boston Pledge and NABC 2006 Award for the most eco-friendly innovation "Conversion of plastics & jute waste to wealth" in the ESP/50K Business Plan Competition in Houston, Texas, USA. In addition, he holds patents on waste plastic processing technology and high-speed jute ribboning technology for preventing water wastage and occupational hazards.

Contribution of Optimal Equipment for ELV Recycling to the Sustainable Development of Serbia and the Region

M. Pavlović, D. Tadić, A. Aleksić, N. Komatina and B. H. Hosam

Abstract The "3R" (reduce, reuse and recycle) principle has been significantly employed in processing ELVs (end-of-life vehicles), particularly ELV parts and methods to promote sustainable development. Motivated by legislation such as Directive (2000/53/EC), recently, ELV problems have seriously been treated in developing countries as well. This paper analyses the impact of newly developed equipment for ELV recycling on the sustainable development of the Republic of Serbia through environmental protection, resource exploitation and socio-economic factors. Research is carried out on the national and regional level. The proposed research and analysis show the significant contribution of newly developed equipment on the sustainable development of the Republic of Serbia and the region.

Keywords Sustainable development · ELV recycling

M. Pavlović (✉)
Faculty of Economics and Engineering Management, University Business Academy, Novi Sad, Serbia
e-mail: milanpavlovic50@gmail.com

D. Tadić · A. Aleksić · N. Komatina
Faculty of Engineering, University of Kragujevac, Kragujevac, Serbia
e-mail: galovic@kg.ac.rs

A. Aleksić
e-mail: aalekisic@kg.ac.rs

N. Komatina
e-mail: nkomatina@kg.ac.rs

B. H. Hosam
Rejto Sandor Faculty of Light Industry and Environmental Protection Engineering, Obuda University, Budapest, Hungary
e-mail: bayoumi.hosam@rkk.uni-obuda.hu

S. K. Ghosh (ed.), *Urban Mining and Sustainable Waste Management*,
https://doi.org/10.1007/978-981-15-0532-4_1

1 Introduction

The term sustainable development has many different meanings and therefore provokes many different responses. In broad terms, the concept of sustainable development is an attempt to combine growing concerns about a range of environmental issues with socio-economic issues (Hopwood et al. 2005). In compliance with that is proposed the classification of different trends on sustainable development, their political and policy frameworks and their attitudes towards change and means of change. The concept of sustainable development should present a socio-economic environmental system as a moving target, which is continuously evolving. Nonaka and Toyama (2005) suggest that sustainability has no clear end to it, and it requires relentless efforts to achieve it. It keeps driving the organization towards unattainable perfection. The sustainability development can be defined as a regulative function of the organization by preventing it from contending the imperfect realities. The sustainability concept includes an environmental concern, but also incorporates economic (Bagheri and Hjorth 2007) and social dimensions (Dempsey et al. 2011). It can be said that this concept strongly links environmental and socio-economic issues making an impact on the quality of life (Milivojević et al. 2011).

In 2005, the European Union adopted the Draft Declaration on Guiding Principles, which emphasizes that a sustainable development should be a key policy goal that can be achieved through the fulfilment of the following objectives: (1) environmental protection (prevention and reduction of environmental pollution and promotion of sustainable models of production and consumption in order to break the link between economic growth and environmental degradation), (2) social equity and cohesion, (3) economic prosperity (promotion of prosperities, innovative, knowledge-rich, competitive and eco-efficient economy that provides a high standard of living and total employment in the European Union) and (4) fulfilling international liabilities.

Respecting all mentioned above, it can be said that the goals of the sustainable development strategy are almost fully achieved by realizing the explained objectives. For designers of recycling equipment, very important information in designing is to what extent this equipment will contribute to sustainable development. The answer to this question is very difficult to be determined precisely. In the literature, there are no papers which treat this problem. In this paper, it is assumed that by using the regression analysis and evidence data, the amount of ELV in subsequent time periods can be predicted.

It may be considered that one of the very important tasks for managers of reverse supply chains is to select recycling equipment. Nikolić et al. (2017) suggest benchmarking analysis of the specialized pieces of equipment of the internationally known manufacturers as well as those on the national level. Evaluation criteria are current market situation and needs, as well as the unit prices. It can be said that choosing of the recycling equipment can be stated as multi-criteria optimization task (by analogy Arsovski et al. 2015; Pavlović et al. 2016).

The paper is organized in the following way: the problem statement is presented in Sect. 2, Sect. 3 presents the results of case study, and conclusion is set in Sect. 4.

2 The Proposed Approach

This research deals with the problem of determining the influence of recycling press and detoxification devices developed in the scope of activities of technological development project TR 35033 for sustainable development of the Republic of Serbia.

Using the data obtained from the Republic Statistical Bureau (http://www.stat.gov.rs/) can present the number of registered motor vehicles in the previous period. By using the regression analysis, the number of registered vehicles can be predicted in the future years. Respecting the results of the best practice from recycling domain of developed countries, it is known that minimum, maximum and average number of ELVs present 4%, 6.7% or 5.35% of total number of registered motor vehicles. With respect to all the mentioned facts, the number of ELV in considered time period is determined. For each recycling level α, $\alpha \in [0 - 1]$, the number of recycled ELV can be determined. The impact of the recycled material on sustainable development is assessed with respect to three aspects: economic, environmental protection and social that are considered from the perspective of the number of employees. The impact values are determined according to the assessment of decision-makers. In this case, the decision-makers are representatives of Ministry of Environmental Protection.

It can be assumed that one ELV, in the process of recycling, may provide 700 kg of steel. This kind of recyclate may be used as a raw material in many production companies of the metalworking sector. While selling this kind of recyclate, recycling centres generate profit which may be calculated as multiplication of obtained recyclate and unit selling prices. In this paper, the impact of treated recycling equipment on the economic aspect of sustainable development is determined as the total profit achieved through the sale of treated recyclate. In practice, the economic aspect of sustainable development should be determined by respecting many factors. The considered factor is just one of the factors that can be considered.

It is assumed that the impact of proposed recycling equipment, with respect to environmental protection, can be determined according to the estimates of decision-makers. They form their assessments on their knowledge and the results of developed countries. In this paper, evaluations of decision-makers are based on common scale measurement (Saaty 1990). Value 1 indicates that environmental protection is very low, and value 9 indicates that environmental protection has been totally achieved, respectively.

The social aspect of sustainable development can be determined by respecting many of the factors defined in the National Strategy for Sustainable Development. One of the factors is the number of employees that can be considered to be the most important factor in developing countries. In this paper, only the number of workers that should be employed in the recycling centres, which is determined in compliance with the available capacity of the considered recycling equipment, is considered.

On the basis of the obtained results, it is possible to conclude that there is a necessity of increasing the level of recycling in the Republic of Serbia. The management of each recycling centre should select recycling equipment that will be used for achieving the set objectives. The obtained results may be very useful, both for recycling centres management and decision-makers who bring and implement a national sustainable development strategy. The proposed algorithm can be realized through the following steps:

Step 1. Collect data for number of registered vehicles in Serbia from Republic Statistical Bureau.
Step 2. Set the regression model. Determine the adequacy of the model.
Step 3. Determine the amount of ELV which may be recycled as well as the amount of motor oil for different levels of recycling.
Step 4. Assess the impact of the obtained results from Step 3 on economic, environmental and social aspects of the sustainable development if pressing machine and detoxification device constructed in the scope of technological development project TR 35033 are used.

3 Case Study in the Republic of Serbia

The proposed approach is tested on the data from the Republic of Serbia. The number of registered vehicles in the Republic of Serbia in the period from 2001 to 2014 is taken from official reports of Republic Statistical Bureau (http://www.stat.gov.rs/en-US/oblasti/saobracaj-i-telekomunikacije/registrovana-vozila) and presented in Table 1 (Step 1 of the proposed algorithm).

The change in the number of registered vehicles ($\hat{y}$) in dependency of time period (x_i) is given in compliance with the data from Table 1 (Step 2 of the proposed algorithm):

$$\hat{y}_i = 1.33804 + 0.02825 \cdot x_i$$

By applying the technique of variance analysis at the risk level of 5%, the existence of linear dependency is tested. The statistic of decision-making is

Table 1 Number of registered passenger cars in the Republic of Serbia in the period (2001–2014)

Year	Number of registered passenger cars		Number of registered passenger cars		Number of registered passenger cars
2001	1,382,396	2006	1,511,837	2011	1,677,510
2002	1,343,658	2007	1,476,642	2012	1,726,190
2003	1,388,109	2008	1,486,608	2013	1,770,162
2004	1,449,843	2009	1,637,002	2014	1,797,252
2005	1,481,498	2010	1,565,550		

$$F_0 = \frac{\sum_{i=1,\ldots,14} (\hat{y}_i - \overline{y})}{\sum_{i=1,\ldots,14} (\hat{y}_i - y_i)/n - 2} = \frac{0.1826}{0.0301/10} = 60.69.$$

The table value of Fisher's distribution is $F_{0.05,1,10} = 4.96$. As the value of statistical decision-making is greater than the value in table, it can be concluded that analysed hypothesis is true: the number of registered vehicles is linearly increased compared to the time period. The measure of scattering around the regression is determined with

$$R^2 = 1 - \frac{\sum_{i=1,\ldots,14}\left(\hat{y}_i - y_i\right)}{\sum_{i=1,\ldots,14}\left(\hat{y}_i - \overline{y}\right)} = \frac{0.0301}{0.1826} = 0.835.$$

This means that 83.5% of the points are grouped around the regression. Based on the obtained test results, it can be clearly concluded that it is quite adequate that the number of registered vehicles in the Republic of Serbia, depending on the period, is described as a linear function.

The issue of assessment of two equipment types for recycling has already been investigated in the literature (Nikolić et al. 2017). This research was applied to different equipment manufacturers with the following evaluation criteria: price, box material, box length, bale size, engine power, working pressure and weight. The weights vector is given, so that it is (0.25, 0.05, 0.125, 0.125, 0.2, 0.15, 0.1). Respecting all criteria and their weights, it can be assumed that for recycling centre in Serbia, it is optimal to purchase the equipment constructed in the scope of national technology development project TR 35033.

According to the procedure (Step 3 of the proposed algorithm), the amount of ELV, the amount of metal and the amount of motor oil, which may be obtained in the process of recycling in 2020, 2025 and 2030, are calculated, and it is presented in Figs. 1, 2 and 3. It is assumed that the level of recycling in 2020 is 15%, in 2025 it is 50%, and in 2030 it is 80%.

Fig. 1 Amount of ELV which can be recycled at the different levels of recycling

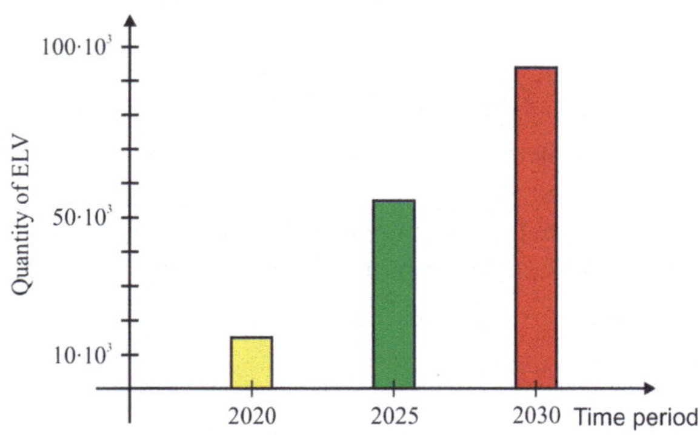

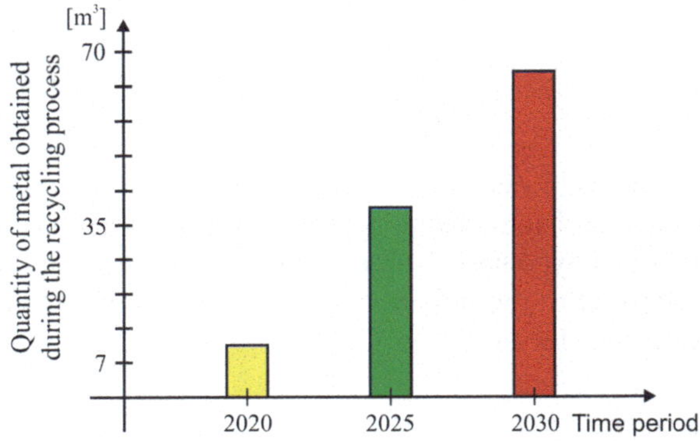

Fig. 2 Amount of metal materials that can be recycled at the different levels of recycling

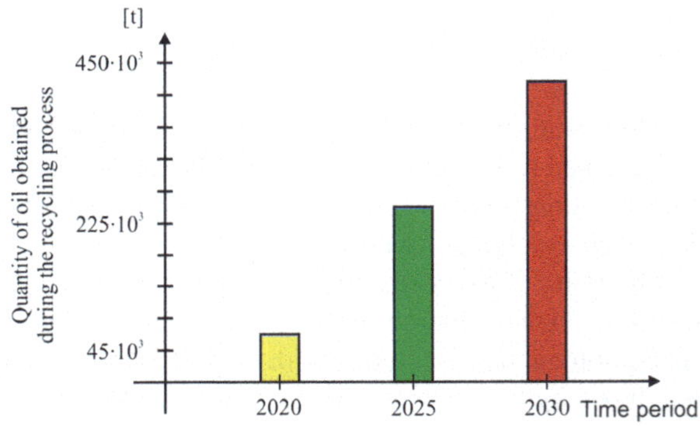

Fig. 3 Amount of motor oil that can be recycled at the different levels of recycling

In the Republic of Serbia, the unit selling price of metal materials is 0.14 €/kg, and for the motor oil it is 2 €/l. Motor oil which may be obtained in the recycling process is sold to companies that produce motor oils where certain chemicals and additives are mixed in. The operating capacity of one pressing machine is delivered in the range of 15–20 min. The operating capacity of device for detoxification is 25 min. By assuming that work is organized only in one shift, with the 242 of working days per year, the needed number of pressing machines may be calculated. By applying the calculation of proportion, the pressing machine may treat 6636.85 ELV/year. In the similar way may be calculated the volume of recycling related to detoxification device, as well as the number of workers that need to be employed for working on these devices. For different amount of ELV, the needed number of pressing machines is calculated. Taking into account that each detoxification device implies the need for engagement of two workers, the number of new employees may be calculated (Tables 2 and 3).

The impact of developed recycling equipment on various aspects of sustainable development, at different levels of recycling, is estimated and presented in Tables 2 and 3.

Table 2 Impact of ELV on sustainable development

The year	Economic aspect	Environmental protection aspect	Social aspect (the number of new employees in recycling centres)
2020	1.46 million euros	2	6
2030	5.36 million euros	5	18
2050	9.16 million euros	8	28

Table 3 Impact of motor oil as waste on sustainable development

The year	Economic aspect	Environmental protection aspect	Social aspect (the number of new employees in recycling centres)
2020	0.23 million euros	1	8
2030	0.49 million euros	4	24
2050	0.85 million euros	7	42

3.1 Discussion

If the recycling centres that exist in the Republic of Serbia use two recycling devices constructed in the scope of technological development project TR 35033 in 2020, they would achieve an income of 1.69 million euros. With increasing the level of recycling, which is one of the goals of the National Recycling Strategy, the revenue generated in recycling centres could be significantly increased. For example, in 2025, the revenue would be 5.85 million euros; in 2030, the revenue would be 10.01 million euros. It can be clearly concluded that the usage of the considered recycling equipment leads to a significant improvement in the economic aspect of sustainable development.

If the level of recycling keeps around 15%, as it is now in the Republic of Serbia, it can be concluded that almost no environmental protection has been achieved. Environmental protection is increasing with increasing the level of recycling. If the same level of recycling remains, the number of employees in recycling centres would increase for new 14 employees in 2020. If it is assumed that the required level of recycling is achieved, which is 80%, the number of employees would increase for new 70 employees.

4 Conclusion

This research deals and treats the impact of optimal equipment for ELV recycling to the sustainable development of Serbia and the region. In relation to that and taking into account the current trends in Serbia and the region, it is assumed that the current level of recycling should be enhanced. The research indicates that recycling centres should increase their attention when the process of equipment selection for recycling is carried on. The equipment selection is a very significant task since it directly corresponds to the achievement of strategy goals.

The obtained results from the first and the second step of the proposed approach may be very from the recycling centres as well as for the other stakeholders that are implementing national strategy related to sustainable development. The main contribution of the research clearly indicates the benefits of exploitation of equipment constructed in the scope of technological development project TR 35033. The new equipment has significant impact on the sustainable development and recycling in the Republic of Serbia.

The future research should examine the impact of non-economic factors to sustainable development of the region.

Acknowledgement Research presented in this paper was supported by Ministry of Science and Technological Development of the Republic of Serbia, Grant TR 35033.

References

Arsovski, S., Putnik, G., Arsovski, Z., Tadic, D., Aleksic, A., Djordjevic, A., et al. (2015). Modelling and enhancement of organizational resilience potential in process industry SMEs. *Sustainability, 7*(12), 16483–16497.

Bagheri, A., & Hjorth, P. (2007). Planning for sustainable development: a paradigm shift towards a process-based approach. *Sustanable Development, 15,* 83–96.

Dempsey, N., Bramley, G., Power, S., & Brown, C. (2011). The social dimension of sustainable development: defining urban social sustainability. *Sustainable Development, 19,* 289–300.

Hopwood, B., Mellor, M., & O'Brien, G. (2005). Sustainable development: mapping different approaches. *Sustainable Development, 13,* 38–52.

Milivojević, J., Grubor, S., Kokić-Arsić, A., Đokić, S., Tonić, N. (2011). Reciklaža motornih vozila na kraju životnog ciklusa u funkciji kvaliteta života, 38. Nacionalna konferencija o kvalitetu, Fakultet inženjerskih nauka, Univerzitet u Kragujevcu.

Nikolić, M., Vulić, M., Pavlović, M., Desnica, E. (2017). Benchmarking of technologies of baling presses machines in the elv recycling process. In: VII International Conference Industrial Engineering and Environmental Protection 2017 (IIZS 2017), 12–13 Oct 2017, Zrenjanin, Serbia

Nonaka, I., & Toyama, R. (2005). The theory of the knowledge-creating firm: subjectivity, objectivity and synthesis. *Industrial and Corporate Change, 14*(3), 419–436.

Pavlović, A., Tadić, D., Arsovski, S., Jevtić, D., & Pavlović, M. (2016). Evaluation and choosing of recycling technologies by using FAHP. *Acta Polytechnica Hungarica, 13*(7), 143–157.

Saaty, T. (1990). How to make a decision: the analytic hierarchy process. *European Journal Operational Research, 48*(1), 9–26.

A Study on Biomedical Waste Management in Chittoor District

N. Hanumaiah, P. Lakshmi Narayana Prasad and J. Karthikeyan

Abstract Chittoor District is famous for renowned temples like Sri Venkateswara Swamy temple at Tirumala, Tiruchnoor, Sriklalahasti and Kanipakam. In Chittoor district is having around 9800 beds serving the medical needs of the people. The biomedical waste is segregated and temporally stored in blue, yellow and red color-coded bags. The non-infectious waste such as glove papers, plastic papers, covers and suture covers are put in blue bags yellow bag contains contaminated cotton swabs, gloves, mops, gauze, specimens, patient's wastage, catheters, I.V. and blood set, suction tubes, urine bags and all draining tubes which are solid and liquid in nature. Human or animal tissues, organs or body parts that are highly infectious are collected and in red bags. The quantity of biomedical waste generated from the hospitals is about 1300 kg/day. Daily all these wastes are collected from all the hospitals and transported to the biomedical waste management plant which is located 25 km away from Tirupati. All these waste is incinerated in a dual-chamber incinerator of 100 kg/h capacity working for 8–10 h/day. The incinerator consists of primary chamber operated at temperature of 800–850 °C and secondary chamber operated at 1000–1100 °C. The incinerator is attached with attendant air pollution control facilities like a cyclone followed by a scrubber. The ash from the incinerator is buried in the landfill. The stack emissions and ash from the incinerator were analyzed for its leaching potential. The problem encountered during incineration, especially of waste components like placenta, etc., are also discussed in the paper.

Keywords Biomedical waste · Moisture content · Incinerator

1 Introduction

Chittoor District is famous for renowned temples like Sri Venkateswara Swamy temple at Tirumala, Tiruchnoor, Sriklalahasti and Kanipakam. Chittoor district has

N. Hanumaiah
AMD Consulting Ltd, Tirupati, India

P. Lakshmi Narayana Prasad (✉) · J. Karthikeyan
Department of Civil Engineering, S.V.U. College of Engineering, Tirupati 517502, India
e-mail: prasadpln10@gmail.com

S. K. Ghosh (ed.), *Urban Mining and Sustainable Waste Management*,
https://doi.org/10.1007/978-981-15-0532-4_2

several regional hospitals in towns like Chittoor, Tirupati, Madanapalli, etc., Tirupati has health-care facilities like SVRRGG hospital, one government maternity hospital, one ayurvedic hospital, super speciality (SVIMS) hospital, 77 nursing homes, 51 clinics, etc., that cater to the requirements of neighboring districts also. The total capacity of all hospitals in Chittoor District in terms of number of beds is around 10,000.

The biomedical waste is the waste that is generated during the diagnosis, treatment or immunization of human beings or animals or in research activities pertaining thereto or in the production or testing of biological components. As per CPCB the biomedical waste is segregated and temporally stored in blue, yellow and red color-coded bags. The non-infectious waste such as glove papers, plastic papers, covers and suture covers are put in blue bags yellow bag contains contaminated cotton swabs, gloves, mops, gauze, specimens, patient's wastage, catheters, I.V. and blood set, suction tubes, urine bags and all draining tubes which are solid and liquid in nature. Human or animal tissues, organs or body parts that are highly infectious are collected and in red bags. Approximately 1300 kg/day of infectious biomedical wastes are generated from these hospitals. Daily all these wastes are collected from all the hospitals and transported to the biomedical waste management plant by a private agency to their BMW treatment facility which is located near Pachikapalam village 25 km away from Tirupati. All these wastes are incinerated in a dual-chamber incinerator of 100 kg/h capacity working for 8–10 h/day.

2 Generation of Biomedical Waste from SVIMS Super Speciality Hospital, Tirupati—A Case Study

Sri Venkateswara Institute of Medical Science (SVIMS) is a super speciality hospital with 15 medical specialties, 6 surgical specialties, 5 diagnostic departments, 25 wards and 5 operation theaters and has a bed capacity of 450. Every day around 1500 outpatients visit the hospital for their medical needs. The total staff working in this super speciality hospital including doctors, nursing, medical and non-medical is 1436. The hospital sanitary staff clears the waste from wards, OTs and laboratories twice in a day, at 7.00 AM and 2.00 PM.

Every day around 1500 patients are visiting SVIMS hospital for their medical needs from various places. In addition to this, there are several in-patients in different wards under treatment. From different departments' clinical waste generated during the treatment, which is collected in three separate color-coded bags depending on level of contamination. The non-infectious waste such as glove papers, plastic papers, covers and suture covers are put in blue bags. Yellow bag contains contaminated cotton swabs, gloves, mops, gauze, specimens, patient's wastage, catheters, I.V. and blood set, suction tubes, urine bags and all draining tubes which are solid and liquid in nature. Human or animal tissues, organs or body parts that are highly infectious are collected and in red bags.

All these bags are deposited by the hospital staff in a protected area for collection by the personal of the biomedical management agency. The color-code bags are collected and placed in a designated vehicle for transportation of biomedical waste. In order to find the quantity of biomedical waste generated in SVIMS, each and every bag was weighed separately. Further, the waste was classified into incinerable and autoclavable. The incinerable waste was obtained as 110–120 kg/day and the autoclavable waste 30–40 kg/day.

3 Transportation

The specially designated vehicles with biohazard symbol collect daily the biomedical waste from all the hospitals located in Chittoor District and are transported to the BMW processing (incineration) and disposal site located at Pachikapalam village 25 km away from Tirupati. Every day the vehicles start from the disposal site in the morning and collect the waste from various hospitals in Chittoor District up to the evening and then return back to the BMW treatment facility and treated.

4 Treatment and Disposal of Biomedical Waste

The biomedical waste thus collected every day is first sorted into autoclavable and incinerable manually in a designated area. Autoclavable waste includes plastic syringes, catheters, I.V. and blood set, suction tubes, urine bags, draining tubes, etc., which are plastics and can be recycled. Before autoclaving, the waste is soaked in a RCC tank filled with 1:3 sodium hypochlorite solutions for 5–10 min. The waste is then placed in autoclave working at a temperature of 110–120 °C and sterilized for 45 min. After that the recyclables are taken out from the autoclave then all the recyclables are sent for shredding. Sterilized recyclable plastic wastes from autoclave are fed into a shredder.

In this shredder, the recyclable items are shred into small pieces of 12–18 mm size, the shredded pieces are sent to an authorized agency of pollution control board for further processing; recyclables are graded and subjected to grinding in a mechanical grinder of 400–500 kg capacity. After grinding, the pellets are washed with detergent solution, dried and made into granules. The granules are sent for reusing and casting.

5 Incineration of Biomedical Waste

Incinerable waste that contains human or animal tissues, organs or body parts, placenta, viscera, etc., which are highly infectious is incinerated in a dual-chamber incinerator. The incinerator consists of primary chamber and attains a temperature

Table 1 Stack monitoring

Parameters	Value
Stack temperature	82 °C
Exit velocity of the flue gasses	5.75 m/s
Flow rate	3291.30 m^3/h
Suspended particulate matter	91 mg/N m^3
Sulfur dioxide	102 mg/N m^3
Oxides of nitrogen	48 mg/N m^3

of 800–850 °C and a secondary chamber where in a maximum temperature of 1000–1100 °C. The capacity of incinerator is 100 kg/h. The incinerator is attached with a cyclone separator followed by a scrubber for treatment of flue gasses with induced draft. The exit of the scrubber is attached to a stack of 45 cm diameter and height of 30 m for venting of emission gasses. The waste from the scrubbing fluid is collected in a sedimentation tank and the clarified water is recycled in scrubber. Sludge from the sedimentation tank is disposed along with ash by land burial. The incinerable waste is directly fed into primary chamber and incinerated for 45–60 min. The draft from the primary chamber is led into the secondary chamber maintained at a temperature at 1000–1100 °C for further destruction of gaseous components. The emissions from the secondary chamber are sucked by induced draft and are first fed into cyclone separator and then to the scrubber where water is sprayed. Ashes are collected from the bottom through the gratings and disposed by land burial.

6 Emission and Ash Analysis

Stack monitoring was performed to determine the pollution levels in the stack gasses and the results of stack emission analysis are shown in Table 1.

The results of stack emission analysis indicate that all regulatory parameters are within permissible limits and the overall working of incinerator is satisfactory.

7 Ash Analysis

The quantity of ash generated is 5–6 kg/100 kg waste incinerated. Thus, a total of 65–80 kg of ash is generated per day of operation. The ash was subjected to further analysis and the chemical analysis is shown in Table 2.

Expectedly ash is rich in phosphorus and nitrogen and may, in fact, be used to supplement vegetative growth. Ash is disposed by land burial in lined pit of size 5 × 7 × 7 m in layers of 30 cm thick with a covering material of locally excavated soils of 10–15 cm thick. The sharps and blades are blunted and cut into pieces and disposed along with ash and sludge by land burial.

Table 2 Ash analysis

Parameters	Value (%)
Nitrogen	0.43
Sulfur	1.80
Phosphorus	2.10
Carbon	3.00
C/N ratio	6.9

8 Conclusion

A total of 1300 kg/day of BMW is generated from all the health-care facilities in the district. Biomedical wastes are treated by incineration of infectious waste and autoclaving of recyclable plastic waste and the ash generated by land burial. The ash and emission analysis are satisfactory.

Treatment of biomedical waste by plasma gasification and verification process may avoid the process of waste segregation, autoclaving and incineration and may further reduce possible pollution from emission and ashes. With this process, the operation and maintenance problems can be minimized and syngasses can be used for generation of power, and therefore the process of plasma gasification and verification is recommended for treatment of biomedical waste.

Reference

Central Pollution Control Board of India (1998) Bio-medical waste (management & handling) rules (Amended in 2000 and 2003).

Knowledge and Practice of Biomedical Waste Management and Awareness of 3 'R's Concept Among Staff Nurses in the Hospital—A Cross-sectional Study

Bonam Soujanya Kumar, Md. Sameer, S. B. Vishnu murthy and G. S. C. N. V. Prasad

Abstract *Background*: 75–85% of healthcare waste is domestic waste, and the rest 15–25% is hazardous which affects the health of staff and community at large. The studies indicate 4–5 kg/bed/day is generated. The knowledge among employees especially doctors, nurses and paramedics involved in direct patient care is essential in proper management. The high turnover of staff and students in a teaching hospital poses constant challenge in training and orienting the staff. There is a need to continuously evaluate the present level of knowledge of the statutory requirements, amendments in policies and guidelines. The theory and practice depend on the attitude and behaviour in implementation within the hospital. *Aim*: To study the knowledge and practice of biomedical waste management and awareness of 3 'R's concept. *Methods*: Cross-sectional questionnaire study to evaluate the knowledge and practice among staff. Nurses working in different areas of the hospital ($n - 86$). *Results*: 81.40% have knowledge about categories, 74.40% are knowing about hazards and infections spread by biomedical waste, 92% are practising segregation, 90.5% know how to handle needlestick injuries, 40% are aware of administrative aspects and amendments, and 55.80% are aware of 3 'R's concept and opined that it is feasible and applies to solids and liquids. *Recommendation*: Strict implementation of guidelines through regular training of staff nurses is essential and constant vigilance and monitoring of the practice for effective management. The behaviour and culture need to be changed through goal setting and support from administration.

Keywords Healthcare waste · Knowledge · Practice · Strategies

1 Introduction

Biomedical waste (BMW) means any solid and/or liquid waste which is generated during diagnosis, treatment or immunisation or in research pertaining to or in the production or testing of biologics, including categories mentioned in Schedule I of

B. S. Kumar (✉) · Md. Sameer · S. B. Vishnu murthy · G. S. C. N. V. Prasad
Department of Hospital Administration, Narayana Medical College and Hospital, Nellore, AP, India
e-mail: bskumar96@gmail.com

S. K. Ghosh (ed.), *Urban Mining and Sustainable Waste Management*,
https://doi.org/10.1007/978-981-15-0532-4_3

BMW rules of 1998 (Reddy et al. 2014; Kumar et al. 2014). It is estimated that 10–25% of healthcare waste is hazardous, with the potential for creating a variety of health problems (Mathur et al. 2011; Sharma et al. 2013). Every concerned health personnel should have the proper knowledge, practice and capacity to guide others in management and proper handling techniques (Jiang 2006).

2 Review of the Literature

2.1 Classification of Hospital Waste

'Biomedical waste' consists of the following types (Kirkby 1993):

(a) 'Animal waste' is waste animal carcasses and body parts.
(b) 'Biosafety level 4 waste' is contaminated with blood and excretions, exudates or secretions from humans or animals.
(c) 'Cultures and stocks' are wastes infectious to humans and include specimen cultures, cultures and stocks of etiologic agents, wastes from production of biological and serums, discarded live and attenuated vaccines.
(d) 'Human blood and blood products'.
(e) 'Pathological waste' biopsy materials, tissues and anatomical parts that emanate from surgery, obstetrical procedures and autopsy. It does not include teeth, human corpses, remains and anatomical parts that are intended for interment or cremation.
(f) 'Sharps waste' is all hypodermic needles, syringes with needles attached, IV tubing with needles attached, scalpel blades and lancets that have been removed from the original sterile package.

2.2 3R's Principle and Waste Hierarchy (Jiang 2006)

The three R's principle of waste otherwise known as the waste hierarchy represented as (1) waste prevention by reduction of usage of any material which generates waste and reuse, (2) recycling and recovery of energy from waste generated and (3) proper disposal has been very widely adopted.

Reduce: The first step towards the effective waste management is to consume only a limited number of resources, by utilising natural resources, and reduce pollution.
Reuse: Reusing old things for a little longer can play a major role in reducing a lot of waste in the surroundings. It helps in reducing waste that needs to be recycled or sent to incinerators and landfills.
Recycle: The process usually involves collecting and processing those materials, which would otherwise be discarded as trash and turning them into new products.

This process will always be of great benefit to surroundings and environment in the following ways.
Aim: To study the knowledge and practice of biomedical waste management and awareness of 3 'R's concept. **Objectives**: (1) To collect and assess the experience and qualifications in the field in patient care, (2) to assess the knowledge in various aspects of biomedical waste management and (3) to evaluate the practice or application of this knowledge in day-to-day patient care.
Inclusion Criteria: (1) Subjects willing to participate in the present study, (2) subjects who were present at the institute when the present study was conducted and (3) subjects who completely filled the questionnaire.
Exclusion Criteria: (1) Subjects who were not willing to participate in the present study, (2) subjects who were absent or on leave for any reason when the present study was conducted and (3) subjects who did not fill the complete questionnaire.
Study Setting: Study was conducted in a 1000+ beds medical college teaching hospital at Nellore, Andhra Pradesh state. The mean age of staff nurses was 22.73 years ranging from 20 to 28 years with std. deviation of 1.514 ($n = 86$). 97.7% are females, and 2.3% are males. The average experience was 5.5 years.
Methods: Cross-sectional questionnaire study to evaluate the knowledge and practice among staff nurses working in different areas of the hospital ($n = 86$). The study does not involve any invasive tests or procedures and consists of questionnaire only. The different domains in the questionnaire related to the biomedical waste management act, classification, disinfection, segregation, dealing with needlestick injuries, colour coding, risks associated and records to be maintained and reporting system have been explained to the participants.

The sample of staff nurses even though are already working with experience and practicing the guidelines are sensitized about the methodology of answering questions. The team of post graduates of hospital administration who are having knowledge of environment protection act and involved in day to day operations of the hospital are trained in research methodology were selected to collect data from a group of staff nurses. Informed consent was taken from all the participants before collecting the data. The questionnaire was distributed in the wards and then collected after ample time. The response rate is 86%.

Regarding 3 'R's, it is important to ask participants to define what they understood by the terms reduce, reuse and recycle and these terms were generally understood. And also to give suggestions the items in their experience which can be reduced, recycled or reused in a hospital.

2.3 *Statistical Analysis*

The data values have been entered into MS Excel, and statistical analysis has been done by using IBM SPSS Version 20.0. For categorical variables, the data values are represented as number and percentages. Inferences were drawn, and frequencies and percentages were calculated for set of questions. The analysis was based and categorised on the domains described in the methodology to evaluate the knowledge and practice.

2.4 *Data Analysis*

NMCH biomedical waste 2017 data

Months	Solid waste in kg	Yellow in kg	Red in kg	Sharps	Total (infectious waste with sharps)
January 2017	90,000	5024	1260	228.5	6512.5
February 2017	84,000	4634	1125	188.5	5947.5
March 2017	99,000	5168	1143	198	5509
April 2017	84,000	4951	1190	190.9	6331.9
May 2017	90,000	5191	1230	195.9	6616.9
June 2017	78,000	6500	1350	250	8100
July 2017	99,000	3849	7552	815	12,216
August 2017	90,000	4952	2925	1580	9457
September 2017	90,000	6194	1425	2936	10,555
October 2017	90,000	7621	1622	3901	13,154
November 2017	90,000	7050	1490	3799	12,339
December 2017	90,000	6350	1150	3243	10,743
Total	1,074,000	67,484 kg	23,462 kg	17,525.8 kg	108,481.8 kg

The solid waste per month is about 89,500 kg and average of 3000 kg/day. Infectious waste is about 7578 kg/month, and it was around 300 kg/day. Sharps constitute around 1460 kg/month and 50 kg/day.

3 Knowledge and Practices of Biomedical Waste Management Practices in Patient Care Areas ($n = 86$)

Q1: Biomedical waste management handling rules were enacted in which year?
Q2: When was the act revised?

Pie Chart 1: Depicting Q1 and Q2.

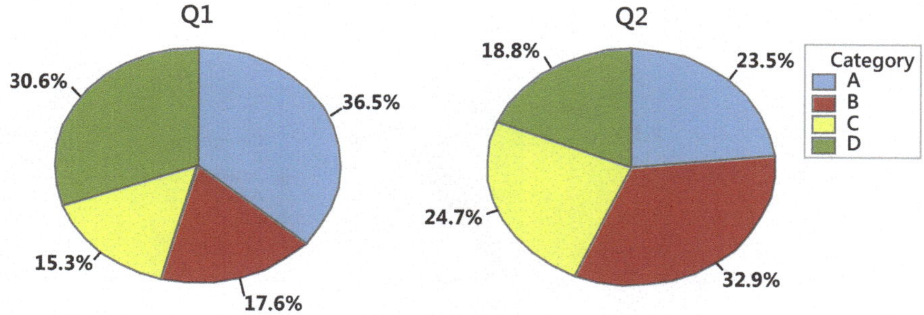

Only 1/3 of staff are aware of the year of enactment of act and amendments to the act.

Pie Chart 2: Indicating Q3, Q4, Q5 and Q6.

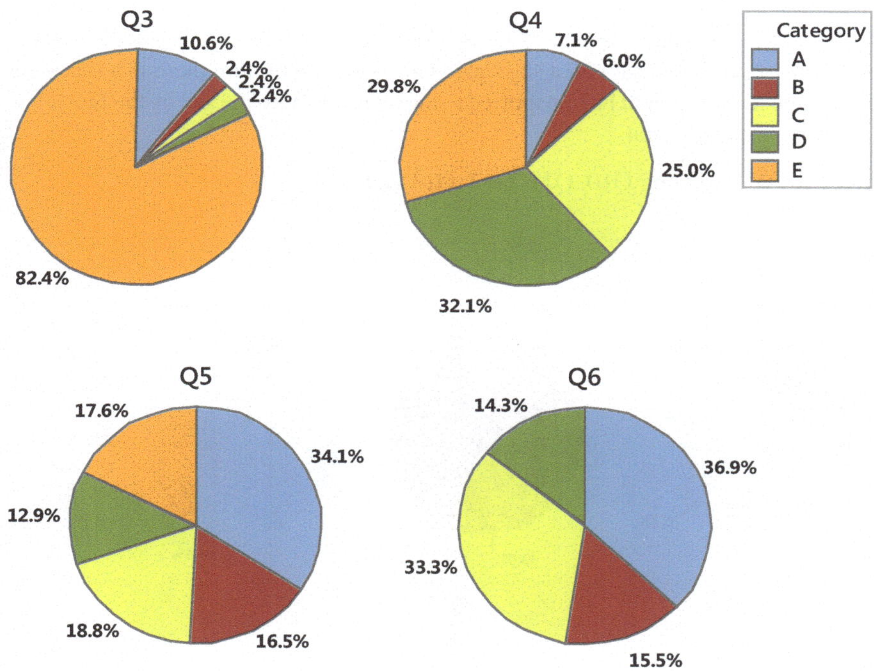

Q3: Definition of biomedical waste; 82.40% are aware of definition and categorisation of waste.
Q4: How much is the percentage of BMW; they opined 10–20%.
Q5: Who is an occupier; 34.10% opined it is the operator and rest opined generator.

Q6: How many categories of waste; 33.60% opined eight categories and another 36.90% opined 10 categories.

Pie Chart 3: Showing Q7.

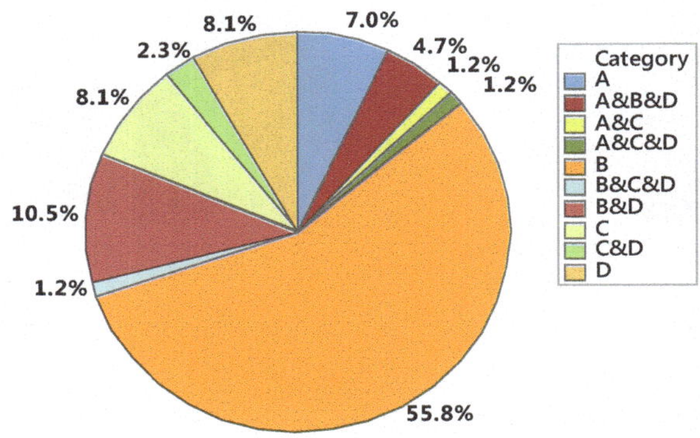

Q7: Segregation of waste should be done at which area?

Segregation at source should be ideal and is the critical step in biomedical waste management. If the waste is not segregated at source, definitely there is a scope for mixing and the same shall be transported to the common garbage site by the bio-waste transports in the hospital.

Pie Chart 4: Depicting Q10, Q11, Q12, Q13 and Q15.

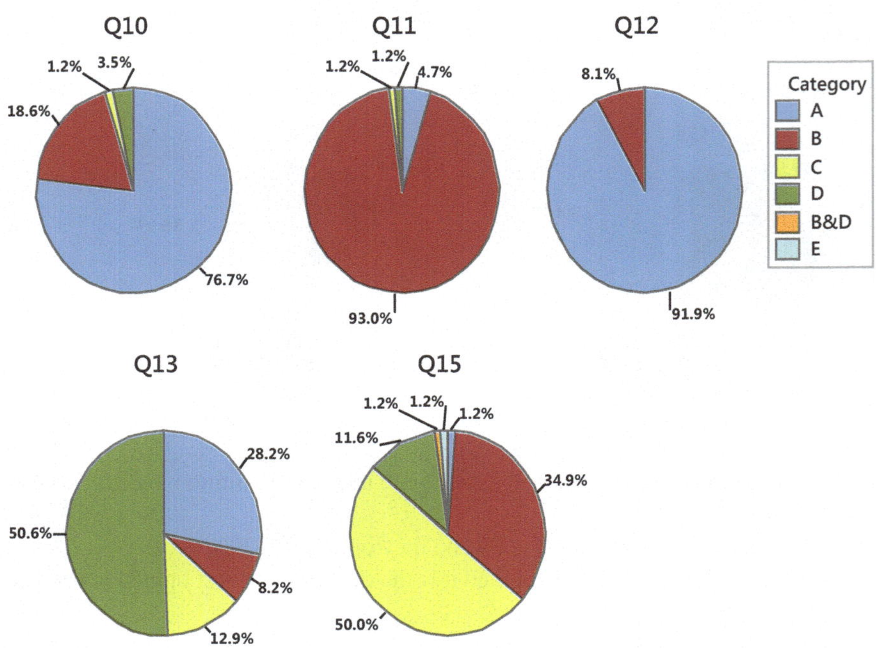

Q10: Placenta should be discarded in which coloured bin?
Q11: IV tubes and catheters should be discarded in which bin?
Q12: Dressings should be discarded in which bin?
Q13: Discarded linen should be discarded in which bin?
Q15: Needles and syringes should be discarded in which bin?

Even though the staff are aware of the issue but while practicing or disposing of the waste of different categories usually at Bed side the commonly used are the IV sets, needles, syringes, dressings needs to disposed in yellow and red coloured bins The general waste like water bottles, paper or food waste in Green or Black bin, 76.70, 93.0, 91.90% of the infectious waste is segregated and disposed correctly in yellow and red bins.

Pie Chart 5: Showing Q21.

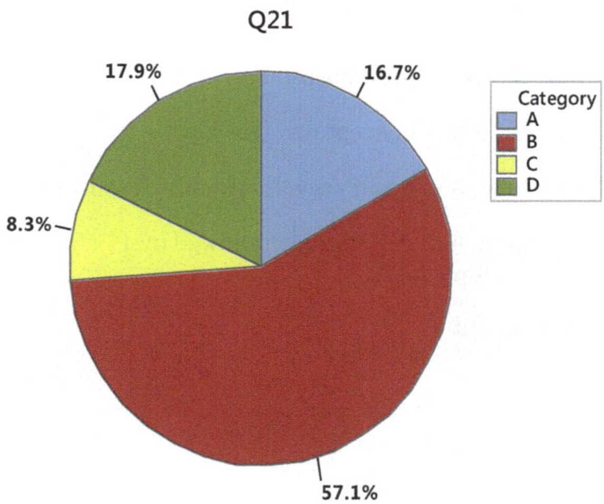

Q21: What is the concept of 3 'R's reduce, recycle and reuse?

57.1% are aware of 3 'R's concept.

Opinion on 3 'R's concept

Bar Chart 1: Representing Q1

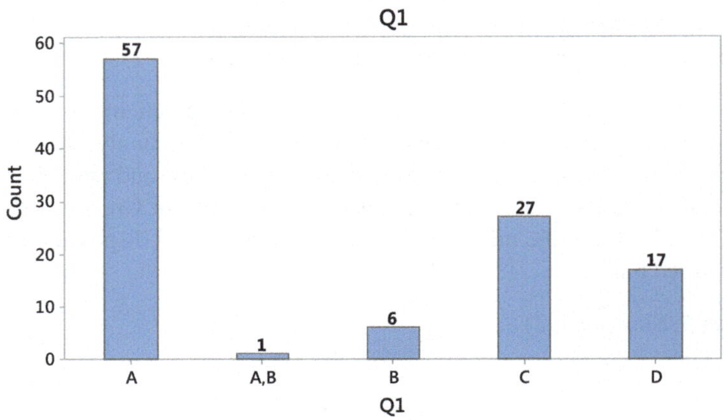

Q1: Feasibility of 3 'R's concept in hospital setting

52.80% of nurses opined 3 'R's concept was feasible.

Q2: Out of 3 'R's reduce, reuse and recycle which is feasible in hospital?

70.80% of nurses opined reduce was feasible followed by 12.50% reuse and 10.40% recycle, and rest gave multiple answers.

Bar Chart 2: Depicting Q2

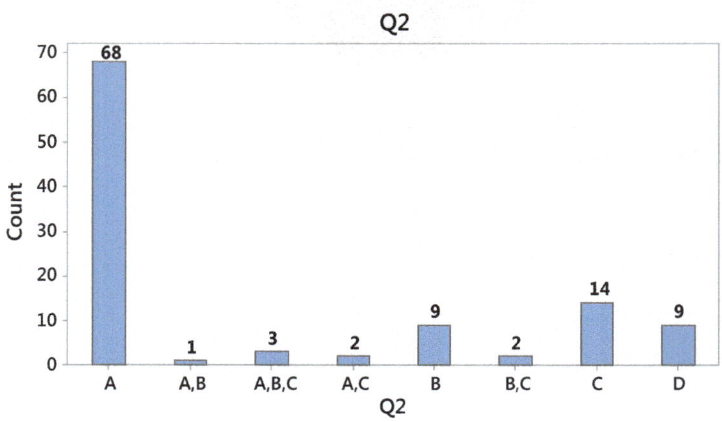

Q3: Feasibility of 3 'R's for solids and liquids or gases

39.60% of staff nurses opined it applies to solids, 25.50% opined it applies to all forms and 33.30% told cannot say.

Bar Chart 3: Depicting Q3

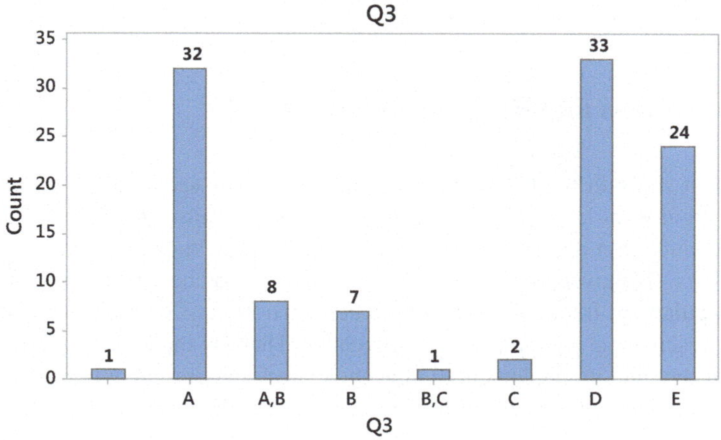

4 Results and Discussion

81.40% have knowledge about definition and categories of waste; 36.0% are able to know the year of enactment of the act; only 55.8% know about the percentage of general versus biomedical waste percentages of the total waste generated in a hospital.

90.50% clearly know about the process to deal with 'needlestick injury' by asking for investigation, consulting doctor, cleaning the wound with soap and water and prophylaxis treatment, and 74.40% are knowing about hazards and infections spread by biomedical waste.

The questions on the disposal of placenta, IV tubes, needles and syringes, dressings and discarded linen distinguish and evaluate the practice of segregation and background knowledge on segregation and colour coding of bins and bags.

92.45% are practising segregation, and 55.80% felt that the waste can be stored for 8 h as many of them working for 8 h and focussing on clearance of waste once they complete their shifts.

60.50% opined that the 'hypo solution' has to be changed once in a day. 40% are aware of administrative aspects and amendments.

55.80% are aware of 3 'R's concept and opined that it is feasible and applies to solids and liquids. They are not aware of gaseous component.

Abah and Ohimain (2011): This study approach involved the estimation of the quantity of HCW achieved, estimation of the waste segregation practices and determination of the knowledge of hospital workers in relation to HCW management.

Daily waste stock of each ward was estimated. A judgement status of the waste management practice in the hospital was created using the following norms: waste management; waste transport; and waste recycling and reuse (Baum 1992).

5 Conclusions and Recommendations

The present study in the hospital setting among a cross-sectional group of nurses revealed significant knowledge and practice in the domains of segregation, colour coding, disinfection and disposal. There is in adequate knowledge in acts and its amendments. Turnover of nurses and entry of new batches have been observed. Hence, regular and frequent training with demonstration of biomedical waste management practices is essential for effectiveness. There is inadequate awareness on 3 'R's concept and its implementation as most are focussing on only biomedical or hazardous component not on solid waste generated in the hospital.

Acknowledgement We thank the support of the management in performing our study. We acknowledge the support of faculty and postgraduates of hospital administration, nursing superintendent, supervisors and cross section of staff nurses who participated in the study. We thank the institute statistician for assisting data analysis.

Conflict of Interest There is **no conflict of interest** among or between authors of the study.

Annexure

Questionnaire

consent

We request your valuable opinion about the following quality attributes about patient care in the hospital .This information is for academic purpose only and for improvement of quality . The personal information is kept confidential .we request your consent for filling and extending the opinion . Thank you for your cooperation

Name .... Sex/__ Age ...... Department .... Designation ....

Mobile number .... signature ....

++

1. Bio medical waste management handling (Rules and Regulations) of Government of India were enacted in the year

   a) 1998 b) 2005 c) 2016 d) 2000

2. When was the act revised

   a) 2000 b) 2005 c) 2011 d) 2016

3. Bio medical waste means waste generated during

a) Treatment B) diagnosis C) surgery D) , research , E) all

4. How much percentage of bio medical waste is hazardous

a) 10 b) 20 C) 50 d) 80 e) 100

5. who is an " Occupier " as per act

a) Generator of BMW b) operator of BMW c) Government d) transporter E) Collector of BMW

6. According to recent guidelines – How many categories of health care waste ?

a) 6 b) 8 c) 10 d) 12

7. What do you do if needle stick injury occurs?

a. Ask for investigations
b. Clean the wound with soap and water solution
c. Clean the wound with spirit or savlon
d. Consult the emergency doctor or physician

8. How much time that the BMW can be stored at source

a) 8 hrs b) 12 hrs c) 24 hrs d) 48 hrs

9. Placenta should be collected in which colour bag

a) yellow b) red c0 white d) black/blue

10. IV tubes and catheters should be collected in which colour bin

a) yellow b) red c0 white d) black/blue

11. dressings done in the ward should be collected in which colour bin

a) yellow b) red c0 white d) black/blue

12. discarded linen should be collected in which colour bin

a) yellow b) red c0 white d) black/blue

13. what is the disinfectant is recommended for sharps

a) savlon b) spirt c) Hypo d) Dettol

15. needles and syringes and should be collected in

a) yellow b) red c0 white d) black/blue

16. which of the following doesn't covered under Bio Medical waste Act

a) Radioactive waste b) Municipal waste c) Lead acid batteries d) All

17 . Risks associated with hospital waste include

a) HIV b) Malaria c) Hepatitis d) Injury e) a and c and d

18. How frequently you change hypo solution?

a) Once in a day b.) Twice in a day c). As and when required

19. How frequent the hospital required to submit report to the authorities

a) Monthly b) quarterly c) half yearly d) yearly

20 How many years the records to be maintained by the hospital

a) one year b) 3 yrs c) 5 yrs d) 2 yrs

21. what is the concept of 3 " R ' s in waste management

a) Research, Recommend ,Reduce

b) Reduce, Reuse , Recycle

c) Recovery, Restrict,Reuse

d) Records, Recovery, Recycle

References

Abah, S. O., & Ohimain, E. I. (2011). Healthcare waste management in Nigeria—A case study. *Journal of Public Health and Epidemiology, 3*(3), 99–110.

Baum, D. L. (1992). The three R's: reducing, reusing & recycling. Cut costs while improving community image. *Southern Hospitals, 58*(4):21.

Jiang, S. (2006). The Practical Implementation of the 3 R's of sustainable waste management in the Norfolk and Norwich University Hospital.

Kirkby, G. (1993) Waste management: three R's (reduce, reuse, recycle) reduce waste, save money. *Leadership in Health Services*, 2(2):30–33. https://www.ncbi.nlm.nih.gov/pubmed/10125210.

Kumar, R., Gupta, A. K., Aggarwal, A. K., Kumar, A. (2014). A descriptive study on evaluation of bio-medical waste management in a tertiary care public hospital of North India. *Journal of Environmental Health Science and Engineering*, *12*, 69.

Mathur, V., Dwivedi, S., Hassan, M. A., Misra, R. P. (2011). Assessment of knowledge, attitude, and practices about biomedical waste management among healthcare personnel: a cross-sectional study. *Indian Journal of Community Medicine: Official Publication of Indian Association of Preventive & Social Medicine, 36*(2), 143–145. https://doi.org/10.4103/0970-0218.84135.

Reddy, B. S., Rao, J. N., & Subrahmanyam, B. V. (2014). Awareness and knowledge practices about the bio medical waste management at tertiary care teaching hospital. *International Journal of Scientific and Research Publications, 4*(5).

Sharma, A., Sharma, V., Sharma, S., & Singh, P. (2013). Awareness of biomedical waste management among health care personnel in Jaipur, India, March 2013. *Oral Health and Dental Management, 12*(1):32–40.

Utilization of Construction and Demolition (C&D) Waste and Industrial Inorganic Wastes in Cement Manufacturing

Parmanand Ojha, Pinky Pandey, Kåre Helge Karstensen and Palash Kumar Saha

Abstract Annual generation of C&D waste and industrial inorganic wastes in India is estimated to be 170 million tonnes and >300 million tonnes, respectively, which is projected to increase with the growing urbanization and economic development. Considering the constant expansion of the cement industry and its need to sustain, waste valourization has emerged as a great opportunity for the industry. Bureau of Indian Standard (BIS) has prescribed standards for use of C&D waste as coarse and fine aggregates, fly ash and granulated blast furnace slag (GBFS) as blending materials in cement manufacturing and other inorganic wastes as performance improvers in cement manufacturing. The physical, chemical and mineralogical characteristics of inorganic wastes suggest their potential to be utilized as alternative raw materials in cement manufacturing. The addition of red mud can improve the strength properties of cement and resistance to sulphate attack. Spent Pot Liner (SPL) can be gainfully utilized as a mineralizer in cement raw mix. NCCBM has explored use of Jarosite as set controller in place of gypsum, use of copper slag and lead–zinc slag as alternative raw materials and as blending materials in Portland Slag Cement (PSC) and use of marble dust as decarbonated raw material and as blending material in Portland Limestone Cement (PLC). There are, however, limited number of studies for use of recycled concrete aggregate (RCA), recycled masonry aggregate (RMA) and fines extracted from recycled concrete, as alternative raw materials in cement manufacturing. As a part of the Indo-Norwegian project on co-processing, NCCBM is currently exploring the possibility of utilization of fines extracted from recycled concrete aggregate and recycled aggregate as alternative raw materials in cement manufacturing. This paper contains potential utilization and further opportunities for increased utilization of C&D waste and industrial inorganic wastes in cement manufacturing process.

P. Ojha (✉) · P. Pandey
NCCBM, Ballabgarh, India
e-mail: pnojhaji@gmail.com

K. H. Karstensen
Asian Institute of Technology, Khlong Luang, Thailand

K. H. Karstensen · P. K. Saha
SINTEF, Trondheim, Norway
e-mail: palashkumar.saha@sintef.no

S. K. Ghosh (ed.), *Urban Mining and Sustainable Waste Management*,
https://doi.org/10.1007/978-981-15-0532-4_4

Keyword C&D waste · Fly ash · Slag · Red mud · Jarosite · Cement

1 Introduction

The rapidly growing economy, urbanization and industrialization are leading to the increasing generation of C&D wastes and industrial wastes. Estimated annual generation of C&D waste in India is in the range of 165–175 million tonnes (Mt) projected to increase further in coming years. Current generation of industrial inorganic wastes is more than 300 Mt. The major industrial inorganic wastes are coal combustion residues, red mud, Spent Pot Liner (SPL), spent catalyst, marble dust, soda ash, lime sludge, jarosite, slags (lead–zinc, copper, steel), etc. It is evident from the physical, chemical and mineralogical characteristics of these wastes, generated from different processes, that they have good potentials for recycling and gainful utilization in cement manufacturing process with 100% material recovery.

2 Construction and Demolition (C&D) Waste

C&D waste is the debris generated from construction, repair, renovation and demolition of structures. It is a non-hazardous solid waste comprising the range of materials, e.g. bricks, mortar, concrete, wood, steel, tiles, metal, masonry, asphalt, stone (marble, granite, sandstone), etc. After processing of C&D waste, generally two types of materials are obtained, i.e. recycled aggregate (RA) consisting of brick masonry, cement mortar, tiles, etc., and recycled concrete aggregate (RCA) consisting of concrete debris. Out of total C&D waste generation about 90–92% is RA and 8–10% is RCA. RA is being processed through wet process, and RCA is being processed through dry process. The treatment and processing of C&D waste in India are at a very nascent stage.

2.1 Current Generation of C&D Waste

The estimated C&D waste generation is in the range of 165–175 Mt/year (BMTPC 2016). Total C&D waste generation from megacities such as Mumbai, Delhi, Bangalore, Hyderabad, Chennai, Ahmedabad, Pune, Surat and Kolkata is 17,600 tonnes/day and in 1 million plus population cities is 4320 tonnes/day.

2.2 Current C&D Waste Processing Plants Operational in India

The processing and recycling of C&D waste in India are limited to four operational plants—three in Delhi located in Burari (IEISL 2017), East Kidwai Nagar (NBCC 2013) and Shastri Park and one in Ahmedabad with capacities of 500 tonnes/day, 150 tonnes/day, 500 tonnes/day and 100 tonnes/hour, respectively. The Municipal Corporations of Jaipur, Bengaluru, Pune and Bhopal are planning to set up processing plants of capacities in the range of 300–750 tonnes/day.

2.3 Current Utilization of C&D Waste in Construction Industry

The C&D waste is processed to produce usable building materials such as fine aggregate, coarse aggregate, bricks/blocks, tiles, paver blocks, kerbstones and prefab slabs. Based on the research conducted in India, BIS has allowed the usage of C&D waste in IS: 383-2016 as coarse and fine aggregates in concrete (BIS 2016). C&D waste may be used in RCA up to 25% in plain concrete, 20% in reinforced concrete of M25 or lower grade and up to 100% in lean concretes of grade less than M15. Liu and Wang (2013) have provided the model for calculating costs of disposal routes for CDW. They have found that on equal conditions, the costs of recycling and reuse are least (Liu and Wang 2013).

2.4 Utilization of C&D Waste in Cement Industry

Limited research has been conducted internationally (Galbenis and Tsimas 2004; Puertas et al. 2008; Gastaldi et al. 2015; Schoon et al. 2015; Kara et al. 2017; Galbenis and Tsimas 2006) and in India on use of C&D waste in the cement industry. Substitution of natural raw material by recycled concrete aggregates and rubbles obtained from C&D waste in clinker production was explored by Galbenis and Tsimas (2004); whereas, Puertas et al. (2008) used ceramic tiles in cement production. Gastaldi et al. (2015) used hydrated cement separated from C&D waste to replace 30% of raw material, while in some other study, Schoon et al. (2015) used C&D waste fines as alternate raw material. Kara et al. (2017) evaluated the suitability of C&D waste components, e.g. tiles, bricks, fire bricks, plaster and concrete as secondary raw material. Galbenis and Tsimas (2006) investigated the use of RCA and RMA with OPC raw meal in different proportions. Currently, C&D waste is not utilized in the Indian cement industry. National Council for Cement and Building Materials (NCCBM) analysed the fines extracted from two samples of RCA and two samples of RA. The results indicate that the fines extracted from RCA and RA are siliceous

Table 1 Chemical analysis of CDW fractions from Delhi conducted by NCCBM

Chemical composition (%)	RCA-1 (%)	RCA-2 (%)	RA-1 (%)	RA-2 (%)
Silica	51.75	52.4	68.7	50.4
Iron oxide	9.93	4.22	3.69	5.25
Alumina	10.63	8.22	8.38	12.11
Calcium oxide	11.66	16.71	7.04	11.25
Loss on ignition	7.6	12.2	6.07	12.31

in nature. Their chemical composition and their conformity with the composition of Ordinary Portland Cement (OPC) raw meal indicate that these materials can be used as an additive in the production of clinker. The use of C&D waste in cement process should be studied on a case to case basis. The analysis is given in Table 1.

3 Fly Ash

The thermal power plants in India are primarily dependent on the combustion of high ash bituminous coal in pulverized fuel-fired systems. Hence, the low-lime fly ash (like Class F of ASTM C 618) is the prime variety generated in India, although significantly smaller volumes of high-lime fly ash (comparable to ASTM Class C) are available in the country.

3.1 Fly Ash Generation Versus Utilization in India Including Cement Industry

As per the data from Central Electricity Authority (CEA), in 2016–17, 169.25 Mt of fly ash was generated in India from coal-fired power plants, almost 2.5 times compared to 1996–97 generation. The fly ash utilization stands at 107.1 Mt in 2016–17 (63.3% of generation) compared to 6.6 Mt in 1996–97. In 2016–17, 40.6 Mt was utilized by the Indian cement industry (CEA 2017).

NCCBM has conducted pioneering studies to enable standardization and commercialization of fly ash in the cement and construction sector. NCCBM has assisted in standardizing and commercializing composite cement recommending 15–35% fly ash and granulated blast furnace slag (GBFS) in the range 20–50% (IS 16415-2015) (BIS 2015a); whereas, Portland Pozzolana Cement (PPC) has already got lions share in the Indian cement market (BIS 2015b). NCCBM is currently investigating production of geo-polymeric bricks using chemically activated fly ash.

4 Red Mud

Red mud is a by-product of the production of alumina from bauxite in the Bayer process which involves reaction with NaOH at high temperature and pressure. Its composition, property and phase vary with the origin of the bauxite and the alumina production process.

4.1 Red Mud Generation and Composition

The total red mud generation in different industries, namely NALCO, HINDALCO, VEDANTA, UTKAL, RAYKAL, Aditya and JSW is 13.7 Mt/year (Samal et al. 2013).

Red mud is alkaline in nature and contains oxides and salts of iron (30–60% by weight), aluminium (10–20% by weight), silica (3–50% by weight), sodium (2–10% by weight) and calcium (2–8% by weight) and a variety of trace elements such as titanium (trace—25% by weight), potassium, chromium, vanadium, manganese, lead, zinc, phosphorus, fluorine, sulphur and arsenic. Mineralogical phases of red mud are haematite Fe_2O_3, goethite FeO(OH), gibbsite $AlOH_3$, diaspore, AlO(OH), quartz SiO_2, cancrinite $Na_6Ca_2[(CO_3)_2|Al_6Si_6O_{24}]\cdot 2H_2O$, kaolinite $Al_2Si_2O_5(OH)_4$ and calcite ($CaCO_3$) (Deelwal et al. 2014).

4.2 Red Mud Utilization and Ongoing R&D Work

Red mud and siliceous additive have been used together to make bricks (Mustafa 2000). CBRI Roorkee has produced clay bricks by partially replacing clay with red mud and fly ash. HINDALCO has sponsored projects for utilization of red mud as additive for production of special cement and mortar. Apart from construction applications, red mud can be used for glass ceramics, water treatment, as catalyst, as geo-technical material (Deelwal et al. 2014) and in recovery of metals from red mud. Red mud can be used as an alternative raw material in cement production (Mishra et al. 2011).

5 Spent Pot Liner

Spent Pot Liner (SPL) is a by-product generated when the carbon and refractory lining of aluminium electrolytic cell, known as a pot, reaches the end of its useful life. Life of aluminium smelter pot is 2500–3000 days, and the generation of SPL per pot is around 60–85 tonnes (Tiwari 2017). SPL has been classified as hazardous waste. SPL generation in India is estimated to be more than 45,000 tonnes/year @ 20 kg/tonnes of aluminium production.

5.1 Utilization of SPL

SPL could be used as an energy source (Parhi 2014) and as raw material component (Chaturvedi et al. 2011; Alka and Kumar 2011). A demonstration trial of SPL carbon fraction has been carried out in a cement plant in Chhattisgarh. SPL has also been used in captive power plants of HINDALCO. Central Pollution Control Board (CPCB) has published standard operating procedures and checklist of minimal requisite facilities for utilization of SPL generated from primary aluminium smelting industries (CPCB 2017).

5.2 Spent Catalyst

Spent catalysts are generated from the catalytic cracking process when catalysts are withdrawn regularly to maintain desired cracking and product yield. Annually, 32,000 tonne of spent catalyst is generated in India. NCCBM has investigated the utilization of spent catalysts as alternative raw material in cement manufacturing process. Spent catalyst has also been used as an asphalt filler. Lin et al. (2017) have also reported the use of 4% spent catalyst as cement raw material; whereas, Al Dhamri et al. (2011) have reported 3.5% use in Portland cement clinker.

6 Lime Sludges

The main sources of the lime sludge are sugar, paper, acetylene, fertilizer, sodium chromate and soda ash industries. It is estimated that approximately 4.5 Mt of lime sludges are generated in these industries (Sengupta 2014).

The following sludges could be used as alternative raw materials in cement manufacturing process: up to 74% of lime sludge on dry basis from paper industry, up to 30% of carbide sludge, approximately 8% of phosphor–chalk limited by P_2O_5, and SO_3; and up to 5% of chrome sludge (as mineralizer) limited by chromium oxide. The lime sludge from paper industry has been found suitable as a blending material for the manufacture of masonry cement in the proportion of up to 30% conforming the Indian Standard Specification of IS: 3466-1988.

7 Marble Dust/Slurry

Rajasthan state has 95% of India's marble deposits (Report on disposal options of marble slurry in Rajasthan 2011) generating 5–6 Mt of slurry every year. The marble dust and slurry from marble stone processing can be gainfully utilized as limestone substitute (up to 5–15%) in cement manufacturing process. However, the presence of MgO and moisture restricts the use of dolomitic marble slurry. Few cement plants in Rajasthan are using 'Makrana' marbles slurry as the replacement of limestone because of the calcined nature of Makrana rock and low moisture percentage.

The technical possibilities for converting marble powder to gypsum are being explored by Centre for Development of Stones (CDOS), Jaipur. Marble dust can be used in road construction, as filler materials and in brick manufacturing. The marble bricks with more than 80% marble slurry have compressive strength that is 2.5 times more than the traditional red bricks. Options for marble slurry management include use in—cement manufacturing, producing synthetic gypsum, road construction, preparing low-cost binder, brick manufacturing and mineral grinding plants (Indian minerals yearbook 2016).

8 Jarosite

Jarosite is a waste generated when iron sulphate from reaction of insoluble zinc ferrite ($ZnO.Fe_2O_3$) and sulphuric acid reacts with ammonium sulphate. Jarosite, due to higher SO_3 content (up to 30%) could act as a gypsum substitute. NCCBM has worked on 3% jarosite replacement for gypsum. Utilization up to 1.5% jarosite as raw mix component has also been reported by NCCBM.

The presence of mineral phases bearing iron, sulphur, sodium, potassium and zinc in jarosite showed its suitability for use as mineralizer in development of clinker mineral phases. Burnability studies showed the mineralizing effect of jarosite in terms of rapid lime assimilation and improved clinker mineral phase formation as compared to control mix prepared without using jarosite.

9 Steel Slag

During steel production, depending on the cooling process, three types of slags are generated, namely air-cooled slag, granulated slag and expanded slag (Ministry of Mines 2018). Steel slag, especially, GBFS is being used as performance improver in cement and as mineral admixture in mortar and concrete (Tiwari et al. 2016). LD slag has calcium and iron and can be used as AR after proper size classification and removal of the magnetic particles. NCCBM is investigating on the production of geo-polymeric cement from slag.

10 Copper Slag

Approximately 3 Mt of copper slag is generated in India; for every tonne of copper, 2.2 tonne of slag is generated. Copper slag is used by the Indian cement industry as a performance improver (IS 269: 2015). Copper slag as an alternative iron corrective reduces calcination temperature and therefore reduces the need for mineralizers. The use of copper slag as pozzolan and in mortar has been reported in many studies.

Up to 15% of copper slag can be used as a cement replacement with constant water: cement ratio of 0.4. This gives higher compressive strength than ordinary cement (Tixier et al. 1997). The dynamic compressive strength of copper slag reinforced concrete generally improved with the increase in amounts of copper slag used as a sand replacement up to 20%, compared with the control concrete (Wu et al. 2010). Since copper slag has higher shear strength and density, it can be used as backfill material in retaining walls to reduce seismic earth pressure (Sathya and Shanmugavalli 2014).

NCCBM and Sterlite Industries (I) Ltd have done joint study that aims to utilize the copper slag as raw mix component in the manufacture of Ordinary Portland Cement (OPC) and as a blending material for Portland Slag Cement (PSC). Leaching behaviour was also studied. The outcome of the study was that copper slag up to 2.5% can be used in raw mix for OPC and up to 35% in manufacture of PSC without any adverse effect on the performance.

11 Lead–Zinc Slag or Imperial Smelting Furnace (ISF) Slag

Imperial smelting furnace (ISF) slag or lead–zinc slag is produced during the pyrometallurgical refining of sulphide metal. Use of lead–zinc slag as a performance improver in OPC (IS: 269) has already been approved by Bureau of Indian Standards (Indian Standard 2015). Morrison et al. (2003) reported increase in density of concrete when sand was substituted with ISF slag due to high specific gravity of ISF slags.

Rajasthan State Pollution Control Board has given consent to Hindustan Zinc Limited for sale of lead–zinc slag for cement manufacturing, road construction, bricks and tiles manufacturing; more than 0.4 Mt of slag has been sold to cement industries in Rajasthan.

12 Conclusion

With the construction sector growing exponentially, C&D waste recycling and recovery will be an emerging sector to work for in India, as negligible amount is being

Table 2 Potential utilization of wastes as AR in cement industry and other uses

Wastes	Generation in India (Mt/year)	AR value	Potential utilization as AR (% of raw mix)	Other uses
C&D wastes	165–175	Iron, aluminium, calcium		Recycled aggregates (RA) and recycled concrete aggregates (RCA) in construction, blocks, tiles and bricks
Fly ash	169	Calcium, silica	1–3%,	Pozzolan in cement, bricks, blocks, tiles, roads and embankments, reclamation of low lying areas, agriculture, geo-polymeric cement
Red mud	13.7	Iron, aluminium	2–3%,	Bricks, tiles
Spent Pot Liner (SPL)	0.04	Mineralizer, fluoride, carbon, sodium	2%	Carbon fraction as alternative fuel in cement
Spent catalyst	0.03	Silica, alumina	RFCC-3.5%, SAC-2.07%	Asphalt filler, metal recovery, sand replacement
Lime sludge	.4.5	Calcium	>70% on dry basis	In Portland limestone cement, cement replacement in mortar and concrete, bricks, lime, composite filler, FGD
Carbide sludge	0.2	Calcium	30%	
Marble slurry/dust	5–6 (slurry)	Calcium, sulphur (gypsum)	5–15%	Bricks, filler materials, production of gypsum, road construction, low-cost binder, mineral grinding

(continued)

Table 2 (continued)

Wastes	Generation in India (Mt/year)	AR value	Potential utilization as AR (% of raw mix)	Other uses
Jarosite (Hindustan Zinc Ltd)	0.045–0.06	Sulphur (gypsum)	1.5%	Set retarder, metal recovery
Blast furnace slag	24	Calcium, mineralizer	10–15%	Pozzolan in cement
Steel/LD slag	12	Calcium, iron corrective, silica	15%	Aggregate, road construction, ceramic tiles
Lead–zinc slag	1	Calcium, silica, iron oxide, alumina	5–6%	Bricks, tiles, road construction, aggregate, Geo-polymeric cement
Copper slag	3	Iron, mineralizer, silica	1.5–2.5%	Sand substitute, filler materials, Pozzolan in cement, recovery of iron, mine backfill

RFCC: Resid Fluidized Catalytic Cracking; SAC: Spent Alumina Catalyst; FGD: Flue Gas Desulphurization

processed and utilized today. For enhancing the utilization of inorganic wastes such as steel slag, lead–zinc slag, copper slag, jarosite, red mud and SPL, by the Indian cement industry, technical guidance is required to be developed and commercial viability needs to be worked out on a case to case basis (Table 2).

References

AL-Dhamri, H., Melghit, K., Taha, R., & Ram, G. (2011) Utilization of by-product from petroleum refinery in Portland cement clinker and cement manufacturing. In *13th International Congress on the Chemistry of Cement*, Madrid, 3–8 July, 2011.

Singh A. K., Alka, M., & Kumar, S. (2011). Utilization of Spent Pot Liner (SPL) as a raw mix component in cement manufacturing. In *XIII ICCC, International Congress on the Chemistry of Cement*, Madrid, 3–8 July 2011.

BIS (Bureau of Indian Standards). (2015a). Composite cement—Specification, IS 16415: 2015.

BIS (Bureau of Indian Standards). (2015b). Portland Pozzolana cement—Specification, IS 1489 (Part 1): 2015.

BIS (Bureau of Indian Standards). (2016). Coarse and Fine Aggregate for Concrete, IS:383, New Delhi, January 2016.

BMTPC (Building Materials & Technological Promotion Council). (2016). Guideline for Utilization of C&D waste.

CEA (Central Electricity Authority). (2017). Report on fly ash generation at coal/lignite based thermal power stations and its utilization in the country for the year 2016–17, New Delhi, December 2017.

Chaturvedi, S. K., Yadav, D., Ali, M. M., Nanda, S., & Das, S. N. (2011). Potential of utilizing spent pot refractory lining waste from alumina smelter in cement manufacture. In *XIII ICCC, International Congress on the Chemistry of Cement*, Madrid, 3–8 July 2011.

CPCB. (2017). SOP and checklist of minimal requisite facilities for utilization of SPL generated from primary aluminium smelting industries. Ministry of Environment, Forest and Climate Change, Government of India, March 2017.

Deelwal, K., Dharavath, K., & Kulshreshtha, M. (2014). Evaluation of characteristic properties of red mud for possible use as a geo technical material in civil construction. *International Journal of Advances in Engineering and Technology*, *7*(3), 1053–1059.

Galbenis, C. T., & Tsimas, S. (2004). Applicability study of recycled aggregates as raw materials in cement clinker manufacturing. In *Advances in Mineral Resources Management and Environmental Geotechnology*, Hania, Greece.

Galbenis, C. T., & Tsimas, S. (2006). Use of construction and demolition wastes as raw materials in cement clinker production. *China Particuology, 04*(02), 83–85.

Gastaldi, D., Canonico, F., Capelli, L., Buzzi, L., Boccaleri, E., & Irico, S. (2015). An investigation on the recycling of hydrated cement from concrete demolition waste. *Cement and Concrete Composites*, (61), 29–35.

IEISL (IL&FS Environmental Infrastructure & Services Ltd.). (2017). C&D waste Project. Retrieved from http://www.cips.org.in. Accessed December 15, 2017.

Indian minerals yearbook. (2016). (Part III) Mineral Reviews 55th Edition, Marble, Govt. Of India, Ministry of Mines, Indian Bureau of Mines.

Indian Standard, Ordinary portland cement—Specification, IS 269: 2015.

Indian Standard, Specification for Masonry Cement, IS 3466: 1988.

Kara, M., Kilic, Y., & Erenoglu, T. (2017). An experimental study on construction and demolition waste usage as secondary raw material for cement production. *World Journal of Innovative Research, 2*(04), 01–07.

Lin, K. L., Lo, K.-W., Hung, M.-J., Cheng, T.-W., & Chang, Y.-M. (2017). Recycling of spent catalyst and waste sludge from industry to substitute raw materials in the preparation of Portland cement clinker. *Sustainable Environmental Research*, *27*, 251–257.

Liu, J., & Wang, Y. (2013). Cost analysis of construction and demolition waste management. *The Open Construction and Building Technology Journal*, 251–263.

Ministry of Mines. (2018). Indian Minerals Yearbook 2017 (Part—II: Metals & Alloys) (56th edn) Slag-Iron and Steel (Advance Release), Government of India, Indian Bureau of Mines, Indira Bhavan, Civil Lines, Nagpur-440 001, March 2018.

Mishra, C. R., Yadav, D., Sharma, P. S., & Ali, M. M. (2011). Production of Ordinary Portland Cement (OPC) from NALCO red mud. In Stephen J. Lindsay (Ed.), *Light Metals 2011, TMS (The Minerals, Metals & Materials Society)*.

Morrison, C., Hooper, R., & Lardner, K. (2003). The use of ferro-silicate slag from ISF zinc production as a sand replacement in concrete. *Cement and Concrete Research*, *33*(12), 2085–2089.

Mustafa, K. (2000). BCR-from byproduct to brick: Using red mud waste as a construction material. Retrieved from https://www.ceramicindustry.com/articles/88630-bcr-from-byproduct-to-brick-using-red-mud-waste-as-a-construction-material. Accessed July 30, 2018.

NBCC (National Buildings Construction Corporation Limited). (2013). Press Release, May 3, 2013.

Parhi, S. S. (2014). *Gainful utilization of spent pot lining—A hazardous waste from aluminium industry* (M. Tech., thesis). Chem Engineering, NIT, Rourkela, Odisha, August 2014.

Puertas, F., Díaz, I. G., Barba, A., Gazulla, M. F., Palacios, M., Gómez, M. P., & Ramírez, S. M. (2008). Ceramic wastes as alternative raw materials for Portland cement clinker production. *Cement and Concrete Composites,* (30), 798–805.

Report on disposal options of marble slurry in Rajasthan, 2011–12, CPCB Zonal Office (Central) Bhopal.

Samal, S., Ray, A. K., & Bandopadhyay, A. (2013) Proposal for resources, utilization and processes of red mud in India—A review. *International Journal of Mineral Processing, 118*, 43–55.

Sathya, M., & Shanmugavalli, B. (2014). Effect of using copper slag as a backfill in retaining wall. *International Journal of Research in Engineering and Technology, 03*(09).

Schoon, J., Buysser, K. D., Driessche, V., & Belie, N. D. (2015). Fines extracted from recycled concrete as alternative raw material for Portland clinker production. *Cement and Concrete Composites, 58*, 70–80.

Sengupta, B. (2014) Co-processing of waste in cement kilns, status and options to promote co processing for waste management. In *International Conference*, Wastech, 2014 organized by GSPCB at Mahatma Mandir, Gandhi Nagar on November 21, 2014.

Tiwari, N. K. (2017). Spent pot liner utilization in cement industry, presentation, BALCO, Korba, C.G. In *3rd International Conference on Alternate Fuel and Raw Material in Cement Industry*, New Delhi, March 23–24, 2017.

Tiwari, M. K., Bajpai, S., & Dewangan, U. K. (2016). Steel slag utilization—Overview in Indian perspective. *International Journal of Advanced Research, 4*(8), 2232–2246.

Tixier, R., Devaguptapu, R., & Mobasher, B. (1997). The effect of copper slag on the hydration and mechanical properties of cementitious mixtures. *Cement and Concrete Research, 27*(10), 1569–1580.

Wu, W., Zhang, W., & Ma, G. (2010). Optimum content of copper slag as a fine aggregate in high strength concrete. *Materials & Design, 31*(6), 2878–2883.

A Review of Studies on Environmental Performance Analysis of Construction and Demolition Waste Management using Life Cycle Assessment

Kishore C. Kumar, V. G. Ram and Satyanarayana N. Kalidindi

Abstract Construction and demolition (C&D) activities generate a large amount of waste throughout their life cycle. Demolition of buildings is on the rise owing to rapid urbanization and dynamic demands. Disposal of C&D waste leads to faster depletion of available landfill space. The environmental performance of C&D waste management needs to be analysed from a life cycle perspective to identify better waste management principles. A selective list of research papers from C&D waste literature, which applied life cycle assessment (LCA) methodology has been reviewed in this paper. The primary aim of this review article is to identify the commonality among those research papers and understand the environmental impacts of various practices being adopted in C&D waste management. LCA case studies were found to vary widely on several categories, and we inferred that they could be compared by selecting normalized unit processes in the system boundaries, functional unit, inventory data sources and impact categories. The environmental impact generated due to the operation of C&D waste recycling facility was reported to be negligible in most of the papers. Transportation of C&D waste was reported to cause the highest environmental impact in most of the articles. Recycling of C&D waste has been recommended as a sustainable alternative to landfilling, and large benefits emerge from savings in valuable land space. Furthermore, energy savings of up to 50% and emission reduction in the range of 7–10 times occur due to recycling of recovered materials, especially metals.

Keywords C&D waste · Life cycle assessment (LCA) · Waste management · Recycling

K. C. Kumar (✉) · S. N. Kalidindi
Indian Institute of Technology Tirupati, Tirupati, India
e-mail: tcskishore@gmail.com

S. N. Kalidindi
e-mail: satyakn@iittp.ac.in

V. G. Ram
Indian Institute of Technology Madras, Chennai, India
e-mail: ramkrithik@gmail.com

S. K. Ghosh (ed.), *Urban Mining and Sustainable Waste Management*,
https://doi.org/10.1007/978-981-15-0532-4_5

1 Introduction

Construction and demolition waste comprise "building materials, debris and rubble resulting from construction, re-modelling, repair and demolition of any civil structure" (Ministry of Environment, Forest and Climate change 2016). C&D waste constitutes about 36% of total solid waste generated in the city of Chennai in the year 2013, and approximate estimation of C&D debris generated was found to equate to 175 kg per capita/year (Ram and Kalidindi 2017). Frequently, C&D waste mixes with municipal solid waste and enters landfill and being inert in nature, they end up occupying valuable landfill space. Lack of proper guidelines and recycling facilities has led to illegal dumping along roadsides, river bed and low-lying areas.

On the other hand, increasing urbanization and economic development create a huge demand for infrastructure, housing and other goods and services, which aggravates resource depletion. The Government of India under the Pradhan Mantri Awas Yojana (Housing for all by 2022) scheme requires the construction of 2 crores dwelling units which adds on to the already existing burden on construction materials. Since residential buildings comprise 67% coarse and fine aggregates by weight (Devi and Palaniappan 2014), the burden on coarse and fine aggregates will increase substantially. In cities like Bangalore, fine aggregates are transported to distances in the range of 70–100 km (Venkatarama Reddy and Jagadish 2003) due to shortage of raw material. In order to avoid large transportation costs, illegal mining occurs in various parts of the country. Several states in India issued a ban on aggregate mining that caused a severe bottleneck in construction and major infrastructure projects leading to the use of alternate building materials like M-sand extracted from granite.

Efficient use of resources and promotion of secondary materials will help in meeting the country's commitment in Paris global accords (2015), by significantly reducing the emission of greenhouse gas. Recycling of C&D waste and substituting with natural quarried aggregates have the potential to solve multiple problems of society. On 28 June 2012, Indian cities with more than 1 million population were asked to set up C&D waste recycling facilities by the Ministry of Urban Development. Further, as a key objective of Swachh Bharat Mission, the GoI aims to process 100% solid waste (including C&D waste) generated in cities/towns by 2 October 2019. Thus, there is an increasing need by public authorities to pursue sustainable waste management solutions which lessen the environmental burden arising due to waste generation and landfilling.

Life cycle assessment commonly known as LCA is used to analyse environmental impacts, benefits of product, process or system along its complete life cycle from cradle to grave (ISO 14040 2006a; ISO 14044 2006b). LCA studies also provide quantitative data that supports decision-making process while considering alternative scenarios for improving environmental performance (Häkkinen 1994).

The overall aim of this article is to review research papers pertaining to LCA studies on C&D waste management and identify commonality in results among studies conducted across the globe. This will help us understand the application of LCA methodology as well as frame guidelines for future LCA studies.

2 Review of LCA Methodologies Adopted in C&D Waste Management Literature

LCA results vary widely due to subjective boundary conditions, differences in methodology, spatial and temporal variations and due to lack of reliable data (Crawford 2011). To reduce high variation in results and to compare similar study results, the methodology proposed by ISO 14040-14044 (2006a, b) needs to be followed. From C&D waste management literature, only those articles, which analysed the environmental performance of C&D waste management by ISO specified LCA methodology, were chosen in this review article. A total of 15 peer-reviewed research papers were found to qualify the chosen criteria, and the papers included were from journals such as "Building and Environment", "Journal of Cleaner Production", "Resources, Conservation and Recycling", "The International Journal of Life Cycle Assessment", "Waste Management" and "Waste Management and Research".

According to ISO 14040-14044 (2006a, b) framework, LCA involves four phases, namely (1) Goal and scope definition, (2) Life cycle inventory analysis, (3) Life cycle impact assessment and (4) Life cycle interpretation.

2.1 *Goal and Scope Definition*

The goal and scope definition phase aims to identify the objective of study and its application, fix the system boundaries and functional unit, identify sources of data for the study and list the assumptions and limitations (ISO 14040 2006a). Initial LCA studies were aimed to assess recycling potential in building demolition (Blengini 2009), to find the best scenario among recycling, landfilling and incineration (Ortiz et al. 2010), to identify the best C&D waste management principles (Mercante et al. 2012; Vossberg et al. 2014) and to analyse the environmental impact caused due to operation of a recycling facility (Coelho and Brito 2013). Recent studies focus on performing comparative analysis between natural aggregates (NA) and recycled aggregates (RA) (Blengini and Garbarino 2010; Faleschini et al. 2016; Hossain et al. 2016; Mah et al. 2018; Rosado et al. 2017; Penteado and Rosado 2016).

All reviewed articles performed cradle to grave analysis of C&D waste. While the cradle is chosen as C&D waste after demolition excluding demolition energy and emissions, grave for each study varied based on how the end-of-life of C&D waste was assumed. C&D waste is either landfilled (as shown in Fig. 1) or sent to a recycling facility (as shown in Fig. 2). A majority of reviewed articles also

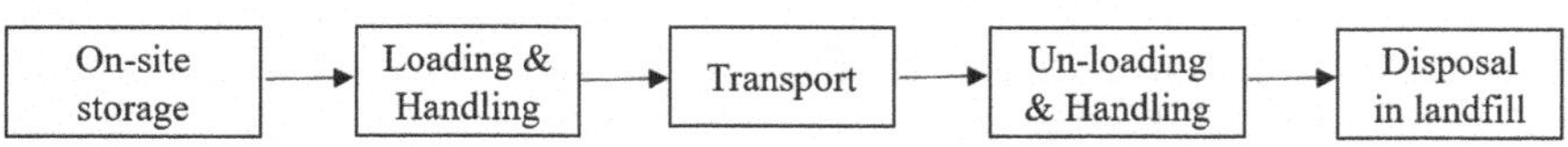

Fig. 1 Flowchart and system boundaries of C&D waste dumping

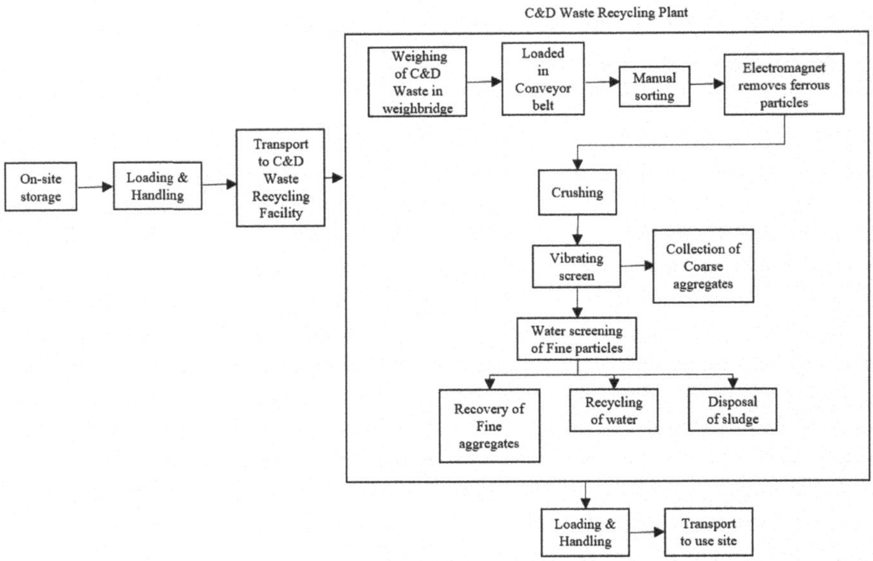

Fig. 2 Flowchart and system boundaries of C&D waste recycling

considered the avoided burdens of landfilling and avoided burdens of production of virgin materials. One ton or 1 m^3 of C&D waste was chosen as the functional unit. While allocation by mass is the most commonly adopted allocation procedure, Blengini and Garbarino (2010) adopted allocation by revenue. Results of LCA studies vary widely due to differences in goal and scope definition phase, and hence, this phase is considered crucial.

2.2 Life Cycle Inventory Analysis (LCI)

The LCI phase quantifies all environmental inputs (energy and raw material) and outputs (emission to air, water and soil) of the unit processes in a complete life cycle (ISO 14040 2006a). Primary data and secondary (or upstream) data are two kinds of data required for forming a complete LCI. Majority of the reviewed articles collected primary data through site visits, questionnaires and interviews. Secondary data is usually obtained from literature, country-specific online database or from commercial LCI inventory databases such as Ecoinvent (2008) and USLCI database (2012).

The commonly collected input and output data include energy and emissions from handling, transportation and dumping the C&D waste as well as rejects/refuse in a sanitary landfill. Further, electricity consumption of recycling facility (which includes steel, wood and aggregate recycling), gasoline required to power engines, oil and other lubricants and water used for dust control and emissions arising from the facility are also collected to form a complete LCI. The emissions arising due to the disposal of rejects/refuse in a sanitary landfill and the corresponding energy required are also considered. While energy embodied in the development of infrastructure and maintenance is rarely included due to lack of reliable data, emission due to leachate production from C&D waste is excluded in most of the reviewed articles owing to its inert nature.

2.3 Life Cycle Impact Assessment (LCIA)

The LCIA phase aims to evaluate the magnitude and significance of all environmental impacts using the results from LCI phase. LCIA phase consists of mandatory and optional stages (ISO 14040 2006a). The mandatory stage includes selection of impact categories, category indicators and characterization models, assigning inventory results to the selected categories (classification) and calculation of each category indicator results (characterization). Each category indicator results are calculated in relative to the reference information (normalization). The impact categories are sorted and ranked (grouping), and the aggregated indicator results are weighed with numerical values based on value choices before mapping them to the impact categories (weighting). Last three stages are optional, according to ISO guidelines.

Two approaches are available in LCIA, the problem-oriented mid-point approach and the damage oriented end point approach. The mid-point LCIA methods chosen in the reviewed articles are CML (Guinée et al. 2002), CML 2 baseline 2000 (CML 2001), Cumulative Energy Demand (CED) (VDI 1997), EDIP (Wenzel et al. 1997), Impact 2002+ (Jolliet et al. 2003) and IPCC (2007). The endpoint LCIA methods include Eco-indicator99 (Goedkoop and Spriensma 2000) and Impact 2002+ (Jolliet et al. 2003). Table 1 shows the impact assessment methods used in the literature.

Since IMPACT 2002+ method by Jolliet et al. (2003) merges the benefits of three LCIA methods (i.e. Eco-indicator 99, IPCC and CML), and recent LCA studies like Blengini and Garbarino (2010), Hossain et al. (2016), Rosado et al. (2017) and Vitale et al. (2017) have adopted this method as the impact assessment method. The Impact 2002+ method consists of 15 mid-point indicators: "human toxicity (carcinogens and non-carcinogens), respiratory effects, ionizing radiation, ozone layer depletion, photochemical oxidation, aquatic ecotoxicity, terrestrial ecotoxicity, terrestrial acidification/nitrification, aquatic acidification, aquatic eutrophication, land occupation, global warming, non-renewable energy, mineral extraction". These are mapped to four damage (end point) categories: "human health, ecosystem quality, climate change and resources" (Jolliet et al. 2003).

Table 1 Impact assessment categories

Author	Impact assessment methodology/category
Blengini (2009)	Eco-indicator 99
Blengini and Garbarino (2010)	IMPACT 2002+ and Eco-Indicator 99
Ortiz et al. (2010)	CML 2 baseline 2000
Mercante et al. (2012)	CML 2002
Coelho and Brito (2013)	Primary energy and CO_2 emission
Simion et al. (2013)	Eco-indicator 99, EDIP/UMIP, CED method.
Vossberg et al. (2014)	CED method and GWP by IPCC 2007
Hossain et al. (2016)	IMPACT 2002+
Penteado and Rosado (2016)	CML 2 baseline 2000
Faleschini et al. (2016)	CML 2002
Rosado et al. (2017)	IMPACT 2002+
Vitale et al. (2017)	IMPACT 2002+
Wang et al. (2018)	BEPAS
Mah et al. (2018)	IPCC GWP 2013 100a
Borghi et al. (2018)	ILCD 2011 and CED method

2.4 Interpretation and Results

The interpretation phase evaluates findings of the inventory analysis and the impact assessment based on the pre-defined goal and scope to find environmental hotspots and to recommend decision-makers (ISO 14040 2006a). This stage also consists of performing completeness, consistency and sensitivity checks to ascertain the reliability of data and reduce uncertainties (ISO 14044 2006b). Sensitivity analysis is the most commonly adopted check in the studied literature. It determines how changes in data, data allocation methods and methodological choices affect the findings from LCIA phase. From the research papers, transportation distance is identified to be the most commonly studied parameter in sensitivity analysis. Table 2 shows the list of parameters considered in sensitivity analysis in the literature.

3 Review of Findings from Process and Comparative LCA Studies

When a complete life cycle assessment is performed, recycling of building waste is found to be feasible from an environment and energy point of view (Blengini 2009).

Table 2 Sensitivity analysis considered in reviewed literature

Author	Parameter considered in sensitivity analysis
Blengini (2009)	Variation in data source of building materials
Blengini and Garbarino (2010), Vossberg et al. (2014), Hossain et al. (2016), Rosado et al. (2017), Mah et al. (2018)	Variation in average transportation distance
Simion et al. (2013)	Application of different end point impact assessment methods
Faleschini et al. (2016)	Application of different transportation scenarios
Penteado and Rosado (2016)	Variation in percentage of waste sent to landfilling and recycling
Vitale et al. (2017)	Variation in percentage of amount of waste, steel and inert material sent to recycling
Borghi et al. (2018)	Variation in transportation distance and replacement co-efficient

Ortiz et al. (2010) found that recycling is an environment-friendly option compared to landfilling and incineration. Onsite recycling showed definitive environmental benefits over landfilling (Vossberg et al. 2014). Recycling of C&D waste, instead of dumping can save valuable landfill space, and there is a substantial increase in end-of-life phase performance of the building under study when the higher quantity of C&D waste is sent to recycling (Vitale et al. 2017). Though most of the studies reported C&D waste recycling to have environmental benefits compared to dumping or virgin aggregate production, it was also reported by Mercante et al. (2012) and Borghi et al. (2018) that C&D waste recycling is not always beneficial. Wang et al. (2018) quantified the environmental benefits using willingness to pay (WTP) approach and recycling 1 ton of demolition waste is found to generate ¥1.21 of environmental benefits.

C&D waste recycling has significant environmental benefits over processing of virgin materials (Faleschini et al. 2016; Hossain et al. 2016; Rosado et al. 2017; Penteado and Rosado 2016; Simion et al. 2013). Replacing virgin aggregates with recycled aggregates leads to energy savings of 17% according to Simion et al. (2013) and 50% according to Hossain et al. (2016). Recycling of C&D waste generates seven times lower CO_2 equivalent emissions compared to crushed stone (Simion et al. 2013), and CO_2 equivalent emissions of almost ten times can be prevented by replacing virgin materials with recycled materials from the C&D waste processing facility (Coelho and Brito 2013).

While transportation caused the highest environmental impact mainly in carbon emissions (Coelho and Brito 2013; Mah et al. 2018; Mercante et al. 2012), highest environmental credits were due to avoided landfill space (Blengini and Garbarino 2010; Wang et al. 2018) and metal recycling (Vitale et al. 2017). Thus, a majority of the sensitivity analysis found in the literature focused on addressing variation in transportation distance. Sensitivity analysis by Borghi et al. (2018) verified that transportation of C&D waste causes the highest environmental impacts. Recycling is preferable to landfilling when the transportation distance from generators of C&D waste and recycling facility is within 30 km (Penteado and Rosado 2016), and if uncertainties in data are considered, the transportation distance reduces to only 6.5 km in order for off-site recycling to have benefits over landfilling (Vossberg et al. 2014). However, recycled aggregates cause lower impacts than natural aggregates (except non-carcinogens) only if RA transportation distance is 20 km more than NA (Rosado et al. 2017). Furthermore, transportation distance can be increased by 45 km without causing additional impacts in GWP category by the use of photovoltaic energy in recycling plants (Faleschini et al. 2016). On the contrary, the benefits of C&D waste recycling were found to be outweighed only if the transportation distance of recycled aggregates increases by a factor of 2 or 3 (Blengini and Garbarino 2010). Also, the net impacts were increased by only 12% when transportation distance increased by 20% (Hossain et al. 2016). Since transportation data plays a vital role, GIS data was used by Blengini and Garbarino (2010) for precise calculation of transportation distance in the city of Turin, Italy and found that 13 out of 14 impact categories according to IMPACT 2002+ LCIA method showed environmental benefits.

4 Concluding Remarks

The articles chosen for review mainly focused on general C&D waste management, comparing various end-of-life scenarios of C&D waste and comparing recycled aggregates with its equivalent obtained from natural sources. Though all the identified papers followed ISO 14040-44 (2006a, b) guidelines, direct comparison of results was not possible. This is due to improper documentation of assumptions, variation in considering the functional unit and system boundaries, differences in primary and secondary data sources and selection of impact assessment categories. Although recycling of C&D waste was found to cause lower environmental impacts in the majority of impact categories, there is a necessity to maintain least transportation distance as transportation was reported to cause the highest environmental impact in most of the articles. Large benefits emerge from savings of valuable land space due to recycling of C&D waste instead of landfilling. Furthermore, the importance of recycling recovered materials, especially metals, was identified as they lead to energy savings of up to 50% and emission reduction in the range of 7–10 times.

References

Blengini, G. A. (2009). Life cycle of buildings, demolition and recycling potential: A case study in Turin, Italy. *Building and Environment, 44,* 319–330.

Blengini, G. A., & Garbarino, E. (2010). Resources and waste management in Turin (Italy): The role of recycled aggregates in the sustainable supply mix. *Journal of Cleaner Production, 18,* 1021–1030.

Borghi, G., Pantini, S., & Rigamonti, L. (2018). Life cycle assessment of non-hazardous construction and demolition waste (CDW) management in Lombardy Region (Italy). *Journal of Cleaner Production, 184,* 815–825.

CML (Centre for Environmental Studies). (2001). University of Leiden, CML 2 baseline method. http://www.leidenuniv.nl/cml/index.html.

Coelho, A., & de Brito, J. (2013). Environmental analysis of a construction and demolition waste recycling plant in Portugal—Part I: Energy consumption and CO_2 emissions. *Waste Management, 33,* 1258–1267.

Crawford, R. H. (2011). *Life cycle assessment in the built environment.* United Kingdom, Abingdon, Oxon: Spon Press.

Devi, P., & Palaniappan, S. (2014). A case study on life cycle energy use of residential building in Southern India. *Energy and Buildings, 80,* 247–259.

Ecoinvent. (2008). The life cycle inventory data version 2. Swiss centre for life cycle inventories.

Faleschini, F., Zanini, M. A., Pellegrino, C., & Pasinato, S. (2016). Sustainable management and supply of natural and recycled aggregates in a medium-size integrated plant. *Waste Management, 49,* 146–155.

Goedkoop, M., & Spriensma, R. (2000). *The Ecoindicator'99: A damage oriented method for life cycle impact assessment: Methodology report.* Amersfoort, The Netherlands: Pré Consultants BV.

Guinée, J. B., Gorrée, M., Heijungs, R., Huppes, G., Kleijn, R., Koning, A., et al. (2002). *Handbook on life cycle assessment. Operational guide to the ISO standards.* Dordrecht, The Netherlands: Kluwer Academic Publishers.

Häkkinen, T. (1994). *Environmental impact of building materials* (Report No. 1590). Espoo: Technical Research Centre of Finland (VTT), p. 38.

Hossain, M. U., Poon, C. S., Lo, I. M. C., & Cheng, J. C. P. (2016). Comparative environmental evaluation of aggregate production from recycled waste materials and virgin sources by LCA. *Resources, Conservation and Recycling, 109,* 67–77.

IPCC. (2007). Intergovernmental panel for climate change. http://www.ipcc.ch.

ISO 14040. (2006a). Environmental management—Life cycle assessment—Principles and framework.

ISO 14044. (2006b). Environmental management—Life cycle assessment—Requirements and guidelines.

Jolliet, O., Margni, M., Charles, R., Humbert, S., Payet, J., Rebitzer, G., et al. (2003). IMPACT 2002+ : A new life cycle impact assessment methodology. *International Journal of Life Cycle Assessment, 8,* 324–330.

Mah, C. M., Fujiwara, T., & Ho, C. S. (2018). Life cycle assessment and life cycle costing toward eco-efficiency concrete waste management in Malaysia. *Journal of Cleaner Production, 172,* 3415–3427.

Mercante, I. T., Bovea, M. D., Ibáñez-Forés, V., & Arena, A. P. (2012). Life cycle assessment of construction and demolition waste management systems: A Spanish case study. *International Journal of Life Cycle Assessment, 17,* 232–241.

Ministry of Environment Forest and Climate Change. (2016). Construction and demolition waste management rules. G.S.R. 317 (E). The Gazette of India, Part-II, Section-3, Sub-section 317.

Ortiz, O., Pasqualino, J. C., & Castells, F. (2010). Environmental performance of construction waste: Comparing three scenarios from a case study in Catalonia, Spain. *Waste Management, 30,* 646–654.

Penteado, C. S. G., & Rosado, L. P. (2016). Comparison of scenarios for the integrated management of construction and demolition waste by life cycle assessment: A case study in Brazil. *Waste Management Research, 34,* 1026–1035.
Ram, V., & Kalidindi, S. N. (2017). Estimation of construction and demolition waste using waste generation rates in Chennai, India. *Waste Management Research, 35,* 610–617.
Rosado, L. P., Vitale, P., Penteado, C. S. G., & Arena, U. (2017). Life cycle assessment of natural and mixed recycled aggregate production in Brazil. *Journal of Cleaner Production, 151,* 634–642.
Simion, I. M., Maria, E. F., Bonoli, A., & Gavrilescu, M. (2013). Comparing environmental impacts of natural inert and recycled construction and demolition waste processing using LCA. *Journal of Environmental Engineering Landscape Management, 21,* 273–287.
USLCI Database. (2012). U.S. Life Cycle Inventory Database. National Renewable Energy Laboratory.
VDI. (1997). Cumulative energy demand—Terms, definitions, methods of calculation. In VDI-Richtlinien 4600. Verein Deutscher Ingenieure, Düsseldorf.
Venkatarama Reddy, B. V., & Jagadish, K. S. (2003). Embodied energy of common and alternative building materials and technologies. *Energy and Buildings, 35,* 129–137.
Vitale, P., Arena, N., Di Gregorio, F., & Arena, U. (2017). Life cycle assessment of the end-of-life phase of a residential building. *Waste Management, 60,* 311–321.
Vossberg, C., Mason-Jones, K., & Cohen, B. (2014). An energetic life cycle assessment of C&D waste and container glass recycling in Cape Town, South Africa. *Resources, Conservation and Recycling, 88,* 39–49.
Wang, T., Wang, J., Wu, P., Wang, J., He, Q., & Wang, X. (2018). Estimating the environmental costs and benefits of demolition waste using life cycle assessment and willingness-to-pay: A case study in Shenzhen. *Journal of Cleaner Production, 172,* 14–26.
Wenzel, H., Hauschild, M., & Alting, L. (1997). *Environmental assessment of products. Methodology, tools and case studies in product development*: *Vol. 1.* United Kingdom: Chapman & Hall; Hingham, MA, USA: Kluwer Academic Publishers.

Extraction of Selected Metals from High-Grade Waste Printed Circuit Board Using Diethylene Triamine Penta-acetic Acid

Auchitya Verma, Amber Trivedi and Subrata Hait

Abstract Electronic waste (e-waste) has emerged as one of the fastest-growing waste streams worldwide. Printed circuit board (PCB), the fundamental component of e-waste, is considered as a secondary resource reservoir due to its rich metallic content including base, toxic, and precious metals. E-waste recycling is necessitated for waste treatment for environmental protection as well as metal recovery for economic development. Owing to the shorter leaching time and high extraction efficiency, the chemical leaching of metals has gained momentum for metal recovery from e-waste. However, various lixiviants including strong inorganic acids and ligands used in the chemical leaching are harmful and persistent in the environment with residual effect. In this context, the present study was carried out to assess the chemical leaching of selected metals, i.e. Cu, Zn, and Ni from high-grade PCB of obsolete computer using diethylene triamine penta-acetic acid (DTPA) as an eco-friendly lixiviant with concentration varying from 0.3 to 0.7 M over a range of pH (5–9) and liquid-to-solid (L/S) ratios (10–100) at a temperature of 20 °C and mixing speed of 450 rpm. At an L/S ratio of 50 and the comminution fines in the particle size range between 0.038 and 1 mm, the maximum leaching of more than 99% each of Cu and Zn and around 81% Ni was observed in five days using 0.5 M DTPA at pH of 9. Efficient metal extraction from waste computer PCB using DTPA can be attributed to its chelating effect. These findings highlight the potential of efficient chemical leaching of metals from e-waste using DTPA as an eco-friendly organic chelator.

Keywords Printed circuit board · Metals · Chemical leaching · DTPA · Chelating effect

A. Verma · A. Trivedi · S. Hait (✉)
Department of Civil and of Environmental Engineering, Indian Institute of Technology Patna, Patna, Bihar 801106, India
e-mail: shait@iitp.ac.in

S. K. Ghosh (ed.), *Urban Mining and Sustainable Waste Management*,
https://doi.org/10.1007/978-981-15-0532-4_6

1 Introduction

With the unremitting advancement in modern technology and lifestyle, the day to day usage of electrical and electronic equipment (EEE) has increased quite significantly. These equipment have penetrated every aspect of our life. With technological innovations taking place so rapidly, EEE becomes obsolete at even greater pace leading to the massive generation of waste electrical and electronic equipment (WEEE) or simply electronic waste (e-waste). Due to the continual consumer demand for the EEE associated with their shorter life cycle, e-waste is being generated at an unprecedented rate worldwide. As per a report published by the United Nations University (UNU) in 2017, about 44.7 million metric tonne (Mt) of e-waste generation was reported worldwide in 2016. The worldwide e-waste generation is estimated to go up by 52 Mt by 2021(Baldé et al. 2017). A large part of the worldwide e-waste generation comes from the developed and the emerging economies. EEE is an assemblage of various components among which printed circuit board (PCB) is an integral part of them. PCBs are the base of the EEE as they mechanically support and electrically connect microelectronic components (Li et al. 2004). PCBs contain valuable as well as toxic metals which require special treatment and management to mitigate the hazardous impacts on the environment and human health (Priya and Hait 2017). A typical composition of PCB consists of around 70–72% non-metallic constituents, viz. polymers, plastics and ceramics and around 28–30% metallic constituents (Li et al. 2004; Zhou and Qiu 2010). Metallic constituents in PCBs include 20–25% Cu, 2–4% Al and other metals like Zn, Ni, and Pb of around 1% along with precious metals like Ag, Au, and Pt and thereby virtually making them as a potential secondary metal reservoir (Creamer et al. 2006; Huang et al. 2009; Li et al. 2007; Priya and Hait 2017; Wang et al. 2005; Zhao et al. 2004). However, the presence of a few elements like mercury, arsenic, lead, and cadmium in the PCB makes it toxic and hazardous. Inappropriate handling and disposal of e-waste poses notable human and environmental health risks and loss of a plethora of valuable metals. Therefore, the recycling of e-waste has become an immediate necessity from the point of view of the environmental hazards as well as the resources present in it.

Several methods based on pyrometallurgy and hydrometallurgy are currently used for the recovery of metals from waste PCBs. Owing to the ease of operation, control over the process, shorter leaching time and high extraction efficiency, the hydrometallurgical process involving chemical leaching is more common for e-waste recycling. PCBs are dismantled and numerous acidic and basic chemical solutions such as sulfuric acid, hydrochloric acid, nitric acid, aqua regia, thiourea, etc., are employed to leach out metals from the PCB (Choubey et al. 2015; Huang et al. 2009; Li et al. 2007; Pant et al. 2012; Silvas et al. 2015). However, these methods release acid and other hazardous chemicals in the form of waste effluent which pollutes groundwater and if not disposed of properly, can cause secondary pollution and severe problems to human health. These impediments of the hydrometallurgical process can be overcome by employing safer and non-toxic chemicals such as organic acids and chelators. Jadhao et al. (2016) used one such organic acid, i.e. ethylene diamine tetra-acetic acid (EDTA) to extract copper from waste computer PCB. However, there is still a scope for research to develop greener chemical leaching technique using other organic acids and chelators to minimize secondary pollution. In this context, the present study explores the feasibility and application of diethylene triamine penta-acetic acid (DTPA) as an organic chelator to leach out selected metals, viz. Cu, Zn, and Ni from waste computer PCB and various process parameters are experimentally optimized.

2 Materials and Methods

2.1 E-waste

E-waste used in this study is in the form of PCB scraps from the used computer procured from the local repair shops and vendors in Bihta, Patna, Bihar, India. For the purpose of sample preparation, mounted electrical components such as capacitors, USB ports, resistors, etc., were manually removed from the PCB by the use of plier, screwdriver, and hammer. The size of PCB was reduced to approximately 2 cm × 2 cm before putting it into the milling machine using pliers and scissors. The PCB sample was then mechanically comminuted for the further size reduction using the cutting mill (SM200, Retsch GmbH, Germany). Fractionation of the sample obtained after comminution was accomplished using the sieve shaker (AS200, Retsch GmbH, Germany) to the size range of 0.038–1 mm. Upon chemical characterization, the comminuted PCB was then subjected to chemical leaching using DTPA at different molar concentrations, pH and L/S (liquid-to-solid) ratio at a temperature of 20 °C and mixing speed of 450 rpm.

2.2 Chemical Characterization of Comminuted PCB

A representative subset of the comminuted and homogenized computer PCB in the particle size range between 0.038 and 1 mm was chemically characterized to obtain the percentage content of selected metals, viz. Cu, Zn, and Ni. The content of selected metals in PCB was determined using the inductively coupled plasma mass spectrometry (ICP-MS) (7800, Agilent, USA) upon microwave-assisted acid digestion following the USEPA 3052 procedure (USEPA 1996). The content of Cu, Zn, and Ni in the waste PCB of computer in the present study was determined as 21.50%, 0.10%, and 0.08%, respectively.

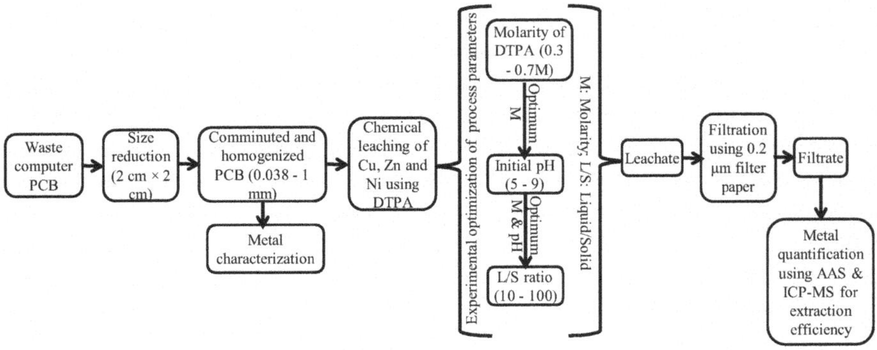

Fig. 1 Experimental scheme for chemical leaching of metals from waste computer PCB using DTPA

2.3 Chemical Leaching

Batch study was conducted for the chemical leaching of selected metals, viz. Cu, Zn, and Ni from the comminuted and homogenized PCB in the particle size range between 0.038 and 1 mm using analytical grade DTPA (Merck, Germany) in a series of 250 mL covered conical flasks (Make: Borosil) employing a temperature-controlled incubated shaker (SIF-5000R, Jeio Tech, South Korea). The schematic flowchart depicting every step of chemical leaching procedure is presented in Fig. 1. In order to investigate the impact of various process parameters, the effects of molar concentration of DTPA (0.3–0.7 M), pH (5–9), and L/S ratio (10–100) on the extraction efficiency of selected metals from waste PCB of computer were sequentially assessed at a temperature of 20 °C and mixing speed of 450 rpm. The leachate samples were withdrawn at different interval till almost asymptotic trend in metal extraction was achieved. All the chemical leaching experiments were conducted in triplicate and the metal extraction results are presented as average values.

2.4 Analytical Determination

Waste PCBs were chemically characterized for the content of selected metals after the sample preparation using microwave-assisted acid digestion procedure as per the USEPA 3052 method (USEPA 1996) using ICP-MS (7800, Agilent, USA). The leaching solutions were analyzed at regular interval for the concentration of the selected metals using AAS (iCE3500, Thermo Fisher, USA) and ICP-MS (7800, Agilent, USA). Distilled, de-ionized water (Millipore) was used for the purpose of reagent preparation and sample dilution. All the analyses were performed in triplicate for better accuracy.

3 Results and Discussion

3.1 *Effect of Molar Concentration of DTPA on the Extraction Efficiency of the Selected Metals*

Extraction of metals using chemical leaching is heavily affected by the concentration of acid used. Due to very less solubility of DTPA (<5 g/L) (Martell and Smith 2003), the concentration of DTPA can be increased up to a certain level only. Molar concentration of DTPA was varied from 0.3 to 0.7 M to assess its impact on the leaching efficiency of selected metals, i.e. Cu, Zn, and Ni from waste computer PCB keeping other process parameters constant at pH of 7, L/S ratio of 50, temperature of 20 °C and mixing speed of 450 rpm. The effect of molar concentration of DTPA on the leaching efficiency of selected metals is depicted in Fig. 2.

It is clear from Fig. 2 that the leaching efficiency obtained with 0.5 M DTPA is better than other molar concentrations. The leaching efficiency of Cu, Zn, and Ni increased from around 68%, 58%, and 38% to 83%, 60%, and 42%, respectively, at 7 days as the molar concentration of DTPA increased from 0.3 to 0.5 M (Fig. 2).

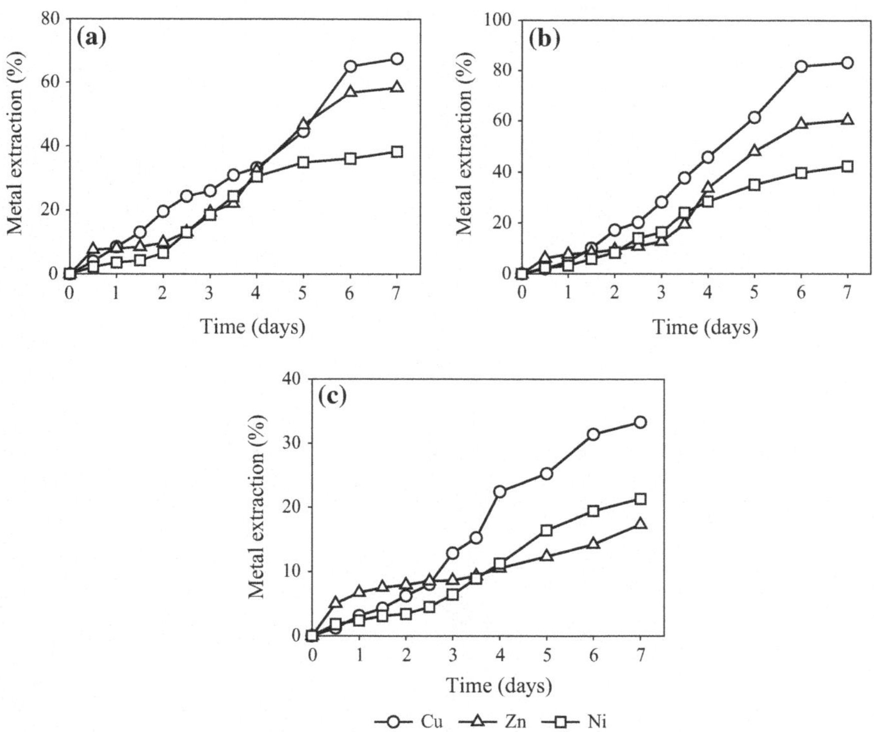

Fig. 2 Effect of molar concentration of DTPA, (**a**) 0.3 M, (**b**) 0.5 M, and (**c**) 0.7 M on the extraction of selected metals from waste computer PCB

However, the leaching efficiency of Cu, Zn, and Ni drastically dropped to 33%, 17%, and 21%, respectively, at 7 days with a further increase in the molar concentration of DTPA to 0.7 M. Therefore, 0.5 M concentration of DTPA was considered as the optimum value to perform subsequent leaching experiments.

3.2 *Effect of Reaction pH on the Extraction Efficiency of the Selected Metals*

Ligands with chelating properties like DTPA are not soluble at its natural pH and that is why the pH plays an important role in the reactions involving these ligands. In this study, NaOH was used to dissolve DTPA and to maintain the desired solution pH. In order to investigate the influence of solution pH on the leaching of selected metals, i.e. Cu, Zn and Ni, the experiments were carried out at acidic (pH = 5), neutral (pH = 7), and alkaline (pH = 9) conditions using 0.5 M DTPA, L/S ratio of 50, temperature of 20 °C, and mixing speed of 450 rpm. Figure 3 presents the effect of reaction pH on the leaching of selected metals from waste computer PCB.

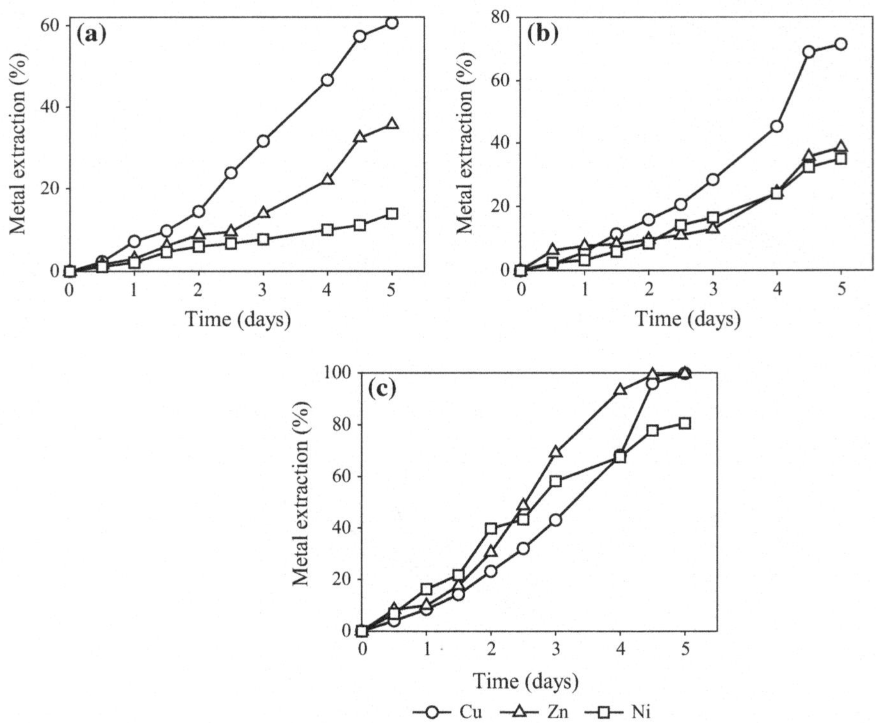

Fig. 3 Effect of reaction pH of (**a**) 5, (**b**) 7, and (**c**) 9 on the extraction of selected metals from waste computer PCB using DTPA

Results showed that DTPA performed better in alkaline condition rather than acidic and neutral conditions. Leaching efficiency of more than 99% each of Cu and Zn and around 81% for Ni was achieved in five days at the alkaline condition (pH = 9) (Fig. 3). Leaching efficiency decreased to around 71% for Cu, 39% for Zn, and 35% for Ni when the reaction pH value was lowered to 7 (neutral condition). At acidic condition (pH = 5), the metal leaching efficiency was found to be the poorest. Hence, the alkaline condition (pH = 9) was considered as the favorable to carry out further metal leaching experiments using DTPA.

3.3 Effect of L/S Ratio on the Extraction Efficiency of the Selected Metals

The liquid-to-solid (L/S) ratio also plays a vital role in the extraction efficiency of metals using chemical leaching as it signifies the chance of solid to interact with the medium. The leaching experiments were conducted at three different L/S ratios i.e. 10, 50, and 100 using 0.5 M DTPA, pH of 9, temperature 20 °C, and mixing speed of 450 rpm. The effect of the L/S ratio on the extraction efficiency of selected metals using DTPA is presented in Fig. 4. The leaching efficiency of Cu, Zn, and Ni increased from around 25, 35, and 19% to more than 99% each for Cu and Zn and around 81% for Ni in five days when the L/S ratio was increased from 10 to 50 (Fig. 4). This increasing trend of metal extraction with the increase in L/S ratio can be contributed to the fact that the viscosity of the slurry is expected to decrease with the increase in L/S ratio facilitating better mixing and thereby reducing the diffusional mass transfer resistance (Rao et al. 2015). However, with the further increase in the L/S ratio to 100, the leaching efficiency of Cu, Zn, and Ni was observed to marginally decrease to around 90%, 83%, and 75%, respectively, at five days. Therefore, an L/S ratio of 50 was found to be the suitable for metal extraction from waste computer PCB using DTPA in the present study.

4 Conclusions

The present feasibility study on chemical leaching of selected metals, viz. Cu, Zn, and Ni from waste computer PCB has demonstrated the possibility of using an organic chelator like DTPA for the efficient metal recovery from e-waste instead of concentrated inorganic acids. Leaching efficiency of more than 99% each of Cu and Zn and around 81% of Ni in five days was achieved using DTPA in the present study. The experimentally optimized conditions for the leaching of the selected metals from waste computer PCB using DTPA were found to be: 0.5 M DTPA, pH of 9, and L/S ratio of 50 at a temperature of 20 °C and mixing speed of 450 rpm. Use of DTPA as an organic chelator can be deliberated as a green chemical process for efficient

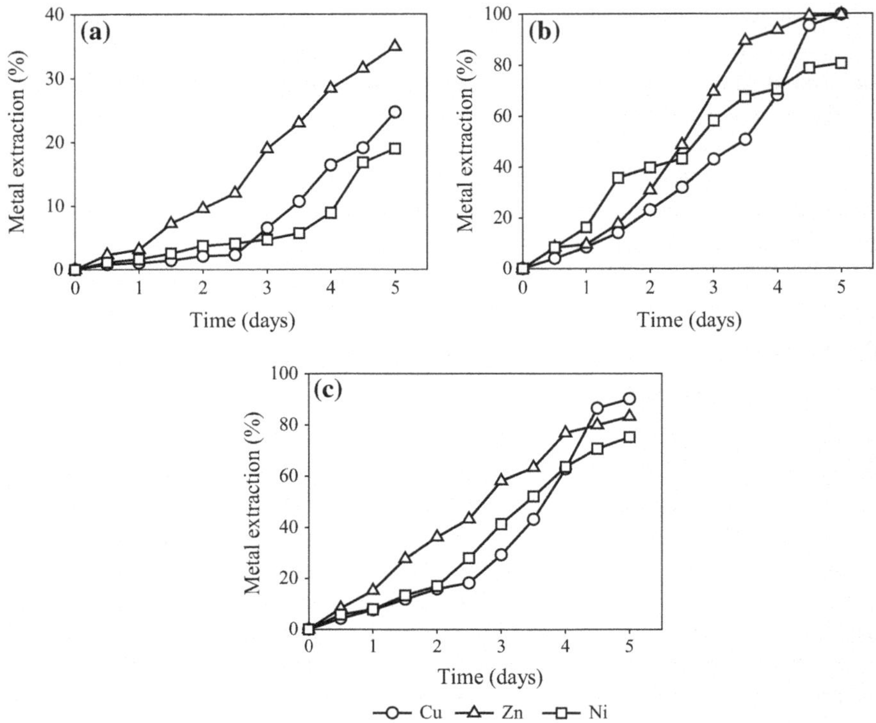

Fig. 4 Effect of L/S ratio of (**a**) 10, (**b**) 50, and (**c**) 100 on the extraction of selected metals from waste computer PCB using DTPA

metal extraction from e-waste. However, the metal leaching time needs to be further improvised by employing oxidant like hydrogen peroxide along with DTPA as an organic chelator.

References

Baldé, C. P., Forti, V., Gray, V., Kuehr, R., & Stegmann, P. (2017). *The global e-waste monitor 2017: Quantities, flows and resources*. United Nations University, International Telecommunication Union, International Solid Waste Association.

Choubey, P. K., Panda, R., Jha, M. K., Lee, J. C., & Pathak, D. D. (2015). Recovery of copper and recycling of acid from the leach liquor of discarded printed circuit boards (PCBs). *Separation and Purification Technology, 156,* 269–275.

Creamer, N. J., Baxter-Plant, V. S., Henderson, J., Potter, M., & Macaskie, L. E. (2006). Palladium and gold removal and recovery from precious metal solutions and electronic scrap leachates by *Desulfovibrio desulfuricans*. *Biotechnology Letters, 28*(18), 1475–1484.

Huang, K., Guo, J., & Xu, Z. (2009). Recycling of waste printed circuit boards: A review of current technologies and treatment status in China. *Journal of Hazardous Materials, 164*(2–3), 399–408.

Jadhao, P., Chauhan, G., Pant, K. K., & Nigam, K. D. P. (2016). Greener approach for the extraction of copper metal from electronic waste. *Waste Management, 57,* 102–112.

Li, J., Lu, H., Guo, J., Xu, Z., & Zhou, Y. (2007). Recycle technology for recovering resources and products from waste printed circuit boards. *Environmental Science and Technology, 41*(6), 1995–2000.

Li, J., Shrivastava, P., Gao, Z., & Zhang, H. C. (2004). Printed circuit board recycling: A state-of-the-art survey. *IEEE Transactions on Electronics Packaging Manufacturing, 27*(1), 33–42.

Martell, A. E., & Smith, R. M. (2003). *NIST standard reference database 46. Critically selected stability constants of metal complexes database, version 7.0.*

Pant, D., Joshi, D., Upreti, M. K., & Kotnala, R. K. (2012). Chemical and biological extraction of metals present in E waste: A hybrid technology. *Waste Management, 32*(5), 979–990.

Priya, A., & Hait, S. (2017). Comparative assessment of metallurgical recovery of metals from electronic waste with special emphasis on bioleaching. *Environmental Science and Pollution Research, 24*(8), 6989–7008.

Rao, S., Yang, T., Zhang, D., Liu, W., Chen, L., Hao, Z., et al. (2015). Leaching of low grade zinc oxide ores in NH_4Cl–NH_3 solutions with nitrilotriacetic acid as complexing agents. *Hydrometallurgy, 158,* 101–106.

Silvas, F. P. C., Correa, M. M. J., Caldas, M. P. K., de Moraes, V. T., Espinosa, D. C. R., & Tenório, J. A. S. (2015). Printed circuit board recycling: Physical processing and copper extraction by selective leaching. *Waste Management, 46,* 503–510.

USEPA. (1996). *Microwave assisted acid digestion of siliceous and organically based matrices USEPA method 3052* (3rd ed.). Washington, DC: United States Environmental Protection Agency.

Wang, H., Gu, G. H., & Qi, Y. F. (2005). Crushing performance and resource characteristic of printed circuit board scrap. *Journal of Central South University of Technology, 12*(5), 552–555.

Zhao, Y., Wen, X., Li, B., & Tao, D. (2004). Recovery of copper from waste printed circuit boards. *Minerals and Metallurgical Processing, 21*(2), 99–102.

Zhou, Y., & Qiu, K. (2010). A new technology for recycling materials from waste printed circuit boards. *Journal of Hazardous Materials, 175*(1–3), 823–828.

A Comparative Study on the Cost–Benefit Analysis on Metal Recovery of WPCB Using Pyrometallurgy with Two Different Thermal Furnaces

R. Balaji and J. Senophiyah-Mary

Abstract E-waste, in particular, PCB is a major component which is the integral part of all the electrical and electronic devices, which will be great source for the natural source deficiency. WPCB has a composition of approximately 28% metals, 23% plastics and remaining percentage as glass and ceramics. Though several methods such as pyrometallurgy, hydrometallurgy and bio-hydrometallurgy have been identified for the recovery of metals and heavy metals, still it has been found with a lot of difficulties. Pyrometallurgical process of recycle helps in immediate recovery of direct metals and it is also one of the conventional and time-consuming process. The emission of toxic gases such as dioxins is opened into the environment, but the emission rate at recycling is much less as compared to the emission during the metallurgical refining process from the ores. This pyrometallurgical process also helps in precious metal recovery like Au, Ag, Pd and some rare earth metals. The metals recovered from the heat treatment process have market value and it can be sold into the market in an easier manner and it reduces the cost for further advanced purification of metals. Cost analysis helps in determining the market value of a scrap PC's and also helps in quantifying and qualifying the treatment processes and commercial values of metal, precious metal recovered. A comparison study on cost–benefit analysis between two different thermal furnaces namely CVD chamber and muffle furnace on metal recovery efficiencies and costwise efficient were discussed. This discussion would help in industrialising the proper treatment technique and thereby reducing the informal sector.

Keywords E-waste · WPCB · Pyrometallurgy · Hydrometallurgy · Bio-hydrometallurgy · Cost–benefit analysis

---

R. Balaji (✉)
Kongunadu College of Engineering and Technology, Trichy, Tamil Nadu, India
e-mail: rbalajicivil@gmail.com

J. Senophiyah-Mary
Government College of Technology, Coimbatore, Tamil Nadu, India

S. K. Ghosh (ed.), *Urban Mining and Sustainable Waste Management*,
https://doi.org/10.1007/978-981-15-0532-4_7

1 Introduction

The world faces a serious problem in protecting our natural sources. We are at the stage of depletion of our natural resources. A day-to-day consumption of natural sources is the major reason for depletion. Steps have been taken for an alternate energy source which will eradicate the depletion problem which we are facing in our world. E-waste plays a major role in eradicating the material deficiency (Shuey and Taylor 2005).

E-waste is nothing but the waste electric and electronic devices in which its value has been considered as zero and it has been thrown out by the user without the intention of reusing it again. Tonnes and tonnes of e-waste have been generated in our country and few only involved in recycling of the e-waste (Huang et al. 2009; Sohaili et al. 2012; Mary and Meenambal 2016). Similarly, a lot of informal sectors emerged in recent years because of its valuable metals, precious metals and some rare earth metals. It is a great platform for the entrepreneurs and business people to look forward to it for its commercial value (Ministry of Environment 2016; Joseph 2007).

E-waste is mainly composed of plastics, metals like steel, aluminium, copper and few more metals are also present in its PCBs (Cui and Zhang 2008; Bizzo et al. 2014). PCB is responsible for the functioning of electronic devices and is found in most of the devices. PCB is mainly composed of plastics, ceramics and metals (Coman et al. 2013; Mary and Meenambal 2015, 2018). E-waste can be used as an alternative energy due to the presence of metals and plastics. The existence of plastics and metals increases the calorific value. Toxic elements removal helps in proper utilisation of the e-waste (Xiu et al. 2017).

E-waste is recycled using any of the following methods. Pyrometallurgy, hydrometallurgy, bio-hydrometallurgy and plasma torch method are generally adopted in the recycling process. Pyrometallurgy is taken as an area of interest and comparison study is done with two different thermal furnaces (Xiu et al. 2017; Cayumil et al. 2014). Pyrometallurgy deals with the metal recovery by the application of heat energy. Similarly, hydrometallurgical process of recycling PCB for its metal is accomplished by the action of chemical solutions. Bio-hydrometallurgical process of recycling is influenced by bacterial action or by the action of micro-organisms (Kumar and Holuszko 2016; He and Duan 2017; Cayumil et al. 2018; Luda 2011; Kavousi et al. 2017).

Chemical vapour deposition (CVD) is a deposition method used to produce high-quality, high-performance, solid materials, typically under vacuum. CVD chamber also known as tubular furnace is used for the separation of metals from the PCB plates as well as PCB in powdered form (Szałatkiewicz 2014; Awasthi and Li 2017; Shuey and Taylor 2005). A muffle furnace is a furnace in which the subject material is isolated from the fuel and all of the products of combustion, including gases and flying ash. Likewise, PCBs are heated at high temperature for the metal recovery process.

To determine the market value of the metals recovered from scrap PC and also to quantify the same, a comparison study on cost benefit analysis was done to industrialise the proper treatment technique and thereby reduce the informal sector.

2 Materials and Methods

2.1 Materials

E-waste is collected from one of the authorised recycler namely Green Era Recyclers, Coimbatore, Tamilnadu. PCB of varied sizes was cut into small pieces of size 2 cm × 2 cm such that each piece was rich in metal content and its approximate weight was found to 2 g. Similarly, PCBs were crushed to form PBC powder where 2 g of powdered PCBs were used as samples.

2.2 Methodology

2.2.1 Chemical Vapour Deposition Chamber

Chemical vapour deposition chamber is an instrument used to heat the samples at inert atmosphere to synthesis the nanoparticles by deposition method. Hydrogen gas is used as a precursor gas and then it is followed by acetylene and argon gas is used for flushing out the gasses present inside the chamber. But in our case, the plate samples and powdered samples are treated in CVD chamber for the direct metal recovery from the samples at the same inert atmosphere. Hydrogen gas and argon gas alone are used for the recovery process (Yu et al. 2009; Kan et al. 2018). The optimisation parameters such as temperature, time period for treatment and gas flow rate can be controlled using this chamber. Optimised values are used for the treatment process since it is directly utilised for the cost analysis comparison with two different furnaces.

2.2.2 Muffle Furnace

Muffle furnace is an instrument used to heat samples at high temperature. Both the samples are heated at 500 °C in furnace and metals get crumbled in this process. More than 90% of the metal gets turned into ashes.

3 Characteristic Study

3.1 Atomic Absorption Spectroscopy

AAS was used to find out the initial and final metal concentration of the samples. For finding out the initial metal concentration, stock solution of different ppm (i.e.) 1–10 ppm was made for different metals like Cu, Zn, Cr, Pb, Ni, Hg and Cd (Hall and Williams 2007; Rajarao et al. 2014). To find out the concentration of metals from the samples, the untreated plate and powdered sample was made into solution by addition of 3 mL of concentrated nitric acid. Then, it was followed by AAS for the concentration of different metals. Similarly, the metallic fractions obtained from the treated plate and powdered samples from both the furnaces were also made into solution for determining its concentration of different metals. Samples with concentration higher than 10 ppm were determined by serial dilution (Khanna et al. 2014; Zhang et al. 2017; Cayumil et al. 2016; Garlapati 2016; Senophiyah-Mary et al. 2018).

3.2 Thermogravimetric Analyser

TGA is a method of heat analysis in which rate of change of sample mass is identified with reference to varying temperature. It helps in identifying the necessary temperature required for setting the temperature of both the thermal furnaces. It can also give the temperature at which the complete degradation of the plastic from the PCBs. From which different temperatures are taken for the study, whereas the degradation was initiated well at 320 °C and it is significant at 500 °C and the complete degradation occurs at 920 °C. For optimising the temperature, the study was carried out at 500 and 900 °C.

3.3 Cost–Benefit Analysis

Weight of PCB from single computer = 250 g					
Recoverable metals	Quantity	Percentage	Percentage to weight (g)	Unit price per kg	Price (Rs.)
Cu (%)	14.2	14.2	35.5	481.84	17.11
Pb (%)	2.5	2.5	6.25	186.38	1.16
Zn (%)	0.18	0.18	0.45	243.45	0.11
Ni (%)	0.41	0.41	1.025	786.63	0.81
Fe (%)	3.08	3.08	7.7	37	0.28

(continued)

(continued)

Weight of PCB from single computer = 250 g					
Recoverable metals	Quantity	Percentage	Percentage to weight (g)	Unit price per kg	Price (Rs.)
Sn (%)	4.79	4.79	11.975	1513.43	18.12
Sb (%)	0.05	0.05	0.125	695	0.09
Na (%)	0.48	0.48	1.2	175	0.21
Ca (%)	1.69	1.69	4.225	240	1.01
Ag (ppm)	317	0.317	0.7925	39,856	31.59
Au (ppm)	142	0.142	0.355	2,789,910	990.42
Cd (ppm)	1183	1.183	2.9575	478	1.41
K (ppm)	180	0.18	0.45	290	0.13
Mn (ppm)	81	0.081	0.2025	175	0.04
Se (ppm)	21	0.021	0.0525	2500	0.13
As (ppm)	11	0.011	0.0275	120	0.003
% of metals	27.6		Total price (Rs.)		1062.62

4 Result and Discussions

4.1 Atomic Absorption Spectroscopy

It is used to determine the percentage of recoverable metals which are obtained after the recovery process. AAS also helps in identifying the percentage of metals in sample before and after the treatment process in terms of concentrations (Ghosh et al. 2015; Wang et al. 2017; Sahajwalla et al. 2015). With the values obtained from initial and final concentration, percentage of metal recovery is obtained.

4.2 Muffle Furnace

The product obtained after the heating process from the muffle furnace was ash and some fractions of metals. These metals were not in its direct form and its recovery percentage also very less when it was tested with atomic absorption spectroscopy.

4.3 Chemical Vapour Deposition Chamber

In the case of CVD chamber, the product obtained after the heating process was rich in metallic fractions as well as they were in its direct form which can be used or commercialised in an easier manner. The slag obtained was also rich in carbon content which can be activated later and this can be used as activated carbon (Joseph 2007; Xiu et al. 2017; Cayumil et al. 2014).

4.4 Cost–Benefit Analysis

The weight of the disassembled PCBs computers, mobile phones and memory card range from 200 to 315 g, 6 to 16 g and 15 to 20 g, respectively, depending on the size, shape and brand of the product (Jessica Hanafi et al.). For our evaluation, we choose 250 g for the weight of PCBs from computers (Senophiyah-Mary et al. 2018; Ghosh 2018).

$$\text{Volume of 600 mL hydrogen gas} = 0.511\ \text{m}^3$$

$$\text{Volume of 300 mL argon gas} = 0.252\ \text{m}^3$$

$$\begin{array}{c}\text{Weight of PCB for Single operation}\\ \text{in CVD chamber}\end{array} = 20\ \text{g}$$

$$\begin{array}{c}\text{Approximate weight of PCB}\\ \text{from a single PC}\end{array} = 200\ \text{g}$$

$$\begin{array}{c}\text{No. of operations required for complete treatment}\\ \text{of a PCB from Single computer}\end{array} = 200/20$$

$$= 10\ \text{operations}$$

4.5 Cost for Single Operation

4.5.1 Cost of Gas for Single Operation

S. no.	Description	Rate per 1 m^3	Quantity (m^3)	Price (Rs.)
1	Hydrogen gas	Rs. 150/-	0.511	76.65
2	Argon gas	Rs. 100/-	0.252	25.2
Total cost of gas				101.85

4.5.2 Cost of Electricity for Single Operation in CVD Chamber

S. no.	Description	Rate per kWh	Quantity (W)	Time (h)	Total consumption (kWh)	Price (Rs.)
1	Furnace	6	2000	5	10	60
2	Temperature controller	6	2000	5	10	60
3	Vacuum system	6	500	0.0167	0.0083	0.050
Total cost of electricity						120.05

$$\text{Total cost for Single operation} = \text{Cost of gas} + \text{Cost of electricity}$$
$$\text{Total cost for Single operation} = \text{Rs. } 221.9/-$$
$$\begin{array}{c}\text{Cost of treatment for PCB}\\ \text{from a single computer}\end{array} = \text{No. of operation} \times \text{Single operation cost}$$
$$\begin{array}{c}\text{Cost of treatment for PCB}\\ \text{from a single computer}\end{array} = 10 \times 221.9$$
$$\begin{array}{c}\text{Cost of treatment for PCB}\\ \text{from a single computer}\end{array} = \text{Rs. } 2219/-$$
$$\text{Total cost from recovered metals} = \text{Rs. } 1062.62/-$$

5 Merits and Demerits

From this comparative study, the metal recovery from CVD chamber is very high when compared with muffle furnace, and here, the metals turns into ashes when treated under muffle furnace. CVD chamber gives the metallic fractions as a thin film, whereas no direct metal is obtained in case of muffle furnace. Metals can be recovered from all the parts of PCBs. Metals of micro-sizes were also recovered in the CVD chamber. Some of the metal elements such as Cu, Al and Ni can directly be identified through physical inspection. Though several other methods had been identified and these methods seems to have added advantages like eco-friendly and economic feasibility, but this method has its unique identity like time reducing process. One of the major disadvantages of this method was toxic gases emission during the heat treatment process. The major toxic gas emission was dioxins and it was freely opened into the atmosphere without reducing its concentration or without any further treatment.

6 Conclusion

It was observed that the chemical vapour deposition chamber (CVD chamber) gave direct metal recovery while comparing with muffle furnace, but it is not the ultimate solution. It would be simple, effective and economical when the chamber has been added with toxic emission control device. The cost required for metal recovery operation was Rs. 1967/-, while the cost we acquire from recovered metals was Rs. 1062.62/-. The major issue with the methodology was emission of dioxin into the environment. Though there were issues with toxic emissions, the chamber has to be modified from laboratory scale to pilot scale which will enhance the heating process for less number of operations with same expenditure. This methodology will function effectively only when the number of operations for complete treatment of PCB from single computer had been reduced to a very large extent and that can be achieved by providing modifications in the size of chamber without affecting the effectiveness of the chemical vapour deposition chamber.

References

Awasthi, A. K., & Li, J. (2017). Management of electrical and electronic waste: A comparative evaluation of China and India. *Renewable and Sustainable Energy Reviews, 76,* 434–447.

Bizzo, W. A., Figueiredo, R. A., & de Andrade, V. F. (2014). Characterization of printed circuit boards for metal and energy recovery after milling and mechanical separation. *Materials, 7*(6), 4555–4566.

Cayumil, R., Khanna, R., Ikram-Ul-Haq, M., Rajarao, R., Hill, A., & Sahajwalla, V. (2014). Generation of copper rich metallic phases from waste printed circuit boards. *Waste Management, 34*(10), 1783–1792.

Cayumil, R., Khanna, R., Rajarao, R., Mukherjee, P., & Sahajwalla, V. (2016). Concentration of precious metals during their recovery from electronic waste. *Waste Management, 57,* 121–130.

Cayumil, R., Ikram-Ul-Haq, M., Khanna, R., Saini, R., Mukherjee, P. S., Mishra, B. K., & Sahajwalla, V. (2018). High temperature investigations on optimising the recovery of copper from waste printed circuit boards. *Waste Management, 73*, pp. 556–565.

Coman, V., Robotin, B., & Ilea, P. (2013). Nickel recovery/removal from industrial wastes: A review. *Resources, Conservation and Recycling, 73,* 229–238.

Cui, J., & Zhang, L. (2008). Metallurgical recovery of metals from electronic waste: A review. *Journal of Hazardous Materials, 158*(2), 228–256.

Garlapati, V. K. (2016). E-waste in India and developed countries: Management, recycling, business and biotechnological initiatives. *Renewable and Sustainable Energy Reviews, 54,* 874–881.

Ghosh, S. K. (2018). *Utilization and management of bioresources*. Springer.

Ghosh, B., Ghosh, M., Parhi, P., Mukherjee, P., & Mishra, B. (2015). Waste printed circuit boards recycling: An extensive assessment of current status. *Journal of Cleaner Production, 94,* 5–19.

Hall, W. J., & Williams, P. T. (2007). Processing waste printed circuit boards for material recovery. *Circuit World, 33*(4), 43–50.

He, J., & Duan, C. (2017). Recovery of metallic concentrations from waste printed circuit boards via reverse floatation. *Waste Management, 60,* 618–628.

Huang, K., Guo, J., & Xu, Z. (2009). Recycling of waste printed circuit boards: A review of current technologies and treatment status in China. *Journal of Hazardous Materials, 164*(2), 399–408.

Joseph, K. (2007). Electronic waste management in India—issues and strategies. In *Eleventh International Waste Management and Landfill Symposium*, Sardinia.

Kan, Y., Yue, Q., Liu, S., & Gao, B. (2018). Effects of Cu and CuO on the preparation of activated carbon from waste circuit boards by H_3PO_4 activation. *Chemical Engineering Journal, 331,* 93–101.

Kavousi, M., Sattari, A., Alamdari, E. K., & Firozi, S. (2017). Selective separation of copper over solder alloy from waste printed circuit boards leach solution. *Waste Management, 60,* 636–642.

Khanna, R., Cayumil, R., Mukherjee, P., & Sahajwalla, V. (2014). A novel recycling approach for transforming waste printed circuit boards into a material resource. *Procedia Environmental Sciences, 21,* 42–54.

Kumar, A., & Holuszko, M. (2016). Electronic waste and existing processing routes: A Canadian perspective. *Resources, 5*(4), 35.

Luda, M. P. (2011). *Integrated waste management* (Vol. 2). Online.

Mary, J. S., & Meenambal, T. (2015). Solubilisation of metals from e-waste using *Penicillium chrysogenum* under optimum conditions. In *Infrastructure Development for Environmental Conservation & Sustenance* (Vol. 2015, p. 285), October 28–30, 2015.

Mary, J. S., & Meenambal, T. (2016). Inventorisation of E-waste and developing a policy-bulk consumer perspective. *Procedia Environmental Sciences, 35,* 643–655.

Mary, J. S., & Meenambal, T. (2018). Removal of copper from bioleachate of electronic waste using banana-activated carbon (BAC) and comparison with commercial-activated carbon (CAC). In *Utilization and management of bioresources* (pp. 233–242). New York: Springer.

Ministry of Environment, F. a. C. C. (2016). *Published in the gazette of India, extraordinary part-II, section-3, sub-section (i).* Government of India.

Rajarao, R., Sahajwalla, V., Cayumil, R., Park, M., & Khanna, R. (2014). Novel approach for processing hazardous electronic waste. *Procedia Environmental Sciences, 21,* 33–41.

Sahajwalla, V., Cayumil, R., Khanna, R., Ikram-Ul-Haq, M., Rajarao, R., Mukherjee, P., et al. (2015). Recycling polymer-rich waste printed circuit boards at high temperatures: Recovery of value-added carbon resources. *Journal of Sustainable Metallurgy, 1*(1), 75–84.

Senophiyah-Mary, J., Loganath, R., & Meenambal, T. (2018). A novel method for the removal of epoxy coating from waste printed circuit board. *Waste Management & Research, 36*(7), pp. 645–652.

Shuey, S., & Taylor, P. (2005). Review of pyrometallurgical treatment of electronic scrap. *Mining Engineering, 57*(4), 67–70.

Sohaili, J., Muniyandi, S. K., & Mohamad, S. S. (2012). A review on printed circuit board recycling technology. *Journal of Emerging Trends in Engineering and Applied Sciences (JETEAS), 3*(1), 12–18.

Szałatkiewicz, J. (2014). Metals content in printed circuit board waste. *Polish Journal of Environmental Studies, 23*(6), 2365–2369.

Wang, H., Zhang, S., Li, B., Pan, D., Wu, Y., & Zuo, T. (2017). Recovery of waste printed circuit boards through pyrometallurgical processing: A review. *Resources, Conservation and Recycling, 126*, 209–218.

Xiu, F.-R., Weng, H., Qi, Y., Yu, G., Zhang, Z., Zhang, F.-S., et al. (2017). A novel recovery method of copper from waste printed circuit boards by supercritical methanol process: Preparation of ultrafine copper materials. *Waste Management, 60,* 643–651.

Yu, J., Williams, E., & Ju, M. (2009). Review and prospects of recycling methods for waste printed circuit boards. In *IEEE International Symposium on Sustainable Systems and Technology (ISSST '09)* (pp. 1–5). IEEE.

Zhang, G., He, Y., Wang, H., Zhang, T., Wang, S., Yang, X., et al. (2017). New technology for recovering residual metals from nonmetallic fractions of waste printed circuit boards. *Waste Management, 64,* 228–235.

Recycling of Polymers from WEEE: Issues, Challenges and Opportunities

Biswajit Debnath, Ranjana Chowdhury and Sadhan Kumar Ghosh

Abstract Waste Electrical and Electronic Equipments (WEEE) contains a nearly 23% polymers including acrylonitrile butadiene styrene, high-impact polystyrene, polystyrene, poly vinyl chloride and polycarbonate. Every year the OEMs manufacture lighter version of Electrical and Electronic Equipments (EEE), and it is only possible by increasing the percentage of polymers and reducing the ferrous metals. The latest End-of-Life (EoL) electronics will have more polymers, and their recycling will become a future concern. However, it is already a problem that the polymers from WEEE are not recycled properly. The primary reason is that the polymer fraction recovered from WEEE is mixed polymer, and their identification is an issue. Different technologies have been developed by the researchers for recycling of WEEE polymers including pyrolysis, gasification and moulding. However, the commercializations of these processes are yet to be realized. Existing literature fails to demonstrate the issues and challenges behind this. Reviews on recycling of WEEE polymers are scant. In this paper, a detailed review of recycling of WEEE polymers has been presented. The issues and challenges pertaining to the recycling technologies of WEEE polymers have been critically discussed. The findings of this paper will be of importance to the research community and the stakeholders of WEEE industry.

Keywords WEEE · Polymers · Issues and challenges

1 Introduction

Electrical and Electronics Equipment (EEE) industry is one of the largest in the world. The electronic industry is expected to reach \$400 billion in 2022 from \$69.6 billion in 2012 (Corporate Catalyst India 2015). The driving force behind the demand of EEE is the technological advancement, short innovation cycles and business strategies

B. Debnath · R. Chowdhury
Department of Chemical Engineering, Jadavpur University, Kolkata, India

S. K. Ghosh (✉)
Department of Mechanical Engineering, Jadavpur University, Kolkata, India
e-mail: sadhankghosh9@gmail.com

S. K. Ghosh (ed.), *Urban Mining and Sustainable Waste Management*,
https://doi.org/10.1007/978-981-15-0532-4_9

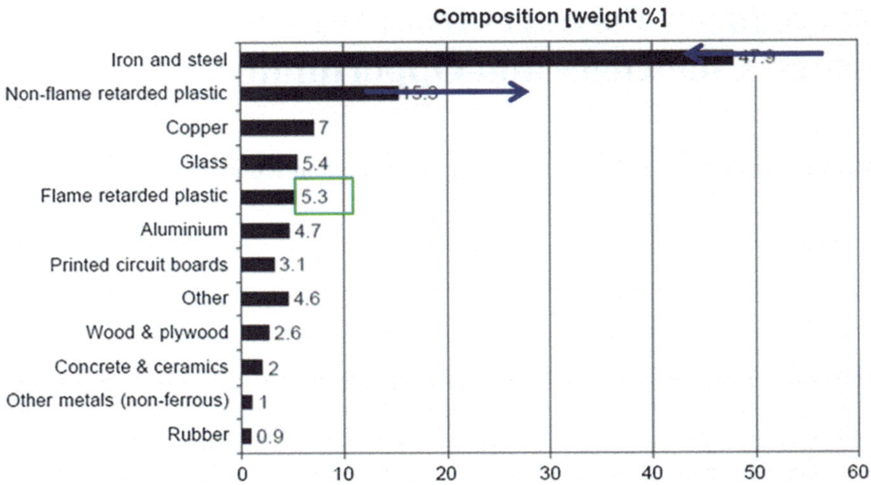

Fig. 1 Composition of WEEE showing decrease in metals and increase in plastics

which shorten the lifespan of the equipments, resulting in e-waste (Ghosh et al. 2016). WEEE generation is increasing exponentially every year (Ghosh et al. 2014). In 2016, nearly 44.7 million metric tons of WEEE was produced globally which is equivalent to 4500 Eiffel Towers (Balde et al. 2017). In 2017, expected WEEE generation will exceed 46 million metric tons and 52.2 MMT by 2021. BRICS nations generated nearly 20% of the total WEEE generated in 2016. China generated 7.2 million metric tons of WEEE in the year 2016 which is highest in both Asia and globally. India generated 2 million tons of WEEE in the year 2016 (Balde et al. 2017).

From Fig. 1, this is quite clear that the electronics are getting lighter because the amount of plastics is increasing and the metals are decreasing. Recent conference on e-waste also raised the question of economical sustainability of metal recycling from WEEE. Polymer fraction of WEEE is often neglected, and they end up in landfill sites. It is high time that the issue with WEEE polymers shall be addressed. It is possible to use these plastics for energy recovery, conversion into nano-materials and even methanol. Interestingly, WEEE plastics recycling open a new door for urban mining. Utilization of these resources will not only enhance the resource efficiency but also ensure circular economy. It has been found that majority of the literature is focused on metal recovery and dehalogenation of plastics from WEEE. A few works on non-metallic fraction of WEEE are available. Studies addressing issues and challenges with WEEE polymers are scant. The research questions that arise are—what are different types of plastics present in WEEE? What are the existing processes available for recycling or treatment of these polymers? What are the issues and challenges associated with the recycling of WEEE polymers? Is there a sustainable way to deal with these polymers? The objectives of this study are to understand the types of plastics present in WEEE, to find out available technologies and the issues for

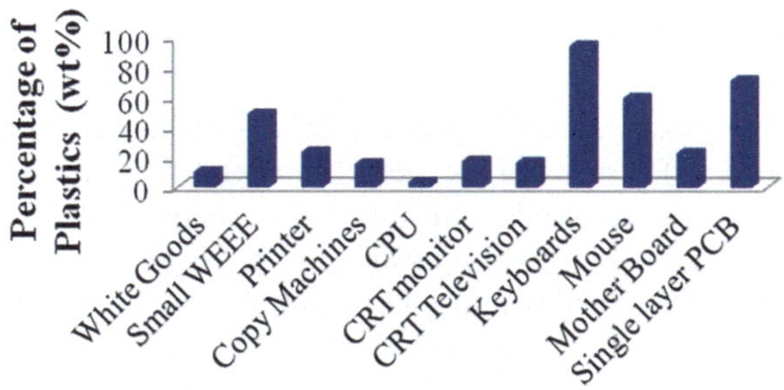

Fig. 2 Amount of plastics in different types of WEEE

recycling of plastics from WEEE and to develop a sustainable supply chain network for treatment of WEEE polymers.

2 Composition of WEEE Plastics

WEEE contains different types of plastics in WEEE which includes acrylonitrile butadiene styrene (ABS), high-impact polystyrene (HIPS), polystyrene (PS), poly vinyl chloride (PVC) and polycarbonate (PC). The amount of these plastics varies widely in different types of WEEE. Figure 2 shows the amount of plastics in different of electronics which reveals that small WEEE has higher amount of plastics. Similar results were reported by Debnath et al. (2016), where using different cases they showed how the amount of plastics varies in different WEEE. In an average, WEEE contains nearly 30% ABS, 25% HIPS and 10% PC, and rest are PP, PVC and other mixtures (Shen et al. 2016). However, some internal study made by Star Plastics Inc. that recycles WEEE plastics reports that the composition handled by them includes 50–60% ABS, 10–20% HIPS, 10–20% PCABS, 2–8% PC and 12–20% various plastics including PVC, PE, PP, PPO, nylon, ABS/PVC and acrylic (https://www.starplastics.com/).

3 Existing Technologies for WEEE Polymer Recycling

The existing technologies for recycling of WEEE polymer include different methods including physical recycling, thermo-chemical recycling and destructive treatments. They are discussed in brief below.

3.1 Pyrolysis

Pyrolysis is a thermo-chemical process, which is primarily used as a pre-treatment process for material recovery from printed circuit boards (Debnath et al. 2018b). Pyrolysis can also be utilized for conversion of plastics recovered from WEEE. The major product is pyro-oil and pyro-char. The process is very efficient for conversion of WEEE plastics into other resources, and it has low environmental food print compared to combustion and land filling (Debnath et al. 2018b).

The contamination of halogens due to the presence of the flame retardants calls for dehalogenation of the feedstock. To achieve clean fuel out of pyrolysis, it is important to have a pre-treatment step. These can be carried out in three different pathways (Fig. 3)—(a) dehalogenation by pre-treatment (mechano-chemical or hydrothermal treatment) followed by thermolysis; (b) in situ dehalogenation during thermolysis by using catalyst or adsorbents and (c) ex situ dehalogenation, i.e. thermolysis followed by further refining (Shen et al. 2016). Debnath et al. have investigated the pyrolysis process of FR-4-type printed circuit boards which contain thermoset polymers. The FR-4-type PCBs are mainly composed of epoxy resin reinforced by glass fibres. They investigated resource recovery opportunities and showed that the polymer part can be converted into pyro-oil via pyrolysis. This process is undoubtedly an excellent process for resource recovery. However, certain issues including corrosive product gas, bad odour during operation and complex kinetics of the process, scale-up issues and economics are the reasons for its large-scale implementation.

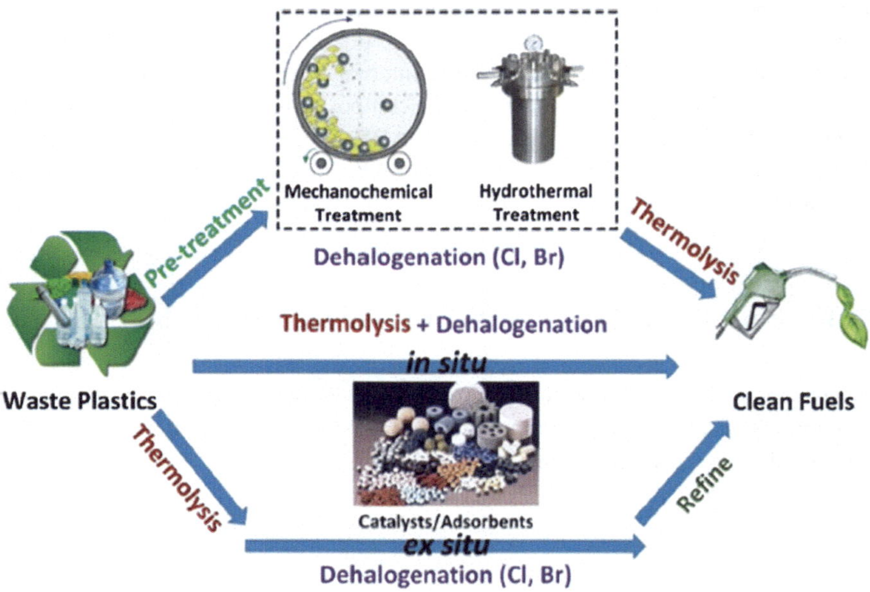

Fig. 3 Pyrolysis pathways for dehalogenation of plastics. *Source* Shen et al. (2016)

3.2 Co-pyrolysis

The presence of halogens in the WEEE plastics often creates nuisance during pyrolysis as they end up contaminating the pyro-oil. To address this problem, bio-mass is partially replaced in the feedstock subjected to pyrolysis. This is known as co-pyrolysis. By definition, co-pyrolysis is a process which involves two or more different materials as a feedstock. In this process, the pyro-oil is free from halogen because the halogen gets trapped in the pyro-char. The pyro-char is hazardous in nature and required additional attention for disposal which incur extra cost (Liu et al. 2013) (Fig. 4).

Co-pyrolysis with catalysts and adsorbents can upgrade the quality of the pyro-oil. However, there exist some issues related to the use of a catalyst in pyrolysis. Some of them are expense of catalyst, short life cycle due to catalyst poisoning/deactivation, the increase of solid residues, the catalysts/sorbents regeneration, etc. The co-pyrolysis of biomass and WEEE plastics can be an optional route. According to Abnisa and Duad (2014), co-pyrolysis is a potential technology for industrialization as it has attractive performance/cost ratio. The key here is the synergistic effect that evolutes due to the reactions of different materials during pyrolysis. The calorific value of the resulting pyro-oil is higher than the pyro-oil of biomass alone. The product oil is enriched with paraffins, isoparaffins, olefins, naphthenes and aromatics (Panda et al. 2010). Co-pyrolysis of WEEE plastics with waste biomass was investigated to produce pyro-oil. The yield of pyro-oil (62.3%) from co-pyrolysis was significantly higher compared to the pyrolysis of WEEEs (53.1%) and biomass

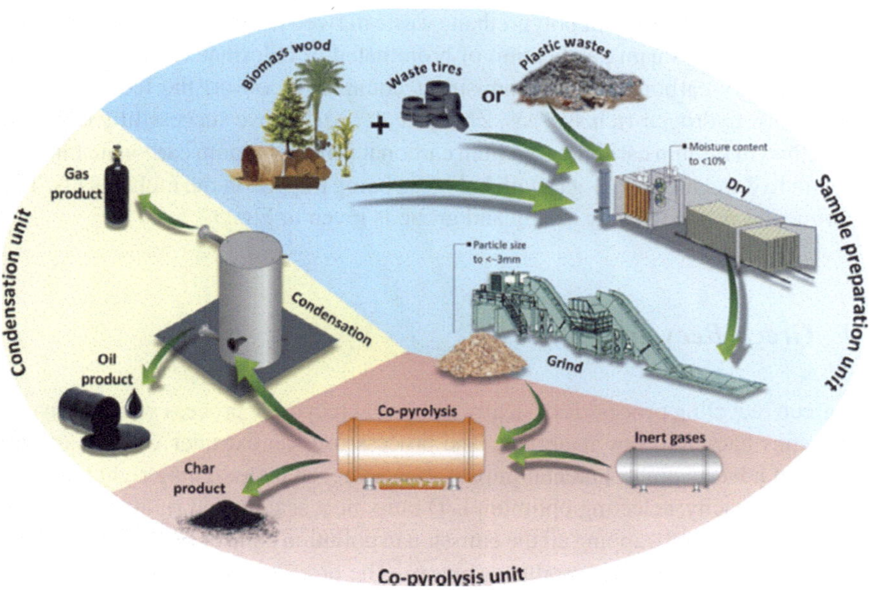

Fig. 4 Schematic of co-pyrolysis of plastics with biomass. *Source* Abnisa and Duad (2014)

(46.3%) alone (Liu et al. 2013). In general, blending of the WEEEs with biomass is in favour of enriching Br to the biochar and decreasing its contents in bio-oil during the pyrolysis process.

3.3 Gasification

Gasification is an established process which converts any feedstock into producer gas with steam or air or both as gasification medium. Using gasification, WEEE plastics can be converted into syn-gas which can be further used for methanol synthesis, paving a way to n number of chemicals. The high temp and the reducing atmosphere suppress the formation of dioxins. Sometimes there is a secondary combustion chamber where dioxins (if any generated in the first step) are decomposed by high temperature and appropriate residence time. Flag and fly ash are recovered as by-product in this process which is further used in construction industry (Shen et al. 2016). High-impact polystyrene resin containing PBDE (polybrominateddiphenyl ethers; mainly decabromodiphenyloxide) as a brominated flame retardant was used as feedstock for a zero-emission recycling process which implements gasification followed by shock cooling. This method successfully prevented the formation of dioxins and furans and also recovered antimony present in the plastics in form of solids. A gaseous by-product called "thin gas" rich in hydrogen was found suitable for use as raw material in the chemical industry (Yamawaki 2003). The effect of steam and sodium hydroxide on hydrogen production using dechlorinated PVC and activated carbon as feedstock has been reported by Kamo et al. Kang et al. (2016) have explored gasification of polyurethane waste derived from WEEE which showed promising results. Steam gasification of brominated or chlorinated plastics in presence of ternary carbonates is a promising technique to convert the feedstock into high-quality, hydrogen-rich syn-gas. Zhang et al. (2013) have successfully explored this method in lithium carbonate, sodium carbonate and potassium carbonate mixture to obtain hydrogen-rich syn-gas with halogens being trapped in the molten salts. The experimental set-up used by Zhang and group is given in Fig. 5.

3.4 Green Recycling Process

The green recycling process is basically a physical recycling process which converts the plastics back into raw materials. The process uses an extruder which has zero significant adverse environmental impact. It is achieved by assigning right motor of minimum capacity, selecting optimum L/D ratio, heat sealing and right temperature for the processes and trapping all the emission in pollution control gadget and treating the pollutant to produce by-products (Fig. 6). The processes followed by extrusion are water cooling and pelletization resulting into plastic granules. PP, PE, PC, ABS

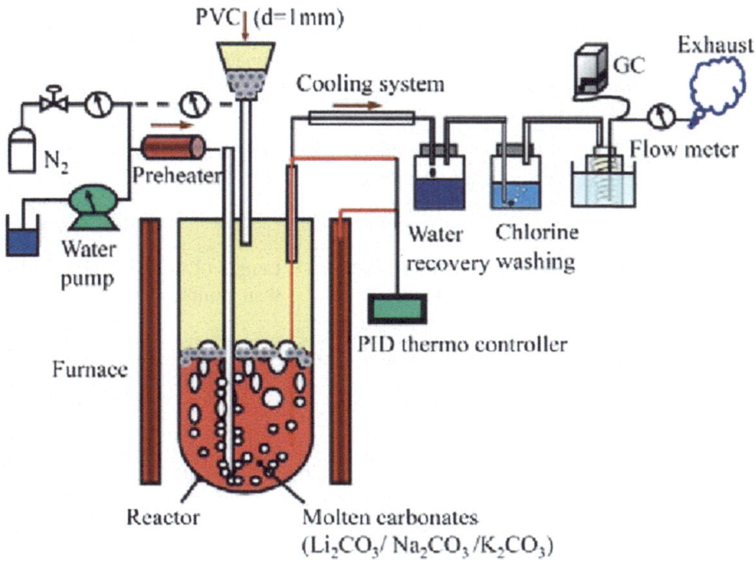

Fig. 5 Schematic of steam gasification set-up as given by Zhang et al. (2013)

Fig. 6 Green recycling of plastics set-up as developed by Ghosh (2004)

and HIPS can be recycled using this method (Ghosh 2004). Currently, a modified version of this process is being carried out by one of the leading e-waste recycler in India.

Table 1 Plastics to CNT from different sources

Feed	Temperature of decomposition (k)	Outer diameter (nm)	Length	Shapes and form
Polyethylene (HDPE, LDPE)	1073	Mean diameter in the range of 30–85 nm	Length typically of 1–5 μm	16 layers of parallel graphene layers, MWCNTs
HDPE AND LDPE (polymer waste)	973	Diameter of 80 nm	Length of more than a micron	MWCNTs
Virgin or recycled polyethylene	973	15–40 nm		Coiled and straight tubes as MWCNTs and multi-walled nano-fibres
Polyethylene	973–1073	Diameter in the range of 10–100 nm		Carbon nano-tubes (MWCNTs)

3.5 Plastics to Carbon Nano-materials

It is possible to convert plastics into carbon nano-materials such as carbon nano-tubes (CNT), graphene and carbon dots. A handful of literature is available on this topic. Mukherjee et al. (2016) have presented a comprehensive review on conversion of plastics to CNT. However, the commercialization of this process is yet to be realized using plastics as feedstock.

Table 1 presents a few examples reported for conversion of CNT from plastics. Though attempts from WEEE polymers have not been taken yet, this area can be explored in the future and could lead to a green way of resource utilization.

3.6 Co-processing

Co-processing is a method which uses waste materials as raw material or energy source or both in industrial process without compromising the quality of the product (de Queiroz Lamas et al. 2013). It is primarily carried out in energy-intensive industry such as cement, limed, glass and steal. It is a preferable alternative for sound and environmental friendly disposal of waste over incineration and other non-scientific methods. Non-recyclable plastics from WEEE can be co-processed in cement kilns (Fig. 7). Undoubtedly, this is an efficient and economically sustainable process which is suitable as a benchmark technology for WEEE plastic recycling.

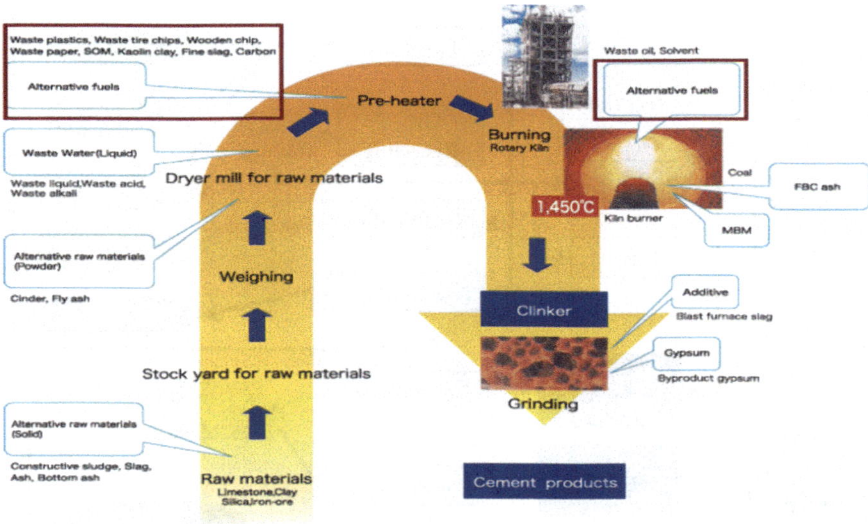

Fig. 7 Co-processing in cement kilns scheme

4 Discussion and Analysis

The previous section has discussed several methods of WEEE polymer recycling. However, there are lots of issues and challenges associated with WEEE plastic recycling. The major concern with recycling of WEEE polymer arises due to the supply chain issues and the lack of awareness and knowledge. In developing countries, the lion's share of WEEE is collected by the informal sector. Their incompetence in identification of polymer, lack of proper knowledge and shrewd awareness restricts the proper utilization of the resource trapped in WEEE plastics. They often end up in landfill. On the other hand, formal WEEE recyclers also face different problems. The primary issue is the identification of polymers for which indecisive attitude prevails. Another major concern is the presence of hazardous substances such as halogenated flame retardants which restrict certain technologies to be useful. In general, non halogenated polymers are subjected to physical recycling, yet the output quality remains questionable owing to the mixture of polymers. Other than this, there are certain environmental and social concerns which should be addressed for sustainable utilization of WEEE plastics.

In order to establish a sustainable pathway for utilization of WEEE polymers, a framework has been proposed. The framework has been developed based on literature findings, brainstorming and the author's previous experience in field surveys. Figure 8 presents the proposed sustainable framework for WEEE plastics recycling. Initially, WEEE is collected by both formal and informal sectors. Then, the collected WEEE is dismantled and plastics are recovered. WEEE plastics are of three types—recyclable, halogenated plastics and non-recyclable plastics. From the recyclable plastics by

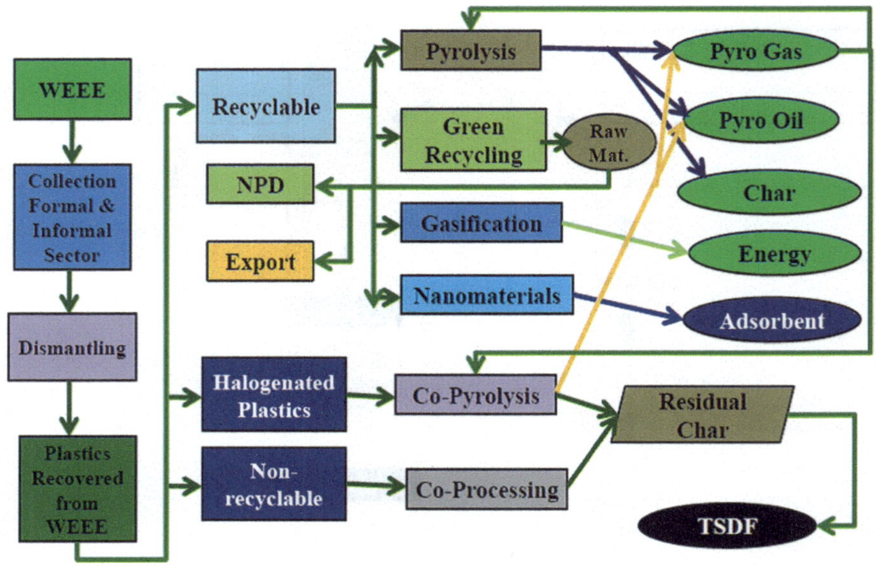

Fig. 8 Sustainable framework for WEEE polymer recycling

pyrolysis, pyro-gas, pyro-oil and pyro-char can be recovered. From this, pyro-gas can again be utilized in the process itself for heating up the system, and the oil and char can be sold to the market for earning revenue. New product development is possible by green recycling process which converts the plastics into raw materials. The raw materials can also be exported to generate revenue. It is also possible to manufacture nano-materials from these plastics which can be used as absorbent. In case of halogenated plastics, the best way is co-pyrolysis with bio-mass. As a result of co-pyrolysis, pyro-oil is produced which is free from halogen has a market value. Pyro-gas can again be utilized in the process to partially substitute the supply of heat. In case of non-recyclable plastics, co-processing in the cement kiln is the only solution. Residues after co-pyrolysis and co-processing of non-recyclable plastics (if there is any) should be sent to the TSDF for further scientific and controlled disposal.

5 Conclusion

In this study, an attempt has been made to look into the possible ways to recycle different types of WEEE polymers. Existing technologies with their drawbacks has been discussed. Issues and challenges associated with WEEE polymer recycling have also been presented to pinpoint the problems. Finally, a sustainable framework has been proposed for effective utilization of WEEE polymers. The findings of the study are expected to be helpful for the researchers, policy-makers and other stakeholders.

Acknowledgements The authors would like to acknowledge the International Society of Waste Management, Air and Water (ISWMAW) for partially funding the work. Additionally, co-operation from Centre for Quality Management Systems (CQMS) Jadavpur University, Kolkata, is gratefully acknowledged.

References

Abnisa, F., & Duad, W. M. A. W. (2014). A review on co-pyrolysis of biomass: An optional technique to obtain a high-grade pyrolysis oil. *Energy Conversion and Management, 87,* 71–85.

Balde, C. P., Forti, V., Gray, V., Kuehr, R., & Stegmann, P. (2017). The global e-waste monitor 2017: Quantities, flows and resources. United Nations University (UNU), International Telecommunication Union (ITU) & International Solid Waste Association (ISWA), Bonn/Geneva/Vienna. ISBN Electronic Version: 978-92-808-9054-9.

Catalyst C (India) Pvt. Ltd. (2015). A brief report on electronics industry in India. Available online http://www.cci.in/pdfs/surveys-reports/Electronics-Industry-in-India.pdf. Accessed January 4, 2016.

de Queiroz Lamas, W., Palau, J. C. F., & de Camargo, J. R. (2013). Waste materials co-processing in cement industry: Ecological efficiency of waste reuse. *Renewable and Sustainable Energy Reviews, 19,* 200–207.

Debnath, B., Chowdhury, R., & Ghosh, S. K. (2018a). Sustainability of metal recovery from e-waste. *Frontiers of Environmental Science & Engineering, 12*(6), 2. https://doi.org/10.1007/s11783-018-1044-9.

Debnath, B., Chowdhury, R., & Ghosh, S. K. (2018b). Studies on sustainable material recovery via pyrolysis of WEEE. In *70th Annual Session of Indian Institute of Chemical Engineers, CHEMCON 2017*, Haldia, West Bengal, India.

Debnath, B., Roychowdhury, P., & Kundu, R. (2016). Electronic Components (EC) reuse and recycling–a new approach towards WEEE management. *Procedia Environmental Sciences, 35,* 656–668.

Ghosh, S. K., Debnath, B., Baidya, R., De, D., Li, J., Ghosh, S. K., Zheng, L. … Tavares, A. N. (2016). Waste electrical and electronic equipment management and basel convention compliance in Brazil, Russia, India, China and South Africa (BRICS) nations. *Waste Management & Research, 34*(8), 693–707.

Ghosh, S. K., Singh, N., Debnath, B., De, D., Baidya, R., Biswas, N. T. … Li, J. (2014). E-waste supply chain management: Findings from pilot studies in India, China, Taiwan (ROC) and the UK. In *Proceedings of the 9th International Conference on Waste Management and Technology*, Beijing, China, 29–31 October, China: Basel Convention Regional Centre for Asia and Pacific (pp. 1131–1140).

Ghosh, S. (2004). Improved eco friendly recycling process of post consumer waste plastics & devices thereof. India. Patent no 202532:(772/KOL/2004).

Kang, J. J., Lee, J. S., Yang, W. S., Park, S. W., Alam, M. T., Back, S. K. … Saravanakumar, A. (2016). A study on environmental assessment of residue from gasification of polyurethane waste in e-waste recycling process. *Procedia Environmental Sciences*, *35*, 639–642.

Liu, W. W., Hu, C. W., Yang, Y., Tong, D. M., Zhu, L. F., Zhang, R. N., et al. (2013). Study on the effect of metal types in (Me)-Al-CM-41 on the mesoporous structure and catalytic behavior during the vapour catalyzed co-pyrolysis of pubescens and LDPE. *Applied Catalysis B, 129,* 202–213.

Mukherjee, A., Debnath, B., & Ghosh, S. K. (2016). Carbon nanotubes as a resourceful product derived from waste plastic—A review. In *Proceedings of the 6th international conference on solid waste management (6th IconSWM)*, Kolkata, India, 26–28 November, Jadavpur University (pp. 1234–1247).

Panda, A. K., Singh, R. K., & Mishra, D. K. (2010). Thermolysis of waste plastics to liquid fuel: A suitable method for plastic waste management and manufacture of value added products—A world prospective. *Renewable and Sustainable Energy Reviews, 14*, 233–248.
Shen, Y., Zhao, R., Wang, J., Chen, X., Ge, X., & Chen, M. (2016). Waste-to-energy: Dehalogenation of plastic-containing wastes. *Waste Management, 49,* 287–303.
Yamawaki, T. (2003). The gasification recycling technology of plastics WEEE containing brominated flame retardants. *Fire and Materials, 27,* 315–319.
Zhang, S., Yoshikawa, K., Nakagome, H., & Kamo, T. (2013). Kinetics of the steam gasification of a phenolic circuit board in the presence of carbonates. *Applied Energy, 101,* 815–821.

Slate Mine Wastewater, the Best Substitute for Cementation

S. Altaf Hussain, S. M. Subhani and S. V. Satyanarayana

Abstract It is the greatness of nature, which has given lot of material required for human being. Lot of mineral matter is extracted from mines; one of such requirements is slate which is used as a note pad for early learners with the help of slate pencil. The written matter is easily erasable and the slate is reusable. Slate rocks are available in the earth crust of Markapur area of Andhra Pradesh. In order to extract slate mine in a slice form, different technologies are being used. After extraction of slate mine the leftover water is found rich in calcium, silica, iron, and alumina which are the essential components in the process of cementation. Wastewater procured from Markapur has been used in making cement cubes and the compressive strength was tested after the curing days of 1, 3, 7 and 28 days, as per Bureau of Indian Standards. Slate mine wastewater has also been used to mix granite waste and ordinary Portland cement, where the results indicated very good compatibility. All the results depict very good compressive strength. It can be concluded that, slate mine water is the best substitute for civil construction, whereas the same water is unsafe for human consumption as well as plants consumption due to excess mineral mater, according to World Health Organization norms.

Keywords Wastewater · Slate mine waste · Granite industry waste · Alternative cementation · Compressive strength · Solid waste management · Powder

1 Introduction

Slate is conventionally used as a mean of preliminary learning, being used to write using slate pencils. Slate is produced from slate mines which is basically quarts,

S. Altaf Hussain (✉)
Lords Institute of Engineering and Technology, Hyderabad, Telangana, India
e-mail: altaf.che@gmail.com

S. M. Subhani
Department of Chemical Engineering, JNTUA College of Engineering, Anantapur, A.P, India

S. V. Satyanarayana
JNTUA University Ananthapuramu, Anantapur, A.P, India

S. K. Ghosh (ed.), *Urban Mining and Sustainable Waste Management*,
https://doi.org/10.1007/978-981-15-0532-4_10

excavated from the mine and then processed to produce a flat plate of around 3 mm thickness. During the process of production of slate lot of water remain as slate mine effluent (SME), where the concentration of different constituents is more than the norms specified by the central pollution control board.

According to the norms iron should be within 0.3 ppm, whereas SME contains 128 ppm, similarly the effluent has aluminium should be within 0.03 ppm, whereas 135 ppm is present in slate mine effluent, likewise sodium and silica to be in the norms are 200 and 370 ppm, whereas the SME contains 432 and 753 ppm to be treated and brought down to the specified norms before it is discharged into the nature, else the effluent will have an effect on aquatic life causing sluggishness in reproductive system as well as causing indirect system to human being through food chain (Jacob et al. 2013). The slate effluent will also affect the plant and animal lives directly in their growth directly as well as harm human being indirectly through food chain because of the impurities present in slate mine effluent. The growth of plants was observed to be sluggish, when effluent was used to water conventional plants, which is because of excess toxic compounds present in effluent (Mohammad et al. 1984). Presence of aluminium concentration in effluent streams demonstrated to be toxic for plants with sluggishness in growth and productivity (Jones et al. 2003). Plants require some amount of salt to be present in the medium, viz. soil, sand and water. Higher concentrations of salts were observed to decay the plant growth (Murphy et al. 2003). The stipulated values are set by central pollution control board as well as state pollution control board towards the biotic life. Concentrations beyond in effluent found to be toxic in fishes leading to depletion in life cycle (Shayne et al. 2005).

The compressive strength and other properties of cement are increasing when underground water is used in comparison with tap water (Su et al. 2002). The compressive strength of cement decreases when wastewater generated in oil fields is used in construction, this could be due to the nature of oil particulates, which are basically organic in nature, whereas cement is an inorganic material (Al Harty et al. 2005). Wastewater coming over from ready mix preparation plants can be the best substitute to be used for mortar preparation; it reduces the capillary nature of cement (Franco et al. 2001). Use of sludge water in concrete mix can be an option for effective utilization of sludge, but it has its positives and negatives, whereby such sludge can be used for constructing boundary wall bricks in ground floor (Chatveera et al. 2006). Use of concrete plant sludge with a maximum of 6% solids is recommended to use in cement mortar, without compromising the quality of cement, in terms of mechanical properties (Chatveera et al. 2009).

Ordinary Portland cement (OPC) is commonly used as the best means of cementing material for construction, with different grades including 33, 43 and 53 grade, available in the market as per Bureau of Indian Standards (IS: 12269: IS 1489:2005). OPC consists of calcium silicates, aluminates and iron. Calcium carbonate is first dissociated at around 900 °C to produce calcium oxide and carbon dioxide, further it is heated at around 1400 °C and made to react with silica, alumina and iron, to form small pebbles called clinker, which is then grinded to approximately 100 μm size to produce OPC cement. An amount of approximately 5% gypsum is added to retard setting time, which serves as a buffer time during construction.

Portland Pozzolana cement (PPC) is produced by mixing a maximum of 35% (BIS: IS 1489:1991) of fly ash in OPC cement. Fly ash is produced at thermal power stations as a waste material, when huge volumes of coal are burnt in boilers. The flue gas is made to pass through electrostatic precipitators and the fine dust (fly ash) is collected. Fly ash consists of silica, calcium oxide and traces of alumina as well as iron, which is cementing compatible material. The present line of work is to replace fly ash with granite waste in OPC cement. Earlier fly ash was considered as waste and was used for landfilling, similarly, granite waste is presently used in landfilling.

In India, about 960 million tons of solid waste is generated annually as by-product from different industrial processes out of which approximately 350 million tones are from organic agricultural sources, 290 million tones inorganic wastage from industrial mining and 4.5 million tones are hazardous waste. Part of the waste from different industries has been recycled in construction industry as well as other industries (Ashokan et al. 2007). The granite sawdust comprises of calcium and iron which has the compatibility to acid soils, granite waste powder is used as a suitable means to neutralize acid soils (Barral et al. 2005). Use of granite and marble rock waste found to be effective in the production of concrete for civil construction (Hanifi et al. 2008). Use of municipal solid waste incinerated fly ash is found to be effective up to 20%, as cement substitute (Lin et al. 2003). Muscovite granite waste has the ability to produce ceramic tiles by adding 20–30% (Mirabbos et al. 2011).

In Brazil, granite waste powder has been effectively studied and proved compatible for making ceramic bricks and tiles and the results have been acknowledged by Brazil standardization for ceramic bricks and tiles (Romualdo et al. 2005). Granite sludge waste is found to be effective filler for pozzolanic motors, where a reddish pigment is produced by calcining at lower temperature (700–900 °C) for short time, therefore, granite sludge can be an effective additive in the preparation of coloured motor (Marmol et al. 2010). The sludge produced from granite cutting polish industries is also found to be effective up to 10% mixing to produce roof tiles with enhance properties (Monteiro et al. 2004) and (Torres et al. 2009). Granite waste is also found to be effective up to 40% (Mg by weight) in producing fly ash magnesium oxychloride cement (Ying et al. 2013). Granite waste is currently used for landfilling at most of the industries in India. A maximum of 35% fly ash is being added to OPC and available in market as PPC (Celik et al. 2008). Current research is intended to mix different proportions of granite sawdust collected from different locations of granite industry and study for the compatibility with OPC.

2 Materials and Methods

Slate mine wastewater was procured from Markapur slate mine area, of Andhra Pradesh, India. Granite samples: Granite industry cuts the granite rock to slabs based on the requirement of the client, using metal saws, and the fine powder is produced during cutting is considered as solid waste and is being sent to dump. During the conveyance from granite cutting zone to waste dump, samples were collected at the first interval, where the density of waste mixture is high in comparison to waste collected from other destinations. Sample 1 is collected just below the cutting zone and the same is blended with different percentages starting from 20 to 40% in ordinary Portland cement (OPC). Similarly, Sample 2 is also collected from the same place but of different granite and the same is blended with different percentages starting from 20 to 40% in OPC. Likewise, another sample is collected from the same place but of different granite. Same is blended with different percentages starting from 20 to 40% in OPC.

3 Preparation of Testing Specimen

Aggregates are prepared with reference to the BIS standards by mixing laboratory-grade sand (supplied by Tamilnadu minerals Ltd, Innore, Chennai) cement and water, in the ratio of 3:1 (sand: cement-mass basis), water is added based on the normal consistency. These composites are thoroughly mixed in pony mixer for 2 min, poured into standard mould of $7.5 \times 7.5 \times 7.5$ cm, then subjected to compression in vibrating machine for 2 min, and then kept in humidity chamber for 24 h, maintaining 27-degree centigrade. The specimens are then taken out and cured for 1, 3, 7 and 28 days in curing chambers of same temperature.

3.1 Testing of Samples

1. Compressive strength: Specimen cubes are subjected to compressive strength test in the standard compressive strength machine, supplied by AIMIL, New Delhi.
2. All the blended samples were analyzed for physical and chemical analysis as per Bureau of Indian Standard norms.

4 Results and Discussion

The slate mine waste from different zones of Markapur slate mines of Andhra Pradesh, India had been procured and mixed with the cement to prepare testing cubes and tested their strength after 1,3,7 and 21 days of curing as per BIS. The 'compressive strength' of samples is shown in Table 6.6. A maximum of 61.61 Mpa for Sample 1, similarly for Sample 2 maximum of 62.11 Mpa, and for Sample 3 62.01 Mpa is observed, when slate mine wastewater is used for preparing the moulds,

Table 1 Compressive strength of cement sample(s) using slate mine effluent at different ages of curing

Cement sample	Day/compressive strength (MPa)							
	Wastewater mix sample 1				Wastewater mix sample 2			
	1	3	7	28	1	3	7	28
Cement sample 1	9.09	15.27	33.95	61.66	10.29	16.29	32.32	62.11
Cement sample 2	8.37	18.22	36.54	56.97	11.42	17.42	34.33	59.86
Cement sample 3	9.42	16.63	34.58	61.72	9.43	15.32	34.42	62.94
Cement sample	Day/compressive strength (MPa)							
	Wastewater mix sample 3				Regular water mix sample 4			
	1	3	7	28	1	3	7	28
Cement sample 1	9.17	18.32	36.50	62.01	8.41	13.87	30.42	55.50
Cement sample 2	8.54	16.34	35.73	59.54	7.31	14.32	31.26	54.43
Cement sample 3	10.42	16.42	35.43	61.93	8.21	12.5	31.34	54.32

whereas 55.5 Mpa for Sample 4 was observed in when conventional water is used for preparing the moulds.

The compatibility of slate mine effluent with the process of cementation may be due to the chemical composition, where silica was observed as 753.8 ppm, alumina 134.4 ppm and iron 122.8 ppm, all these values are much above threshold limits as shown in Table 6.5. The compressive strength is increasing as the number of days of curing increases, in line with the established research.

Granite Sample 1, Sample 2 and Sample 3 are mixed up to 40% in OPC, and its physical properties are shown in Table 1. It has been observed an increase of bulk density and tapped density as the percentage of granite mix increased (Matthews et al. 2011); this could be because of variance in particle size and shape of granite waste and OPC cement (Abdullah et al. 1999).

The critical quality parameter 'compressive strength' of samples is shown in Table 2. A maximum of 54.32 Mpa for 20% granite mix and minimum of 43.23 Mpa for 40% granite mix are observed for Sample 1, similarly for Sample 2 maximum of 48.33 Mpa and minimum of 41.17 Mpa are observed, and for Sample 3, a maximum of 48.86 Mpa and a minimum of 42.96 Mpa are observed, after curing 28 days. The compressive strength is increasing as the number of days of curing increases, in line with the established research (Wig et al. 1915). The values of compressive strengths of Samples 1 and 3 for day 1, 3 are found to be increasing as the percentage of granite wasteis increased. This could be because of initial reaction of iron, as the percentage of iron is higher for all the samples compared to other points in discharge stream of waste. Iron is present in granite waste as eroded powder of iron saw caused during sawing operation. The compressive strengths of 28 days was found to be decreasing as the percentage of granite waste increases for all samples.

Bulk densities of samples were varying; it could be because of morphology of sample and its movement from sawing zone to dumping zone (Tong et al. 2002).

Table 2 Chemical analysis of slate mine wastewater

S. no.	Parameter	Max. permissible limit (ppm)	Effluent from slate industry (ppm)
1	Al	0.03	134.6
2	Ca	200	633.3
3	Fe	0.3	122.8
4	Mg	30	43.9
5	Mn	0.1	2.2
6	Na	200	432.1
7	K	20	110.8
8	P	0.10	2.3
9	Si	370	753.8
10	Ti	1.0	7.2
11	B	0.8	7.5

Table 3 Bulk densities of Sample 1, 2 and 3

Granite %	Bulk density (gm/cc)		
	S-1	S-2	S-3
20	0.93	0.88	0.91
25	0.92	0.92	0.89
30	0.94	0.87	0.90
35	0.89	0.93	0.91
40			

The compressive strength values obtained after mixing with granite shows the same trend as in case of fly ash mix (Celik et al. 2008); therefore, granite waste can be perfect mix as a cementing material (Tables 3 and 4).

The critical quality parameter 'compressive strength' of samples is shown in Table 2. A maximum of 54.32 Mpa for 20% granite mix and minimum of 43.23 Mpa for 40% granite mix are observed for Sample 1, similarly for Sample 2 maximum of 48.33 Mpa and minimum of 41.17 Mpa are observed, and for Sample 3, a maximum of 48.86 Mpa and a minimum of 42.96 Mpa are observed, after curing 28 days. The compressive strength is various chemical properties like, insoluble residue, loss on ignition and SO_3 is measured for all the samples and obtained results are shown in Table 5. The values of insoluble residue found to be increasing, as the percentage of granite increases, this could be because of nature of granite material. The range of values is similar with PPC cement BIS: IS 1489: (1991). The other chemical property 'loss on ignition' found to be approximately 1% which infers organic material, which is similar in case of OPC as well as PPC. Percentage of gypsum as indicated by SO_3 found to be slightly decreasing as the percentage of granite increases. The normal

Table 4 Compressive strength of Sample 1, 2 and 3, at different ages of curing and different percentages of granite saw waste mix

Granite %	Day/compressive strength (MPa)											
	Sample 1				Sample 2				Sample 3			
	1	3	7	28	1	3	7	28	1	3	7	28
20	13.34	22.21	31.27	54.32	14.32	19.76	29.69	48.33	12.88	27.21	36.22	48.86
25	15.22	24.83	33.43	44.83	15.44	28.44	28.90	44.45	14.76	29.87	38.12	44.34
30	16.88	23.33	34.76	39.98	15.83	27.99	27.98	45.39	17.33	29.33	31.11	42.22
35	18/41	25.65	33.33	38.33	16.55	29.33	39.41	41.86	18.43	28.55	33.85	41.22
40	19.22	25.55	34.33	43.23	19.32	28.90	39.88	41.17	19.08	29.93	33.96	42.96

Table 5 Chemical properties of samples

Samples	Granite%	Loss on ignition%	Insoluble residue%	SO_3 %
Sample 1	20	1	20	6
	25	1	21	6
	30	1	25	5
	35	1	27	5
	40	1	28	6
Sample 2	20	1	20	6
	25	1	23	6
	30	1	30	5
	35	1	32	5
	40	1	33	6
Sample 3	20	1	33	6
	25	1	33	6
	30	1	33	5
	35	1	33	5
	40	1	33	6

percentage of gypsum in OPC and PPC as SO_3 should be less than 5%, which is added to clinker externally, for retarding setting time. There has been a slight increment SO_3, values for all samples of granite mix, which could be because of the nature of granite.

5 Conclusions

The objective of present work is to study the compatibility of high-density granite saw waste as a cementing material, and the results found are encouraging for a maximum mix of 40% in OPC. The results were compared with Bureau of Indian Standard (BIS) norms and are the values are well in line within the range, up to addition of 40% granite waste, little above the lines of PPC cement. Presently, granite waste was considered to be a waste product and is being used for landfilling. The waste of granite industry, collected from the first point of sedimentation can effectively be used as a substitute in blending with ordinary Portland cement, for a maximum percentage of 40.

References

Abdullah, E. C., & Geldart, D. (1999). The use of bulk density measurement as flow ability indicators. *Powder Technology, 102,* 151–165.

Al-Harthy, A. S., Taha, R., Abu-Ashour, J., Al-Jabri, K., & Al-Oraimi, S. (2005). *Cement and Concrete Composites, 27,* (1), 33–39.

Ajma, M., & Khan, A. U. (1984). *Ecological and Biological*, 34(4), 367–379.

Binici, H., Shah, T., Aksogan, O., & Kaplan, H. (2008). Durability of concrete made with granite and marble as recycle aggregates. *Journal of Materials Processing Technology, 208,* 299–308.
Bureau of Indian Standards IS 1489(part 1). (1991).
Bureau of Indian Standards IS: 12269: IS 1489. (2005).
Celik, O., Damci, E., & Piskin, S. (2008). Characterization of fly ash and its effects on the compressive strength properties of Portland cement. *Indian Journal of Engineering and Material Sciences, 15,* 433–440.
Chatveera, B., Lertwattanaruk, P., & Makul, N. (2006). *Cement & Concrete Composites, 28*(5), 441–450.
Chatveera, B., & Lertwattanaruk, P. (2009). *Journal of Environmental Management, 90*(5), 1901–1908.
Franco, S., & Elisa, F. (2001). *Cement and Concrete Research, 31*(3), 485–489.
Gad, S. C. (2005). *Encyclopedia of Toxicology*, 233–239.
Hojamberdiev, M., Eminov, A., & Xu, Y. (2011). Utilization of muscovite granite waste in the manufacture of ceramic tiles. *Ceramics International, 37,* 871–876.
Jacob, D. O., Dube, M. G., & Niyogi, S. (2013). *Ecotoxicology and Environmental Safety, 91,* 188–197.
Jones, D. L., & Ryan, P. R. (2003). *Encyclopedia of Applied Plant Science*, 656–664.
Li, Y., Yu, H., Zheng, L., Wen, J., Wu, C., & Tan, Y. (2013). Compressive strength of fly ash magnesium oxychloride cement containing granite wastes. *Construction and Building Materials, 38,* 1–7.
Lin, K. L., Wang, K. S., Tzeng, B. Y., & Lin, C. Y. (2003). The reuse of municipal solid waste incinerator fly ash slag as a cement substitute. *Resources, Conservation and Recycling, 39,* 315–324.
Marmol, I., Ballester, P., Cerro, S., Monros, G., Morales, J., & Sanchez, L. (2010). Use of granite sludge waste for the production of coloured cement based mortars. *Cement & Concrete Composites, 32,* 617–622.
Menezes, R. R., Ferreira, H. S., Neves, G. A., de Lira, L. H., & Ferreira, H. C. (2005). Use of granite sawing wastes in the production of ceramic bricks and tiles. *Journal of the Europian Ceramic Society, 25,* 1149–1158.
Mullarney, M. P., Beach, L. E., Dave, R. N., Langdon, B. A., Polizzi, M., & Blackwood, D. O. (2011). Applying dry powder coatings to pharmaceutical powders using a comil for improving powder flow and bulk density. *Powder Technology, 212,* 397–402.
Murphy, D. J. (2003). *Encyclopedia of Applied Sciences*, 1478–1483.
Monteiro, S. N., Pecanha, L. A., & Vieira, C. M. F. (2004). Reformulation of roofing tiles body with addition of granite waste from sawing operations. *Journal of the European Ceramic Society, 24,* 2349–2356.
Pappu, A., Saxena, M., & Asolekar, S. R. (2007). Solid waste generation in India and their recycling potential in building materials. *Building and Environment, 42*(6), 2311–2320.
Silva, M. T., Hermo, B. S., Garcia-Rodeja, E., & Freire, N. V. (2005). Reutilization of granite powders as an amendment and fertilizer for acid soils. *Chemosphere, 61*(7), 993–1002.
Su, N., Miao, B., & Liu, F. S. (2002). *Cement and Concrete Research, 32*(5), 777–782.
Tong, X. D., & Sun, Y. (2002). Particle size and density distributions of two dense matrices in an expanded bed system. *Journal of Chromatography A, 997,* 173–183.
Torres, P., Fernandes, H. R., Olhero, S., & Ferreira, J. M. F. (2009). Incorporation of wastes from granite rock cutting and polishing industries to produce roof tiles. *Journal of European Ceramic Society, 29,* 23–30.
Wig, R. J., Williams, G. M., & Gates, E. R. (1915). Compressive strength of Portland cement mortars and concretes. *Journal of the Franklin Institute, 180,* 608–613.

Waste Coal Utilization in India: A Review

Krishna Kant Dwivedi, Prabhansu, M. K. Karmakar, A. K. Pramanick and P. K. Chatterjee

Abstract Waste coal production is an important worldwide issue, especially in India. Indian waste coal has great importance due to the increase of production in latest years. With the increasing demand of high-quality coal, waste category coals also can be used for combustion and chemical production. Indian waste coal is usually featured with high moisture content, high ash content and low carbon content, which especially exhibits high volatile and low heating value. The physio-chemical analysis (proximate and ultimate analysis) of Indian waste coal were carried out as per ASTM-D 5373 and results shows that the volatile matter, fixed carbon, ash contents and moisture contents are in the range of (14–16%), (5–8%), (73–79%) and (2–3%), respectively, with a moderate amount of carbon contents (10–15%). The present review article provides a comprehensive overview of the various thermal treatments and advanced technologies, characterization of Indian waste coal (proximate and ultimate analysis) for the possibility of using waste category coal as a fuel. The review starts from basic aspects of the process such as important operating parameters than focus on comparative analysis of the utilization of Indian waste coal with their characterization and the environmental performances of different fluidized-bed gasifiers. The analysis indicates that gasification and pyrolysis are technically viable

K. K. Dwivedi (✉) · A. K. Pramanick
Department of Mechanical Engineering, National Institute of Technology, Durgapur, West-Bengal, India
e-mail: kkdwivedi44@gmail.com

A. K. Pramanick
e-mail: akpramanick@yahoo.com

Prabhansu
Department of Mechanical Engineering, Muzaffarpur Institute of Technology, Muzaffarpur, Bihar, India
e-mail: prabhansu.nitp@gmail.com

M. K. Karmakar · P. K. Chatterjee
Energy Research and Technology Group, CSIR-Central Mechanical Engineering Research Institute, Durgapur, West Bengal, India
e-mail: malaycmeri@gmail.com

P. K. Chatterjee
e-mail: pradipcmeri@gmail.com

S. K. Ghosh (ed.), *Urban Mining and Sustainable Waste Management*,
https://doi.org/10.1007/978-981-15-0532-4_11

option for the waste conversion. The advantages in terms of utilization of Indian waste coal are also covered in detail and shows that Indian waste coal could be an effective energy source that will not only contribute to reuse the waste materials but also reduces waste disposal landfills.

Keywords Waste coal · Pyrolysis · Gasification · Thermo-chemical conversion · Circulating fluidized bed

1 Introduction

Energy demand is a major step of the growing population. Many researchers are trying to find out the alternative forms of energy. There are different ways of alternative energy sources, but Indian waste coal has the unique advantage as an energy source. Large amount of waste coal is reserved underground in Indian coal mines, especially in west zone coalmines. Indian waste coal cannot be conveniently used like fine coal because Indian waste coal has low caloric value and high moisture content. Therefore, the utilization of waste coal is far from satisfying in the world (Tanigaki et al. 2012; Liu et al. 2016; Belgiorno et al. 2003). Many methods have been developed in order to use Indian waste coal as an energy source since the 1920s. There are some technical steps for utilization of low-rank coal-like steam drying, supercritical water drying and hydro-thermal–mechanical compression drying (Koukouzas et al. 2008; Morris and Waldheim 1998). The future energy resources and chemicals supply basically depends on efficient and cost-effective utilization of Indian waste coal. In case of energy source, Indian waste coal is well known because it has high concentration of oxygen-containing functional groups and high proportion of transitional (Pan et al. 2000). Gasification is the process in which any carbonaceous fuel can be converted into a gaseous product and it will be used for energy conversion (Shehzad et al. 2016). Gasification of Indian waste coal as an energy resource is a current domain in all over the world. There are many thermo-chemical conversion methods for conversion of coal into fuel such as pyrolysis can justify the weight loss profile with respect to time and temperature from a variety of sources including waste category coal into liquid, solid and gaseous fuels (Idris et al. 2010; Mondal et al. 2018; Saini and Srivastava 2017).

In the present study, a literature survey has been made to evaluate the status and identify the problems on uses of Indian waste coal. The study also aims at encouraging researchers to work towards the improvement of the present overview on Indian waste coal through suggestions and recommendations.

Table 1 State wise distribution of coal resources

S. no.	Name of the state	Geological resources of coal in million tonnes
1.	Andhra Pradesh	17,715
2.	Arunachal Pradesh	90
3.	Assam	375
4.	Bihar	160
5.	Chhattisgarh	41,450
6.	Jharkhand	74,392
7.	Madhya Pradesh	20,346
8.	Maharashtra	9670
9.	Meghalaya	460
10.	Sikkim	73
11.	West Bengal	28,335

2 Availability of Coal in India

India is the world's sixth-largest energy market in the world. The quantity of different quality of Indian coals depends on different conditions like standard of living, degree of commercial activities and different geological phenomena. In case of waste category coal washing is an important step for efficient industrial use. The production of waste category coal requires less ash content to minimize the formation of energy consuming. Percentage of moisture present in waste coal can affects the gasification process. In Indian waste coal moisture is between 7 and 13% (Dwivedi et al. 2018). The availability of Indian coal in different state listed in Table 1.

3 Characterization and Composition of Indian Waste Coal

3.1 Proximate and Ultimate Analysis

Chemical properties of Indian waste coal are an essential step to evaluate the alternative processing and recovery options of coal samples. It has two major steps first proximate analysis and second is ultimate analysis. The approximate values for proximate and ultimate analysis for Indian waste coal are described below as follows:

3.1.1 Proximate Analysis

With the help of proximate analysis, we can evaluate the amount of moisture, volatile matter, total ash and fixed carbon in any coal samples as shown in Table 2.

Table 2 Proximate analysis for Indian waste coal

Sample	Volatile (%)	Ash (%)	Fixed carbon (%)	Moisture (%)
Indian waste coal	16.63	73.45	6.73	3.19

Table 3 Ultimate analysis for Indian waste coal

Sample	Carbon C (%)	Hydrogen H (%)	Nitrogen N (%)	Sulphur S (%)	Oxygen O (%)
Indian waste coal	15.02	1.13	0.48	0.20	13.61

3.1.2 Ultimate Analysis

Ultimate analysis is the process to calculate the percentage of carbon, hydrogen, oxygen, nitrogen, sulphur and ash in Indian waste coal samples. The approximate values for ultimate analysis for Indian waste coal are shown in Table 3.

4 Thermal Treatment Techniques Applied for Utilization of Indian Waste Coal

4.1 Pyrolysis

Pyrolysis is a thermo-chemical conversion of a sample with respect to time and temperature. During the pyrolysis process, no flue gas is required to utilization. Basically, pyrolysis is a thermo-chemical conversion process that helps to evaluate the physio-chemical properties of samples (coal, biomass, etc.) with respect to time and temperature (Dwivedi et al. 2018). The pyrolysis study of coal mostly shows the thermo-chemical conversion of material with defined time and temperature conditions. The hole process includes different chemical reactions in mainly in three stages with releases of volatile matter second is fast pyrolysis and third is slow pyrolysis process. In the first stage about 70% weight losses in case of waste coal. So the pyrolysis of Indian coal helps to examine the ignition, rate of combustion and particle emission of coal. In the other hand, we can say that it is a method char production to improve its calorific value. During the gasification and co-combustion process, thermo-chemical conversion of coal plays an important role and also helpful for the design and development of boiler and gasifier. Coal pyrolysis passes mainly in two steps first is de-polymerization where vapour, gas and tar are formed and second is re-polymerization where char forming occurs. Mass loss of coal can be measured by using pyrolysis. During the thermo-gravimetric analysis of coal in terms of coal pyrolysis we can calculate weight loss percentage and maximum weight loss at peak of temperature with the help of TG and DTG profiles. The kinetic parameters such as

Table 4 Literature survey for TGA with operating conditions and status

Sample	Operating conditions (temperature) (°C)	Notes	References
Malaysian low-rank coal	25–900	Results show that there are three thermal decomposition evolution profiles	(Saini and Srivastava 2017)
Indian coal	30–950	The characteristic parameters of different particle size coals increased significantly with increasing the heating rate. Corresponding calculated mean value of activation energy for coal is found 241.132 kJ/mol	(Dwivedi et al. 2018)
Low-rank Indonesian coal	30–900	Chemical composition and blending can improve the ability of Indonesian waste coal	(Mondal et al. 2018)
Dezhou (DZ) low-rank coal	30–1000	The reactivity rate of coal with steam/coal ratio of 1:2 is 1.98 times higher	(Idris et al. 2010)

activation energy and pre-exponential factors can be calculated through model-free analysis which may be useful in terms of development of pyrolysis technology and design of reactors (Table 4).

4.2 Gasification Technologies for Indian Waste Coal

Coal gasification technology is one of the most common technologies in India. Different types of solid waste like plastic waste and waste coal are dried and granulated to obtain required particle size for gasification at 700–900 °C. After discharging of gas and char, it may be utilized via combustion in a boiler to raise steam. Gasification can convert carbonaceous fuel into gaseous products.

Gasification of Indian waste coal as an energy recovery process and nowadays it is a current research domain in all over the world. Several researches focused on municipal solid waste gasification technologies by using fluidized-bed system. The fluidized-bed reactor technology has key role in the gasification of solid waste like coal and biomass (Idris et al. 2010; Mondal et al. 2018; Saini and Srivastava 2017). The syngas production and gas-heating value can be helpful to justify the feasibility of fluidized-bed system. Figure 1 shows the common thermo-chemical process and their output for utilization of waste coal (Fig. 2).

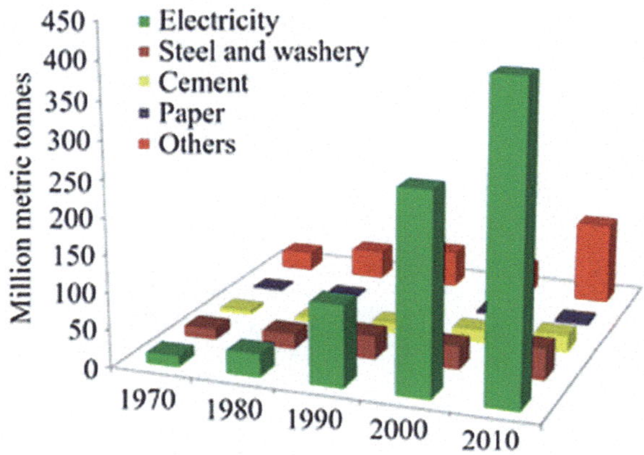

Fig. 1 Consumption of waste coal in India

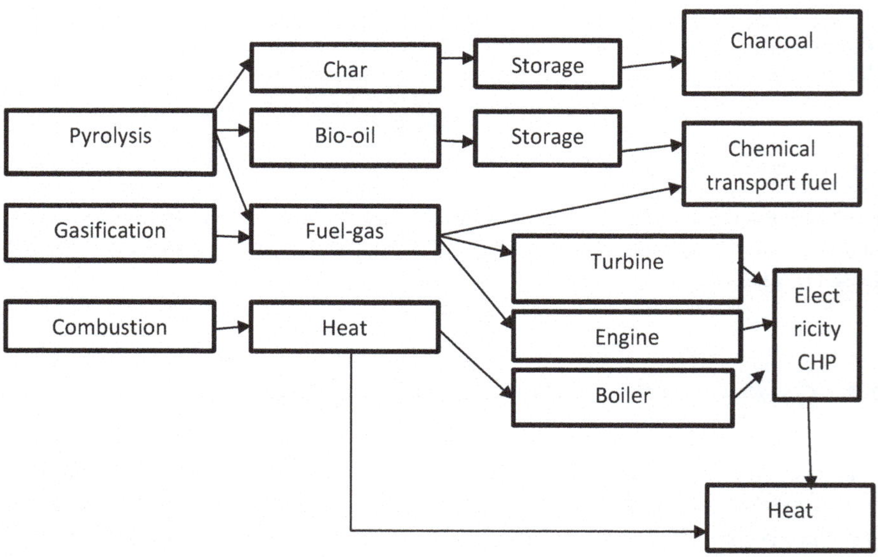

Fig. 2 Thermo-chemical process and their output for utilization of Indian waste coal

5 Some Treatment Strategies to Reduce Environmental Impacts of Indian Waste Coal

Indian coal generally has more ash content and low calorific value, which makes it undesirable for industrial purpose. Indian waste category coal needs washing to make it suitable because it has more ash content. Normal Indian coal generally is more suitable because it has less ash content but it has some another harmfully materials which can reduce gasification efficiency. Since, in India coal mining, regarding uses of waste coal has many issues so it is necessary to manage with good impact its utilization (Pan et al. 2000; Shehzad et al. 2016; Idris et al. 2010). Earlier research also developed some mineral matter and sulphur exhibit harmful effects on utilization of Indian waste coal. So removing of sulphur and high ash content is essential for utilization of waste coal used in different industries (Mondal et al. 2018). Many other ways is on the way to promote the uses of Indian waste coal samples such as some gasification technology-based fluidized-bed gasifier systems (Saini and Srivastava 2017).

6 Conclusion

Continuously improvement on waste management systems is an ever on-going process. In this review article, it is justified that we can use Indian waste coal in the form of pyrolysis and gasification. It is concluded that Indian waste coal utilization can be considered as future energy source with caution. Finally, the study concluded that with some new technologies and with some thermal treatment Indian waste coal can be used as an energy source. The various technologies of gasification and pyrolysis presented in this paper and have contributed greatly to the eco-image of waste management and particularly to Indian waste category uses, treatment and recovery. Using of waste category coal will certainly benefit the current situation. It is very important to consider recycling and energy recovery methods for utilization of Indian waste coal. This study required many further research and development in near future.

Acknowledgements The authors sincerely express their thanks to the Director, NIT Durgapur and Director, CSIR-CMERI, Durgapur for their support.

References

Belgiorno, V., Feo, G., Rocca, C., & Napoli, R. (2003). Energy from gasification of solid wastes. *Waste Management, 23,* 1–15.

Dwivedi, K. K., Chatterjee, P. K., Karmakar, M. K., & Pramanick, A. K. (2018). Experimental study on pyrolysis of coal by thermo gravimetric analysis (TGA) under different temperature conditions. *Journal of Energy and Environmental Sustainability, 5,* 49–52.

Idris, S., Rahman, N. A., Ismail, K., Alias, A. B., Rashid, Z. A., & Aris, M. J. (2010). Investigation on thermochemical behaviour of low rank Malaysian coal, oil palm biomass and their blends during pyrolysis via thermo gravimetric analysis (TGA). *Bioresource Technology, 101,* 4584–4592.

Koukouzas, N., Katsiadakis, A., Karlopoulos, E., & Kakaras, E. (2008). Co-gasification of solid waste and lignite—A case study for Western Macedonia. *Waste Management, 28,* 1263–1275.

Liu, Z., Lin, C., Chang, T., & Weng, W. (2016). Waste-gasification efficiency of a two-stage fluidized-bed gasification system. *Waste Management, 48,* 250–256.

Mondal, S., Mondal, A. K., Chintala, V., Tauseef, S. M., & Pandey, J. K. (2018). Thermochemical pyrolysis of biomass using solar energy for efficient biofuel production: a review. *Biofuels*. https://doi.org/10.1080/17597269.2018.1461512.

Morris, M., Waldheim, L. (1998). Energy recovery from solid waste fuels using advanced gasification technology. *Waste Management 18*, 557±564.

Pan, Y. G., Velo, E., Roca, X., Manya, J. J., & Puigjaner, L. (2000). Fluidized-bed co-gasification of residual biomass/poor coal blends for fuel gas production. *Fuel, 79,* 1317–1326.

Saini, M. K., & Srivastava, P. K. (2017). Effect of coal cleaning on ash composition and its fusion characteristics for a high-ash non-coking coal of India. *International Journal of Coal Preparation and Utilization, 37,* 1–11.

Shehzad, A., Bashir, J. K., & Sethupathi, S. (2016). System analysis for synthesis gas (syngas) production in Pakistan from municipal solid waste gasification using a circulating fluidized bed gasifier. *Renewable and Sustainable Energy Reviews, 60,* 1302–1311.

Tanigaki, N., Manako, K., & Osada, M. (2012). Co-gasification of municipal solid waste and material recovery in a large-scale gasification and melting system. *Waste Management, 32,* 667–675.

Temporal Changes of Solid Waste at Limestone Quarries in and Around Yerraguntla, YSR District, A.P., using Google Earth Images

Y. Sudarshan Reddy, B. Suvarna, M. Prasad, V. Sunitha and M. Ramakrishna Reddy

Abstract In limestone mining and cement industry, solid and liquid wastes are generated in every day and every stage of the operations and are required to mitigate properly. Different types of waste generated from both the industries are cement and limestone quarry. Due to environmental impact and public health and safety, proper efforts must be made to decrease waste generation and hence efficient disposal practices have to be followed. Hence, this study aims at a better understanding of spatial and temporal changes of unplanned dumping sites from 2006 to 2018. Google Earth mapping is one of the most advanced methods for identification of rock-solid waste clearly for collecting on satellite image data. This paper is centered on application of Google Images in assessing the temporal changes of solid waste at limestone quarries in and around Yerraguntla. Time series multi-date Google Earth imageries of 2006–2018 are used to demarcate the evolutionary changes in limestone waste disposal management and to understand the spatial and temporal changes that happened due to the changes by expanding the rock waste dump around Yerraguntla village, YSR district, A.P. Results revealed that the solid waste management in the study area is very poor which need to be properly monitored so as to mitigate the present and future environmental threats.

Keywords Temporal changes · Solid waste · Limestone quarries · Google Earth · Yerraguntla

1 Introduction

Waste is defined as the discarded and discharged material generated during every stage of life causing adverse health and environmental impact (Bringi 2007). Environmental contamination and waste management are the major concerns to earth scientist and form other related fields of science all over the world both in developing

Y. Sudarshan Reddy (✉) · B. Suvarna · V. Sunitha
Department of Geology, Yogi Vemana University, Kadapa 516005, India
e-mail: yenugusudharshan@gmail.com

M. Prasad · M. Ramakrishna Reddy
Department of Earth Sciences, Yogi Vemana University, Kadapa 516005, India

S. K. Ghosh (ed.), *Urban Mining and Sustainable Waste Management*,
https://doi.org/10.1007/978-981-15-0532-4_12

and in developed countries (Soupios et al. 2004). The major factors which contribute solid waste generation are growing population, rising economic growth, and rise in community living standards. Limestone solid wastes represent huge masses of bulk materials that are often deposited without any consideration, causing a lot of difficulties and inviting illegal deposit of other kinds of waste and mine garbage (Kartam et al. 2004). As a result, the building and construction industry play an important role in waste generation. At present, waste management became an unavoidable challenge in regional level, national level, and global level. Mining is one of the major incoming sources for the developing countries like India, and at the same time it is the major source of huge amounts of solid waste generation in terms of overburden, milling, processing, and tail end wastes which need to be monitored and maintain proper mine plan and environmental policy particularly in opencast mining areas. The unplanned mining became a serious impact on the environment by unplanning mining activity causing reduction of forest cover, soil erosion, pollution air, water, land and reduction of biodiversity in earth surface. The waste rock dumps could alter landscape around mining areas (Mondal et al. 2013; Sarma and Kushwaha 2005). Mining activity also results in environmental impact with respect to human settlements, degrading air, water quality, forest land, etc. (Dumitru et al. 2016). In India, limestone mines are operated opencast mining method (Indian Minerals Yearbook 2017). When compared to underground mining, opencast mining activity causes adverse impacts; hence, proper environmental management plan has to be adopted. In order to ensure long-term stability of storage and dispersed facilities and to minimize air, water, and land contamination, proper management technique is needed. If the Mine water is good enough quality it can be used for public supply but when the management failure of control solid waste harmful impacts, such as depletion of water resources by dewatering, or the pollution of surface water courses by poor quality mine waters and mine waste leachates, demand careful inspection (Younger and Wolkersdorfer 2004). The inappropriate or unsafe management of solid rock wastes at quarry working continues to create opposition from local communities and the general public and has contributed to the negative impact on public perception of the mining industry. It can be better monitored by the remote sensing satellite data than the underground mining. To identify suitable sites for mining and also take a remedial measure guidelines approved by the Ministry of Environment Forests and Climate Change (MoEF and CC 2008) and Indian Bureau of Mines (IBM) can reduce the major possible impacts. In a present scenario where the mineral extraction was done on a large scale, it became a highly subject of concern regarding the limestone quarry waste being generated by this limestone mineral extraction, which has a huge impact on the environment; therefore, a regular study should be taken in accordance with the application of latest technology in order to sustain this earth (Das and Choudhury 2013). The Google Earth (GE) shows the area at fine spatial resolution helpful clearly identified waste dumps surrounding the limestone quarries. Using Google Earth images identifies dumps in limestone quarry locations to better understand the effect of mines on vegetation, air, and groundwater disturbances (Jade and Sunitha 2015). If detailed change detection is needed for a study, post-classification will be also needed (Zhang et al. 2017). This research used Google Earth data detection solid waste changes near Yerraguntla town (Wibowo et al. 2016).

2 Study Area

Yerraguntla is an industrial village in YSR district in Andhra Pradesh. Geographically, Yerraguntla is located at 14.6333 °N 78.5333 °E with average elevation of 152 meters (501 ft). It is 37 km away from district head quarter Kadapa and well connected with rail road. The study area is well known for various cement industries like Zuari cement Ltd, Bharathi cement Ltd, and has good number of stone polishes industries, besides Rayalaseema Thermal Power Plant was established. Niduzivi, Koduru, Valasapalli areas of Yerraguntla Mandal are known for Napa slabs/Kadapa slabs are black colored limestone belonging to Koilakuntla Limestones (Sunitha et al. 2014). YSR district occupies lion share of limestone reserves in Andhra Pradesh, which is the major state among the limestone-producing states with 31, 96,000 ton production as per Indian Mineral Yearbook (2017) (Figs. 1 and 2).

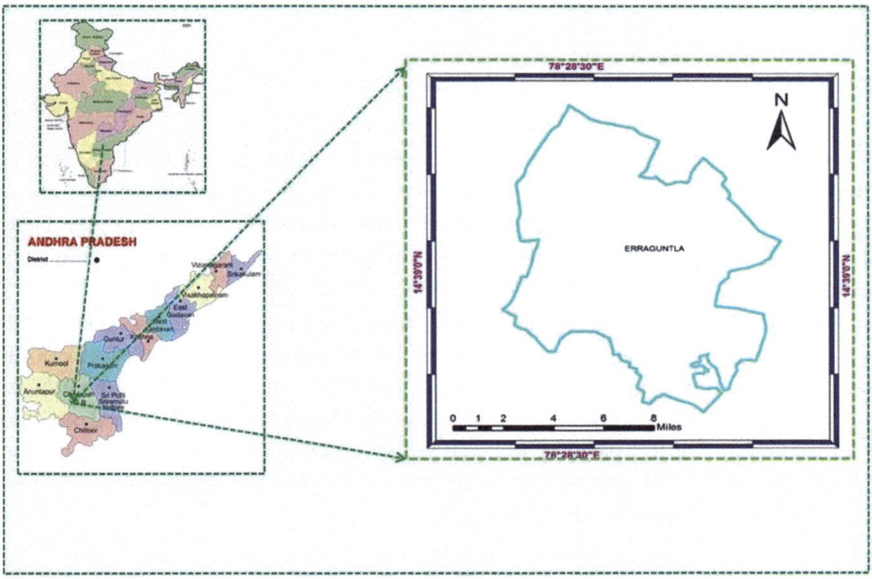

Fig. 1 Location map of the study area

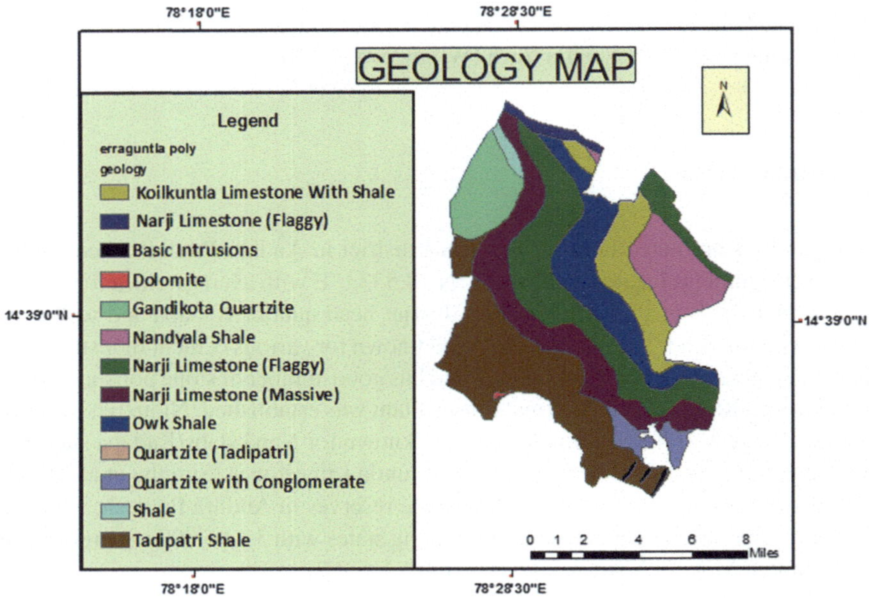

Fig. 2 Geology map of the study area

3 Methodology

A systematic integrated methodology is adopted for the present study to assess the temporal changes in waste disposal. The intensive study of the relevant literature, periodicals, etc., has been carried out; based on the review of the literature, it is clear that remote sensing is the only effective tool to monitor the changes for multi-date scenarios. Thus, Google Earth satellite images were chosen as the data source for the study area and the time series changes have been mapped for the years 2006 and 2018. Field work has been carried out in the study area like waste disposal sites, mines, tailings, cement industries, stone cutting and polishing industries to know to type of waste generation, handling, recycle and disposal. Data generated from the Google Earth was compared and verified ground truth data by GPS survey, topographic maps, field photographs, and questioner from local sources and mine and industrial management (Figs. 3 and 4).

4 Results and Discussion

Limestone mining waste disposal locations were identified on Google Earth imagery, based on the image elements like tone, texture, shape, size, pattern, and association. The active dumps are clearly visible gray tone with medium to coarse texture due

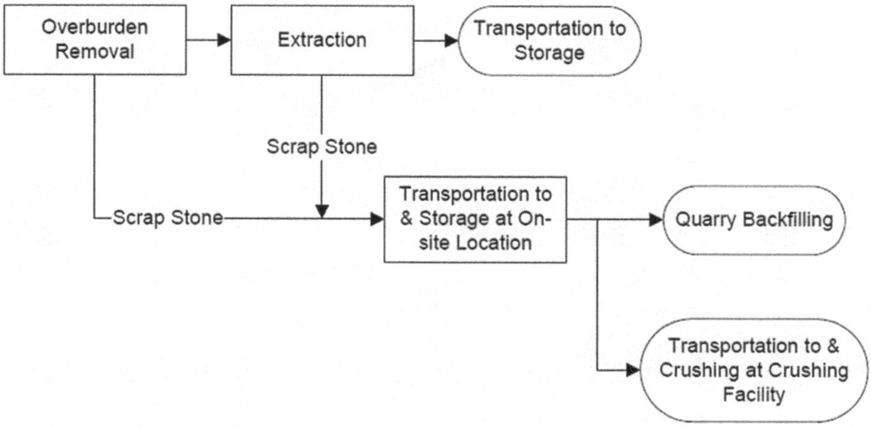

Fig. 3 Flowchart of the limestone quarry processing

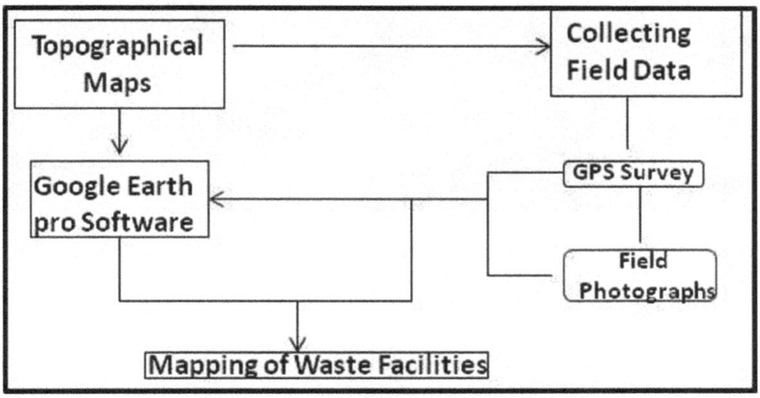

Fig. 4 Flowchart of the methodology

to the presence of debris of waste and overburden and irregular in shape and size, whereas the mine pits are with perfect shape and mainly in dark blue tone due to the presence of water in the pits as shown in Figs. 5 and 6.

In mining and cement industry, wastes are generated in every stage of the operations; in the present study, we have identified two major locations of solid waste disposal: One is near the mine, and the second one is besides the road. Spatiotemporal changes in 2006–2018 are clearly depicted by drawing the red-colored polygon for both the locations (Figs. 5a–f and 6a–f).

By observing the Google Earth temporal images as shown in Figs. 5 and 6 and the field observations as shown in Figs. 7, 8, 9 and 10, it is clear that overburden management and solid waste of these quarries are extremely poor; most of the mining companies are poor in practices on overburden management and fugitive dust from waste pile. They do not bother to take preventive measures against the limestone waste

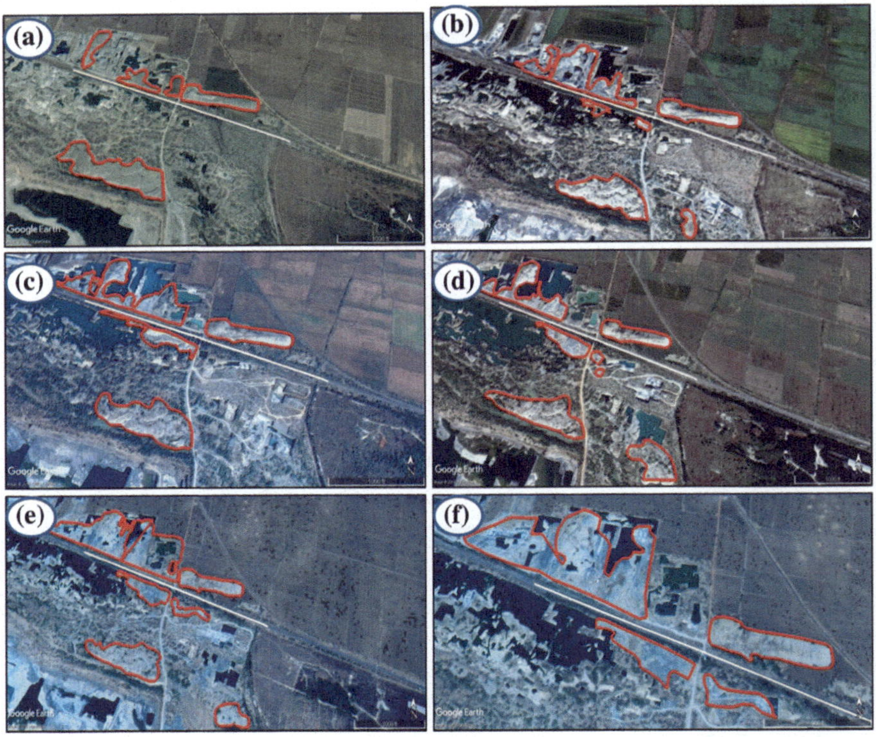

Fig. 5 Temporal and spatial formation and changes during 2006–2018 in location of solid waste beside railway track surrounding Yerraguntla village

management. The best method is to backfill the excavated land, and this is hardly implemental as companies keep opening from time to time different faces of mine open began without exhausting previous one. The opencast mines are characterized by enormous open pits surrounded by huge dump yards surrounded by the croplands. Dust and slurry (Figs. 11 and 12) from these sites are being transported by the air and rainwater to nearby cropland ponds which is major threat to soil porosity, productivity, permeability, and crop yield. During dry summer, these dumps become a key source of air pollution for the surrounding areas.

5 Limestone Waste Properties

Diverse stone wastes like granite, quartzite, limestone, marble, each one has different properties. Limestone waste may have different physical properties depending on the extraction of mineral and method of generation of waste. The properties of limestone waste used as an aggregate in concrete are presented in Table 1 (Lakhani et al. 2018).

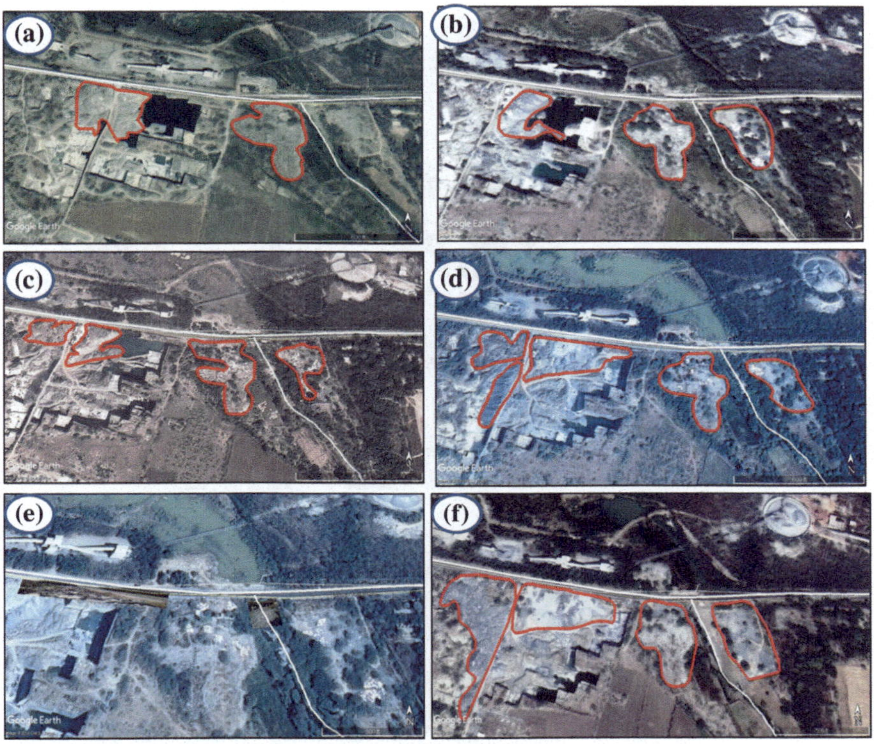

Fig. 6 Temporal and spatial formation and changes of solid waste during 2006–2018 beside highway road near India cement factory surrounding Yerraguntla village

Fig. 7 Limestone dump near India cement factory, Yerraguntla village

Fig. 8 Limestone surrounding Yerraguntla village

Fig. 9 Overburden of Niduzuvvi limestone quarry surrounding the Yerraguntla village

6 Conclusion

The conclusion of this research on the spatial–temporal change detection using the Google Earth data is very important, low cost, and very useful as spatial–temporal detection for solid waste investigation, monitoring, and mapping in mining industries. The present study intended to assess the spatiotemporal changes in the disposal of solid waste derived from the different types and the quantities of mining waste, in and around Yerraguntla village area in Kadapa district by using the Google Earth imagery. Based on the satellite data analysis and ground truth verification, it is concluded that overburden and the tailings from the limestone are extremely poor and there is no strict control on the solid waste disposal and management plan which resulted in the dumping of waste in the nearby streams, roadside, and the banks of the highway.

Fig. 10 Solid waste in a stream near India cement factory, Yerraguntla village

Fig. 11 Water inundation in limestone quarry surrounding Yerraguntla village

Fig. 12 Development of fine dust during transport surrounding Yerraguntla village

Table 1 Physical and mechanical properties of solid and powder limestone waste

S. no.	Property					
	Type of stone waste	Specific gravity	Bulk density (kg/m^3)	Water absorption (%)	Maximum size (m/mm)	Color
1.	Limestone	2.6–2.65	1568–1743	2–4	2 m (solid) 0.2 mm (powder)	Gray

Finer material from Limestone quarries had been carried out by rain water into near by water bodies or lands and pollutes both media. It also observed that dust emissions from the mine and dumps become a key source of air pollution for the surrounding areas.

Acknowledgements My grateful thanks to DST for the financial support and limestone quarry management for providing the permission to carry out the fieldwork in limestone quarries.

References

Bringi, S. D. (2007). *Application of 3D principles to solid waste management on the Asian Institute of Technology (Ait) Campus* (Unpublished M.Sc. thesis). Indonesia.

Das, R., & Choudhury, I. (2013). Waste management in mining industry. *Indian Journal of Science Research, 4*(2), 139–142.

Dumitru, M., Carabis, D., Parvan, L., & Sarbu, C. (2016). Environmental rehabilitation of Mine dumps. *Agriculture and Agricultural Science Procedia, 10*, 3–9. https://doi.org/10.1016/j.aaspro.2016.09.002.

Indian Minerals Yearbook. (2017). Limestone and other calcareous materials. Government of India Ministry of Mines Indian Bureau of Mines.

Jade, R. K., & Sunitha. (2015). Temporal changes due to mining in khetri copper complex, Rajasthan. *Procedia Earth and Planetary Sciences, 11*, 165–172. https://doi.org/10.1016/j.proeps.2015.06.020.

Kartam, N., Al-Mutairi, N., Al-Ghusain, I., & Al-Humoud, J. (2004). Environmental management of construction and demolition waste in Kuwait. *Waste Management, 24*, 1049–1059. https://doi.org/10.1016/j.wasman.2004.06.003.

Lakhani, R., Kumar, R., & Tomar, P. (2018). Utilization of stone waste in the development of value added products: A state of the art review. *Journal of Engineering Science and Technology Review, 7*(3), 180–187.

MoEF, & CC. (2008). Climate change adaptation research climate change, gender and vulnerable groups in Bangladesh.

Mondal, S., et al. (2013). Application of GIS techniques for assessment of changes in land use pattern and environmental impact of mines over a small part of Keonjhar District of Orissa. *IOSR Journal of Research and Method in Education (IOSR-JRME).*

Sarma, K., & Kushwaha, S. P. S. (2005). *Coal mining impact on land use/land cover in Jaintia hills district of Meghalaya, India using remote sensing and GIS technique*. www.csre.iitb.ac.in/~csre/conf/wp-content/uploads/OS5_17.pdf.

Soupios, P., Papadopoulos, I., Kouli, M., Georgaki, I., Vallianatos, F., & Kokkinou, E. (2004). Investigation of waste disposal areas using electrical methods: A case study from Chania, Crete, Greece. *Environmental Geology, 51*(7), 1249–1261. https://doi.org/10.1007/s00254-006-0418-7.

Sunitha, V., Muralidhara Reddy, B., & Ramakrishna Reddy, M. (2014). Mineral resources of Cuddapah Basin. *The Journal of Biological Chemistry, 31* (1), 226–235.

Wibowo, A., Osman Salleh, K., Sitanala Frans, F. T. R., & Mulyo Semedil, J. (2016). Spatial temporal land use change detection using Google earth data. *IOP Conference Series: Earth and Environmental Science, 47*, 012031. https://doi.org/10.1088/1755-1315/47/1/012031.

Younger, P. L., & Wolkersdorfer, C. (Eds.). (2004). Mining impacts on the fresh water environment: Technical and managerial guidelines for catchment scale management. *Mine Water and the Environment, 23*, S2–S80.

Zhang, M., Zhoua, W., & Lib, Y. (2017). The analysis of object-based change detection in mining area: A case study with pingshuo coal mine. *The International Archives of the Photogrammetry, Remote Sensing and Spatial Information Sciences, XLII-2/W7*. https://doi.org/10.5194/isprs-archives-XLII-2-W7-1017-2017.

Waste Is not a Waste—It's Time to Realize

M. Rao Divi

Abstract In the modern world, managing the waste is a crucial challenge for most organizations. Some organizations dispose waste as their main objective, but they ignore to see the value in it. This paper explains the views of Divi's Laboratories Limited (Divis) on waste, i.e., it is a profit-making byproduct. It implies that Divis looks waste as a valuable byproduct and follows the concept of byproduct value recovery management. Waste is nothing but a useful byproduct, which is further managed (6R's strategy) through value additions at the same time ensuring less impact on environment, improved occupational health and safety, conservation of natural resources, sustainable livelihood, and carbon footprint reduction. Some value recovery management strategies that are being practiced in Divis are elaborated in this paper. The byproducts which can't be managed or used within Divis directly due to several reasons like contamination, impurities, and safety concerns are considered as waste. Waste which can add some value to Divis or to other recipient industries with less impact on EH&S is considered as value attaining waste. The waste which can't add value to Divis and to other recipient industries is considered as real waste, and it is managed through safe, eco-friendly waste management system. The practices that are being followed at Divis by using different strategies, techniques and by fulfilling the compliance obligations for waste management are deemed as safe disposal management.

Keywords Byproduct · Value recovery management · Gigantic hidden treasure · Value additions · Value attaining waste · Real waste · Common byproduct value recovery facility

---

M. Rao Divi (✉)
Divis Laboratories Limited, Hyderabad, India
e-mail: raodivi@gmail.com

S. K. Ghosh (ed.), *Urban Mining and Sustainable Waste Management*,
https://doi.org/10.1007/978-981-15-0532-4_13

1 Introduction

1.1 Background

Divis philosophy on 'Waste':

Waste is not a waste!!!

Divis considers **'Waste'** as a **'Valuable Byproduct'** where value is extracted by means of **'VALUE RECOVERY MANAGEMENT' (VRM)**.

Never call waste as 'Waste' because the word 'Waste' psychologically affects the thought process, i.e., many people liquidate waste by disposal. But they don't foresee the value addition attained through 'value recovery management,' which includes sustainability and profits.

Don't ignore the gigantic hidden treasure

Waste is nothing but a useful byproduct, which is further managed (6R's strategy) through value additions at the same time ensuring less impact on environment, improved occupational health and safety, conservation of natural resources, sustainable livelihood, and carbon footprint reduction.

1.2 Management of Waste

It is time industries realized that everything they dub as waste cannot be waste. Most of it is inappropriately managed. Also it is pertinent to call waste management as 'BYPRODUCT VALUE RECOVERY MANAGEMENT.' Finally, waste is the waste that cannot be profitably realized as byproducts.

Byproduct value recovery management prior to waste management

There is nothing secret behind profits of Divis.

Divis enjoy better profits by

Byproduct Value Recovery Management

1.3 Definitions

Byproduct

For Divis, byproduct is a secondary product derived from a manufacturing process or chemical reaction or the non-consumed input which is a part of a chemical process and which cannot be reused directly in the same process or any other process due to several reasons like contamination, impurities, environment, health, and safety concerns.

Waste

Divis definition of waste is the byproduct,

- which cannot be recovered due to less or no value addition to Divis when compared to recovery efforts (recovery cost, more impact on occupational health, safety, and environment, more consumption of natural resources, and more carbon footprint) and which may add more value to others (recipient industries) which is nothing but **'VALUE ATTAINING WASTE.'**
- which cannot be recovered due to no value addition (recovery cost, more impact on occupational health, safety, and environment, more consumption of natural resources, and more carbon foot print) to Divis and to others (recipient industries) is considered as **'REAL WASTE.'**

Value attaining waste is one which needs proper and detailed assessment in terms of identification of appropriate recipient industries, qualifying the recipient industries, safe transfer of waste, value addition (either to Divis or to the recipient industry or both).

Real waste is the waste that needs to be disposed-off in a safe, eco-friendly manner at optimum cost by fulfilling the applicable compliance obligations.

2 Methods

2.1 Value Recovery Management of Byproducts, Value Attaining Waste

The purpose of management of byproducts as well as value attaining waste is to gain benefit from the waste through recovery of byproducts by Divis and facilitate gain to recipient industries through a well-designed system called as **'BYPRODUCT VALUE RECOVERY MANAGEMENT' (BPVRM)**.

Efficient value recovery management will not only yield value to the organization but also it helps in

- Elimination of waste
- Minimization of waste
- Achieving clean workplace environment

- Pollution reduction
- Conservation of natural resources
- Reduction of carbon footprint
- Occupational safety and health benefits
- Achieving sustainable livelihood.

The concept of byproduct value recovery management is presented below in the form of flowchart

Waste management is only the last resort

2.2 *Schematic Flow of Byproduct Value Recovery Management*

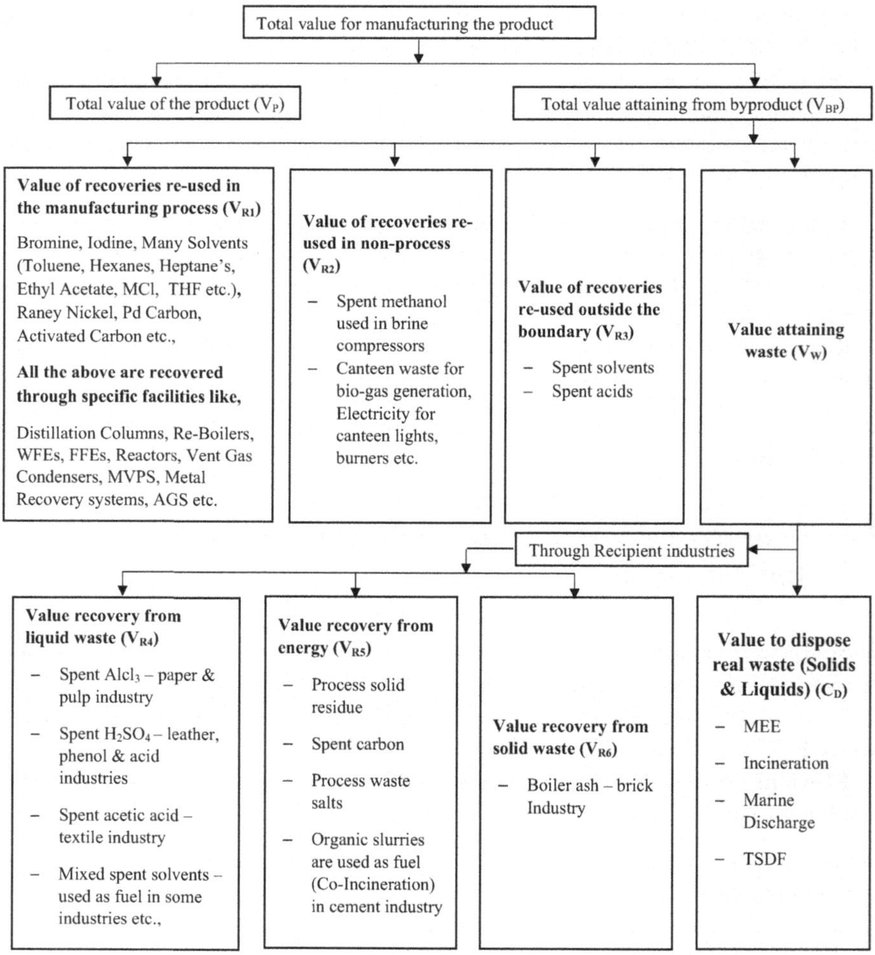

$$\text{Total value attaining from byproduct } (\mathbf{V_{BP}}) = (\mathbf{V_{R1}} + \mathbf{V_{R2}} + \mathbf{V_{R3}})$$

$$\text{Total value attaining from waste } (\mathbf{V_{W}}) = (\mathbf{V_{R4}} + \mathbf{V_{R5}} + \mathbf{V_{R6}})$$

$$\begin{aligned}\text{Total value for manufacturing the Product } (\mathbf{V_T}) &= \text{Total value of the Product } (\mathbf{V_P}) \\ &\quad + \text{Total value attaining from Byproduct } (\mathbf{V_{BP}}) \\ &\quad + \text{Total value attaining Waste } (\mathbf{V_W}) \\ &= (\mathbf{V_P}) + (\mathbf{V_{BP}}) + (\mathbf{V_W}) \\ (\mathbf{V_T}) &= \mathbf{V_P} + (\mathbf{V_{R1}} + \mathbf{V_{R2}} + \mathbf{V_{R3}}) + (\mathbf{V_{R4}} + \mathbf{V_{R5}} + \mathbf{V_{R6}})\end{aligned}$$

2.3 Practices by Divis for Byproduct Value Recovery Management

- Byproduct value recovery management policy within the organization.
- Awareness training to Divis employees and contract personnel on **6R's strategy**.
- Establishing an arm in Research and Development Center where people continuously work on chemical processes by applying Green Chemistry principles.
- Creating Process Development Center where majority of members are directly working on byproduct value recovery management. Byproducts are studied thoroughly through in-house support from Chemical Hazards Evaluation Laboratory (CHEL) and external expertise for evaluation of each byproduct.
- A management information system (MIS) on byproduct value recovery management to the board.

3 Results and Discussions

3.1 Major Byproduct Value Recovery Management Projects in Divis

The best practices are implemented in waste reduction and in collection, segregation, treatment, and disposal of waste in Divis.

Reduction is better than recovery

In Divis, byproducts derived from various processes are managed in three different forms, namely

1. Solid byproduct value recovery management;
2. Liquid byproduct value recovery management;
3. Gaseous/Vapor byproduct value recovery management.

3.1.1 Solid Byproduct Value Recovery Management

(a) ***Product A-extraction***

The byproduct from one of the chemical processes is being efficiently managed through establishment of a wide-range technologies in association with special equipment and instruments from which valuable solid chemical is extracted. The recovered product is reused in other chemical processes.

Cost of the recovery facility for the product-A is Rs. **200 lakhs with an attractive payback period of two months** (Fig. 1).

3.1.2 Liquid Byproduct Value Recovery Management

(a) ***Product B-extraction***

The byproduct from one of the chemical processes is being efficiently managed through establishment of a wide-range technologies in association with special equipment and instruments from which valuable liquid chemical is extracted. The recovered product is reused in other chemical processes.

Cost of the recovery facility for the product-B is Rs. **180 lakhs with an attractive payback period of three months** (Fig. 2).

(b) ***Product C-recovery***

As the specifications of a chemical are very stringent requiring, inter alia, very less moisture content in the solvent, the costly TBD (to be distilled) solvents used to be sold to outside parties for their use.

Fig. 1 Product-A recovery facility

Fig. 2 Product-B recovery facility

Fig. 3 Product-C recovery facility

To overcome this challenge, worldwide technologies were reviewed and an appropriate recovery system was introduced in the plant thereby successfully attaining the desired purity on a consistent basis.

Cost of the recovery facility for the product-C is Rs. **2300 lakhs with an attractive payback period of 1.8 years** (Fig. 3).

3.1.3 Gaseous/Vapor Byproduct Value Recovery Management

(a) ***Product D-recovery: (under execution)***

A facility of liquid nitrogen based modern condensing technology for recovery of a chemical vapor emitted from one of the chemical process is under execution. Once the scheme is put into operation, the chemical will be fully recovered for reuse and the present practice of scrubbing the vapor that is adopted to prevent the release of emissions into environment will be discontinued.

Cost of the recovery facility for the product-D is Rs. **420 lakhs with an attractive payback period of three months** (Fig. 4).

3.2 Other Byproduct Value Recovery Management Projects

3.2.1 Solvent Recovery Systems

Different solvents are used for manufacture of every product during reaction, extraction, and absorption. The solvent used for an operation cannot be reused as the

Fig. 4 Product-D recovery facility

solvent reaches its saturation capacity or it is mixed with other solvents or the product. Many solvents in Divis are being recovered by means of technically assessed, well-established facilities like distillation columns/reactors, WFEs, FFEs, re-boilers, and vent gas condensers. These are the major value recovery projects of Divis with negligible impact on environment (Figs. 5, 6, 7 and 8).

Fig. 5 Distillation columns

Fig. 6 Falling film evaporator

Fig. 7 Re-boiler

Fig. 8 Wiped film evaporator

Fig. 9 Reverse osmosis plant, rejected RO water used for plantation

3.2.2 RO Rejected Water

Water which is rejected from reverse osmosis is used for the gardening/plantation. This is one of the effective conservation measures of natural resources, as the usage of fresh raw water for plantation is eliminated (Fig. 9).

3.2.3 Sewage Water Treatment

Water that is used for kitchen and toilet cleaning is not drained unnecessarily. This water is collected and treated through multiple sewage treatment plants and is used for gardening/plantation. This is one of the effective conservation measures of natural resources, as the usage of fresh raw water for plantation is eliminated (Fig. 10).

Sludge attained from this sewage treatment plant is being used as manure for gardening and plantation. This again becomes a value addition to Divis along with conservation of natural resources (Fig. 11).

Fig. 10 Sewage treatment plant, treated sewage water used for plantation

Fig. 11 Sludge from sewage treatment plant, used as manure for plants

Fig. 12 Canteen waste, bio-gas generation

3.2.4 Utilization of Canteen Miscellaneous

The solid organic canteen waste is treated in 'Bio-Gas Generation Plant' to produce methane gas, which is used as fuel in the kitchen, i.e., cooking and for generation of electricity for use in kitchen (Fig. 12).

3.3 Few Projects on Value Attaining from Waste

3.3.1 Boiler Ash

Ash obtained from the boiler is a byproduct which cannot be used by Divis anymore. Divis sells this waste to brick industry where it is utilized as a raw material for brick making, adding value to the industry. This promotes clean environment at Divis along with less personnel exposure (Fig. 13).

3.3.2 Spent Solvents

A few of different solvents are contaminated, having improper specifications, needs more efforts for recovery in terms of value with more load on environment. These

Fig. 13 Boiler, boiler ash, bricks manufacturing using boiler ash

Fig. 14 Spent solvents transporting to other industries

solvents are seperated, sold and transported (Fig. 14) to other companies at nominal cost. These solvents are used in the recipient companies. This activity promotes clean environment within the premises.

3.3.3 Disposal of Various Items for Recycling

Decontaminated empty drums, iron scrap, used electrical scrap which cannot be re-used further are sold to authorized external parties for recycling (Figs. 15, 16, and 17).

Fig. 15 Empty drums for recycling

Fig. 16 Iron scrap for recycling

Fig. 17 Electrical scrap for recycling

4 Real Waste Management

The waste which can't be recovered, used, or managed by Divis and others (recipient industries) is considered as real waste, and such waste shall be managed through safe, eco-friendly waste management. Few waste management facilities at Divis are shown in Figs. 18, 19, 20, 21, 22, and 23.

Fig. 18 Scrubbing system

Fig. 19 Multiple-effect evaporator

Fig. 20 Incinerator

Fig. 21 Effluent treatment plant

Fig. 22 Guard pond complex

Fig. 23 Jetty—marine discharge facility of Divis

5 Conclusion and Recommendations

5.1 Conclusion

- Once again, it is reminded *'Waste is not a waste'!!!*
- *'Byproduct value recovery management'* ***can be profitably*** *implemented for waste management to attain value addition.*

5.2 Recommendations

- For the benefit of small manufacturing units, it's time to think about having a *'Common byproduct value recovery facility (CBPVRF)'*.

Reference

Divis in-house practices. (2017).

Effective Treatment for COD Removal of Landfill Leachate by Electro-coagulation

P. T. Dhorabe, A. R. Tenpe, V. S. Vairagade, Y. D. Chintanwar, B. R. Gautam and V. R. Agrawal

Abstract Landfill leachate (LL) is the liquid waste generated from solid waste dumping site. Due to the high toxicity and high COD-BOD ratio of leachate, it is not possible to treat it in conventional wastewater treatment plant. There is a special treatment required to treat it. In this study, a complete batch electro-coagulation (EC) process is used to treat the LL. Aluminum (Al) and iron (Fe) electrodes were used for effective removal of COD by varying the treatment parameters such as initial voltage (V), current (A), initial pH, treatment time (t), electrode spacing (s), and number of electrodes (n). The maximum COD removal by Al electrode is found to be 85% at 25 V and by Fe electrode is 45% at 20 V. The optimum number of electrodes is found to be 6 at alkaline pH range.

Keywords COD · Current · Electro-coagulation · Landfill leachate · Voltage

1 Introduction

The technological processes and consumptive processes result in the formation of solid wastes. The solid waste is generated during technological processes involving mining, manufacturing, and packaging. The process of consumption of products also results in the formation of solid waste in urban and rural areas. In the long-term perception, such a production–consumption imbalance damages the environment. The landfill is one of the methods of disposal of solid waste in which the wastes are dumped in low-lying areas or a trench or ramp is prepared for disposal. This method

P. T. Dhorabe · V. S. Vairagade (✉) · V. R. Agrawal
Priyadarshini College of Engineering, Nagpur, India
e-mail: vikrantvairagade@gmail.com

A. R. Tenpe
Visvesvaraya National Institute of Technology, Nagpur, India

Y. D. Chintanwar
Priyadarshini J. L. College of Engineering, Nagpur, India

B. R. Gautam
Priyadarshini Indira Gandhi College of Engineering, Nagpur, India

S. K. Ghosh (ed.), *Urban Mining and Sustainable Waste Management*,
https://doi.org/10.1007/978-981-15-0532-4_14

of solid waste disposal is not maintained properly as it leads to the production of leachate.

Leachate is any liquid passing through matter, extracts solutes, suspended solids, or any material through which it has percolated. The landfill leachate considerably varies in composition, and it depends upon type of waste composition and the age of the landfill (Henry and Heinke 1996). It can typically contain both suspended and dissolved materials. The leachate is generated, caused by precipitation which is percolating through layers of solid waste placed in a landfill. The percolating water comes in contact with decomposing waste and becomes contaminated and pours out of the waste material as leachate. The extra leachate volume is produced during the disintegration of carbonaceous material that produced a large quantity of other materials such as carbon dioxide, methane, and a complex mixture of aldehydes, organic acids, simple sugars, and alcohols. The leachate when emerges from a landfill site is a black, brown, or yellow color, strong odor turbid liquid. The odor is offensive and acidic and may be very persistent due to nitrogen, hydrogen, and mercaptans (sulfur rich organic species). For landfills which do not obtain run-off from outer areas, a very small quantity of leachate generation can be obtained by assuming it to be 25–50% of the precipitation from the active landfill area and as 10–15% of the precipitation from covered areas. This is a thumb rule and can only be used for preliminary design (CPHEEO manual). The toxicity of leachate made it problematic for the disposal. There are certain technologies which are using for the treatment of leachate such as membrane filtration like reverse osmosis (Ushikoshi et al. 2002), biological treatments like rotating biological contactor (Bishop and Kinner 1986), aerobic activated sludge process (Loukidou and Zouboulis 2001), sequencing batch reactor (Yalmaz and Oztürk 2001), anaerobic sequencing batch reactor or up-flow anaerobic sludge blanket (Im et al. 2001), various physicochemical treatments like coagulation flocculation (Amokrane et al. 1997; Tatsi et al. 2003) and adsorption (Morawe et al. 1995; Geenens et al. 2001). All these methods are effective but cost imparting methods. Due to their expensive nature, we cannot say them economical and remedial solution for leachate treatment. In this paper, we have used the very new efficient and economical method for the treatment of leachate 'electro-coagulation' (EC) process. This is nothing but the electrochemical treatment of leachate. In this study, the leachate is treated by EC by using the aluminum and iron electrodes and by varying the treatment parameters such as initial voltage (V), current (A), initial pH, treatment time (t), electrode spacing (s), and number of electrodes (n) to the COD (chemical oxygen demand) removal efficiency of the treatment.

2 Materials and Method

2.1 *Chemicals*

All the chemicals used in the study were of analytical reagent (AR) grade. Ammonium ferrous sulfate, 1-10 phenanthroline, polyacrylamide (PAA), mercuric sulfate (Hg_2SO_4), silver sulfate (Ag_2SO_4), potassium dichromate ($K_2Cr_2O_7$), 98% Conc. sulfuric acid (H_2SO_4), Conc. hydrochloric acid (HCl), and sodium hydroxide (NaOH) were obtained from LOBA Chemicals Ltd., Mumbai (India).

2.2 *Electrodes and Reactor*

The iron (cold rolled) sheets and galvanized aluminum sheets with more than 99% purity were used as electrode materials for treatment. The iron and aluminum electrode plates as procured from the local suppliers were cut into the required shape, degreased by using 15% HCl solution followed by washing with distilled water and then oven-dried the electrodes at 50 °C prior to their use in the EC treatment.

The batch reactor of EC was fabricated of acrylic sheets of 5 mm thick in rectangular shape. The dimensional characteristics of the reactor and the configuration details are as shown in Table 1. A magnetic stirrer was used for continuous stirring. The magnet was placed inside the bottom of ECR. Dual DC power supply (TD3203D, Aplab Co. Ltd) in the range from 0 to 30 V and current range from 0 to 3.0 A are used

Table 1 Characteristics of electrodes and electro-coagulation reactor (ECR)

Electrodes characteristics		
Material (anode and cathode)	Iron	Aluminum
Shape	Rectangular plate	Rectangular plate
Size of each plate (mm)	70 × 115	70 × 115
Thickness (mm)	1.5	2.0
Plate arrangement	Parallel	Parallel
Each plate electrode surface area (cm^2)	80.5	80.5
Submergence (effective area) (cm^2)	42 (7 × 6)	42 (7 × 6)
Reactor characteristics		
Make	Acrylic glass	
Reactor cross section	Square	
Dimensions (mm)	120 × 120 × 150 (length × width × height)	
Volume (L)	1.5	
Thickness of sheet (mm)	5	
Stirring mechanism	Magnetic bar	

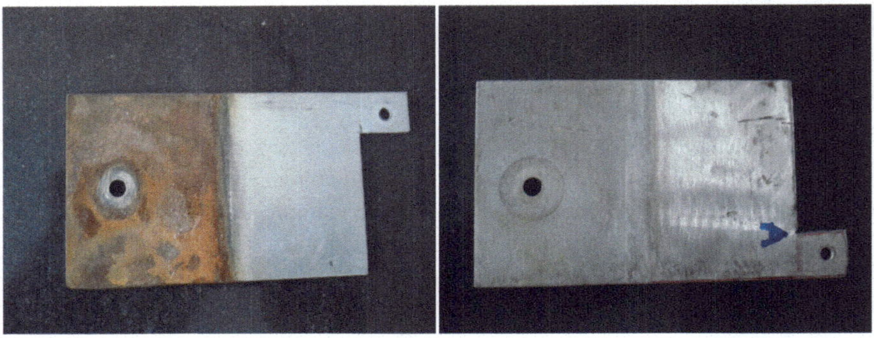

Fig. 1 Iron and aluminum electrodes used in ECT

to supply the required quantity of voltage and current. Magnetic stirrer plus heater (1MLH REMI CORPORATION Ltd.) of maximum stirring speed is of 1200 rpm for 1 L volume and heating capacity 150 W and a digital pH meter (Systronics μ pH system 361) with digital temperature measuring rod were used for the measurements as shown in (Fig. 1).

2.3 Leachate Generation

The leachate was prepared by storing mixed solid waste collected from VNIT dumping site into a 40-L plastic bucket. At the bottom of a bucket, a tap for the collection of leachate was fixed at about 5 cm above the bottom. To prevent the chocking of tap, at the bottom of bucket a 4-cm-thick aggregate layer was placed over which 2 kg of collected solid waste was dumped and covered by 15-cm soil cover. 5 L of water was added in the bucket and closed air-tight cover, and the bucket was placed in the open atmosphere as shown in Fig. 2. The water sample or leachate was collected daily and analyzed for the chemical oxygen demand (COD) and total solids (TS). The leachate of seventh day was collected in sufficient quantity (near about 1.5 L) as a stock solution or the modeled leachate. For the further analysis and for EC treatment (ECT), we were used this stock by diluting it with doubled distilled water. The characteristics of leachate were determined by standard methods.

2.4 Batch Experiment

Experiments were run in a batch reactor consisting of a 1.5-L acrylic rectangular container equipped with a cathode and anode, both made of iron or aluminum with a particular spacing. The electrodes were dipped into the reactor containing leachate

Fig. 2 Leachate-generating bucket filled with solid waste collected from VNIT dumping site

with a 1.0 L working volume and continuously stirred by using magnetic stirrer. The experimental setup used in the present study is shown in Figs. 3 and 4.

The leachate generated was having very high COD; therefore, it was diluted 200 times and then used for all EC batch experiments. The batch study was performed over diluted leachate to study the effects of various parameters, viz. time (0–30 min), voltage (0–30 V), current (0–3.0 A), number of electrodes (2–6 plates), pH (2–12), electrodes spacing (1.0–2.0 cm), and temperature of ECR leachate (30–50 °C) for iron and aluminum electrodes. The solution was kept continuously stirring by using magnetic stirrer at about 150 rpm. It is a monopolar EC treatment so cathode and anode have consisted of same material. After each batch treatment, polarity of electrodes was reversed for 1 min so that the flaky depositions over electrode plates were cleaned and will give better efficiency of EC treatment for next batch study. The environmental parameter considered for removal efficiency is chemical oxygen

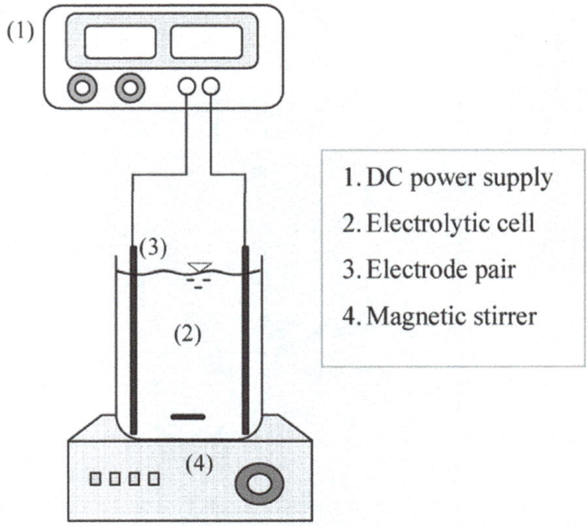

Fig. 3 Schematic experimental setup of ECT

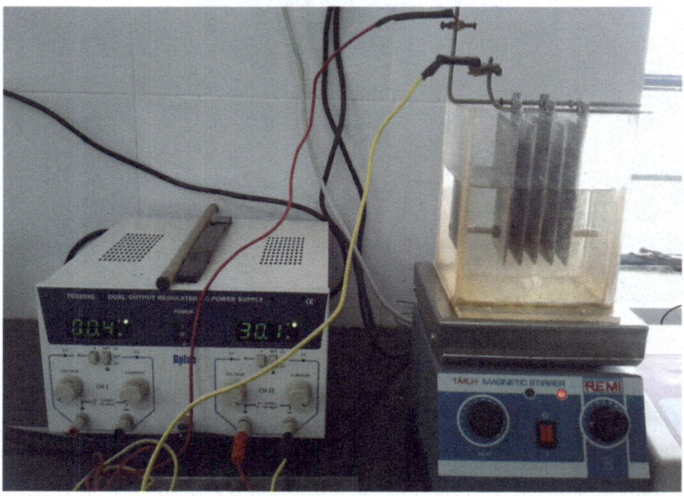

Fig. 4 Experimental setup used for leachate treatment in ECT

demand (COD) which was determined as per the standard method by open reflux method. After running each batch of EC treatment samples were collected from one-third depth of the ECR by using pipette (attached with pipette pump) and were filtered by using filter papers of laboratory grade (to remove the flocs formed during the process), and the filtrate was used for further COD analysis.

3 Results and Discussion

3.1 *Solid Waste Composition and Leachate Characteristics*

The solid waste was collected from VNIT solid waste (SW) dumping site and analyzed for the following characteristics which are shown in Table 2.

The characteristics of leachate generated from solid waste collected from VNIT SW dumping site are shown in Table 3. The daily variations in COD (mg/L) and total solids (TS) (mg/L) of leachate are also shown in Fig. 5.

The graph in Fig. 6 shows that initially the COD increased with time, and after seventh day, it was decreased because of the biological decomposition of leachate. It has been observed that the maximum COD (mg/L) reached at seventh day. Therefore, the seventh-day leachate was collected and stored as stock solution. This stock solution was used for the study by diluting it in a required proportion as and when required. The total solids gradually increase with time even after COD reduction.

3.2 *Batch EC Study Over Iron (Fe) and Aluminum (Al) Electrodes*

3.2.1 Effect of ECT Time on Removal of COD and pH Change

The effect of time on removal of COD by using Fe and Al electrodes was studied by varying voltage and keeping other parameters (viz. pH, no. of electrodes, spacing, temperature, and current) constant and is shown in Figs. 6 and 7.

Table 2 Characteristics of dumped solid waste

Parameter	Value
Total weight (kg)	2.0
Biodegradable (kg)	1.427
Non-biodegradable (kg)	0.573

Table 3 Characteristics of leachate used for the ECT

Parameter	Value
pH	5.10
COD (mg/L)	20,000–22,000
BOD (mg/L)	17,600
Alkalinity (mg/L of $CaCO_3$)	4400
Chloride (mg/L)	808.8
Total solids (g)	1.11

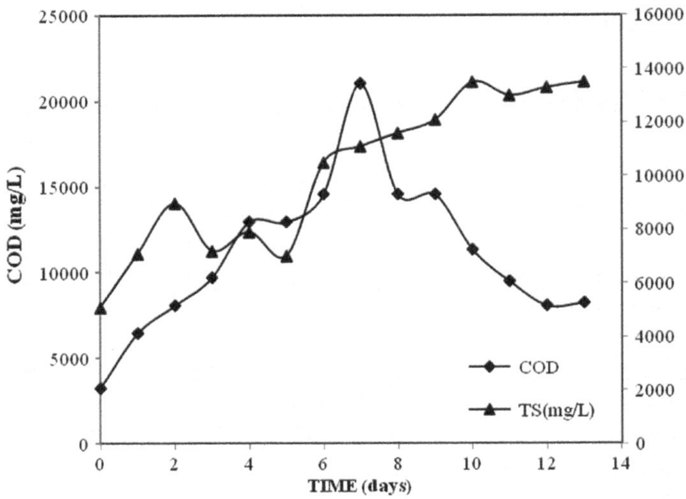

Fig. 5 Daily variations of COD (mg/L) and TS (mg/L) for leachate generated

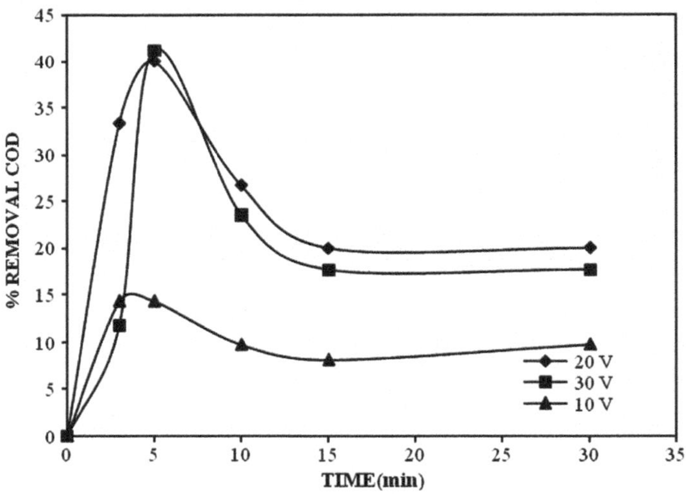

Fig. 6 Effect of time on % removal of COD at various voltages (pH of D.W = 5.81, No. of Electrodes = 6, Current = 3 A, Temp = 30 °C, Spacing = 1 cm) for Fe

For Fe, the maximum COD removal at 10, 20, and 30 V was observed at 5 min as 14.29, 41.17, and 40.00%, respectively (Fig. 6). During the EC treatment, it is observed that with increasing time of treatment, the color changes from light yellow to darkish reddish brown at lower voltages 10 and 20 V but at 30 V color changes from light yellow to dark reddish brown and finally change into dark greenish black due to the formation of various complex compounds of Fe hydroxides with other

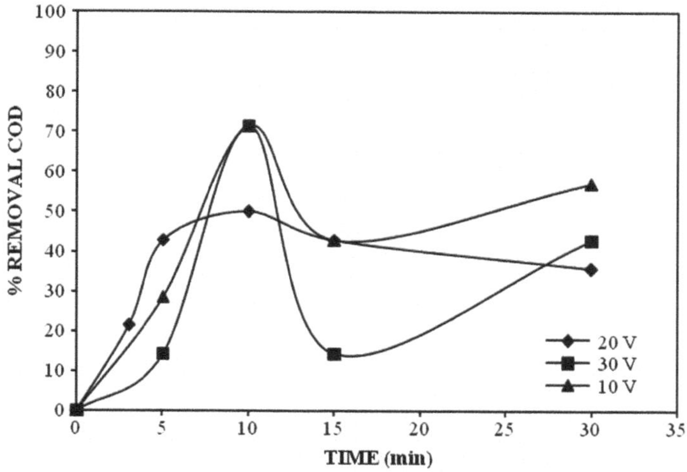

Fig. 7 Effect of time on % removal of COD at various voltages (pH of D.W = 5.37, No. of Electrodes = 6, Current = 3 A, Temp = 30 °C, Spacing = 1 cm) for Al

constituents of leachate (Moreno et al. 2007). The turbidity of leachate solution also increases with time, which may be due to the formation of flocs during ECT.

For Al electrode, the maximum percentage removal of COD at 10 min was found to be 71.43, 50.00, and 71.43%, respectively, at the voltages of 10, 20, and 30 which are shown in Fig. 7. It has been observed that during process of ECT, initially leachate solution is colorless but as time increases, the milkiness of leachate increase because of EC and turbidity also increases. As the time increases, the foam formation on the surface also increases. All the foams were observed on the top having white milky color appearance because foam floats with the flocs formed during EC which are white in color.

From Figs. 8 and 9, it is observed that during EC treatment process, pH of leachate increases for both Fe and Al electrodes at various voltages also.

For Fe at 10 V, initially pH increases from 6.14 to 7.07 up to 10 min and then decreases, and at the end of 30 min, it becomes 6.56. At 20 V, pH gradually increases from 5.38 to 6.74, and at 30 V, initially pH rapidly increases from 5.57 to 7.2 up to 3 min and then gradually decreases, and at the end of 30 min, it becomes 6.38 as shown in Fig. 8. This change in the pH is due to the various ions, hydroxides, and polyhydroxide species of iron formed during EC treatment (Moreno et al. 2007).

For Al electrode at 10, 20, and 30 V, pH increases from 6.29 to 9.11, 6.04 to 9.04, and 6.0 to 9.5, respectively, for the 30 min ECT process as shown in Fig. 9.

3.2.2 Effect of Voltage on Removal of COD

The effect of voltage on the removal of COD was studied by varying voltage from 5 to 30. The other parameters (viz. time, temperature, pH, current, no. of electrodes,

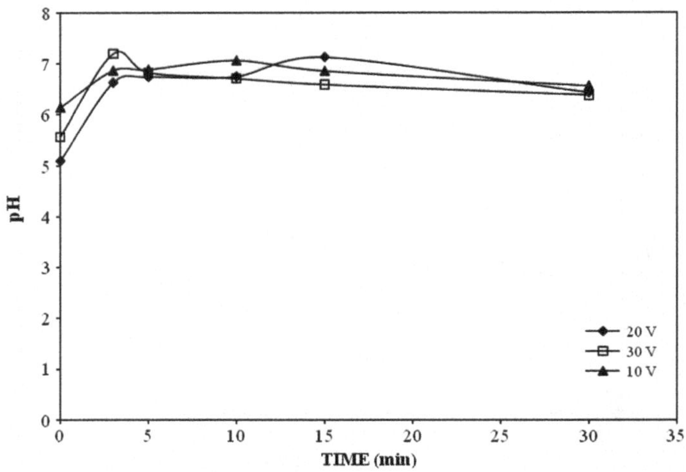

Fig. 8 Effect of time on pH changes of leachate during treatment at various voltages (pH of D.W = 5.81, No. of Electrodes = 6, Current = 3 A, Temp = 30 °C, Spacing = 1 cm) for Fe

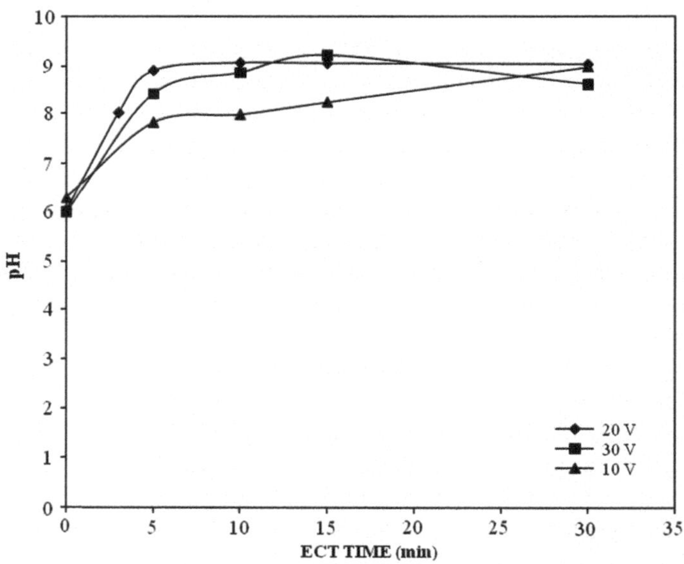

Fig. 9 Effect of time on pH changes of leachate during treatment at various voltages (pH of D.W = 5.37, No. of Electrodes = 6, Current = 3 A, Temp = 30 °C, Spacing = 1 cm) for Al

spacing) were constant for both Fe and Al electrodes. The results are shown in Figs. 10 and 11.

From Fig. 10, it is clear that the % removal of COD increases with increasing voltage of the EC system for iron electrode. % removal of COD was found to be 21.4, 14.28, 21.4, 35.71, 28.5, and 28.5% at the respective voltage of 5, 10, 15, 20, 25, and 30. The optimum COD removal was found to be at voltage 20. At this point, the color of the leachate solution becomes light turbid yellow.

For Al electrode, the COD removal at 5, 10, 15, 20, 25, and 30 V was found to be 12.5, 62.5, 62.5, 75.0, 87.5, and 81.25%, respectively, as shown in Fig. 11. The optimum COD removal was found at voltage of 25. At this point, the color of the leachate solution becomes light milky white and the foam formed at the top containing very fine bubbles.

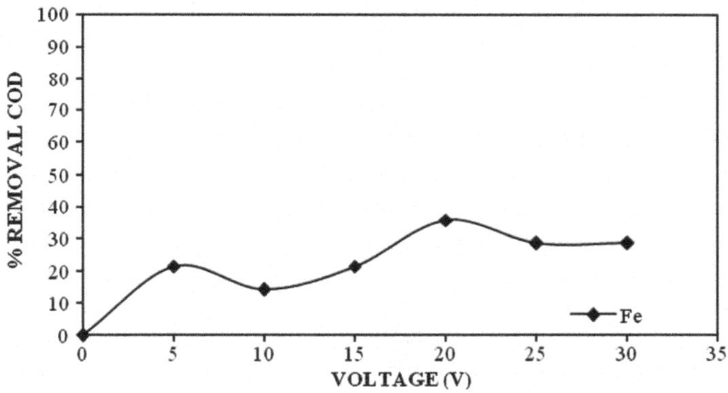

Fig. 10 Effect of voltage on % removal of COD for iron electrodes (No. of Electrodes = 6, Spacing = 1 cm, Current = 3 A, Time = 5 min, and Temp = 30 °C)

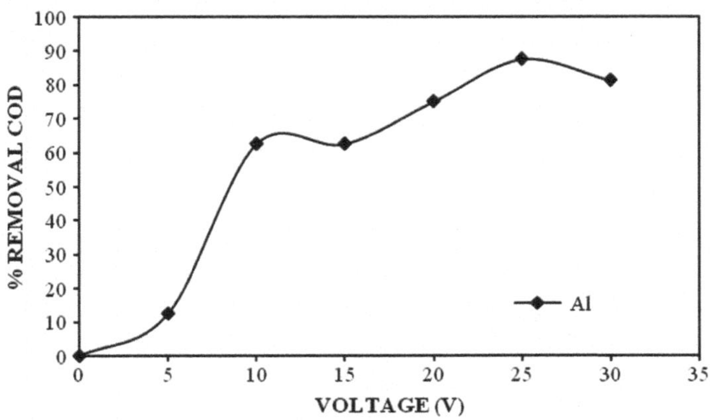

Fig. 11 Effect of voltage on % removal of COD for aluminum electrodes (No. of Electrodes = 6, Spacing = 1 cm, Current = 3 A, Time = 10 min, and Temp = 30 °C)

This increasing of COD removal during the ECT process is due to the formation of Fe and Al hydroxides complexes with leachate compounds like phenols or benzenes or humic acids which are initially before treatment are not oxidized during COD analysis but after ECT, they become oxidized in COD analyzing process (Moreno et al. 2007).

3.2.3 Effect of Applied Current on Removal of COD

The effect of current flowing through the electro-coagulation cell was studied by varying the current from 0.5 to 3.0 A by keeping other parameters constant for both Fe and Al electrodes, and the results are shown in Figs. 12 and 13. During the experiment, it was observed that the current applied to the EC cell was not constant because of the changing of salts concentrations during treatment, some dissolved salts becomes precipitated due to that the resistance of ECR changes during ECT (Mouedhen et al. 2008).

For Fe electrode, it was observed that as the applied current increases, the removal efficiency of COD also increases up to 2.0 A, and it remains constant after 2 A current. For current of 0.5, 1.0, 1.5, 2.0, 2.5, and 3.0 A, the COD removal efficiency was found to be 14.28, 21.42, 28.57, 35.71, 35.71, and 35.71%, respectively. It is maximum at 2.0 A current where leachate solution color observed was reddish yellow.

For Al electrode, it has been observed that as the applied current increases, the removal efficiency of COD also increases up to 2.5 A, and then remains constant after this current. The % removal of COD was found to be 50, 62.5, 62.5, 75.0, 84.37, and 84.37% for the current of 0.5, 1.0, 1.5, 2.0, 2.5, and 3.0 A, respectively. The percentage removal for COD by using aluminum electrode was found to be maximum at 2.5 A current where leachate solution appears to be milky white with

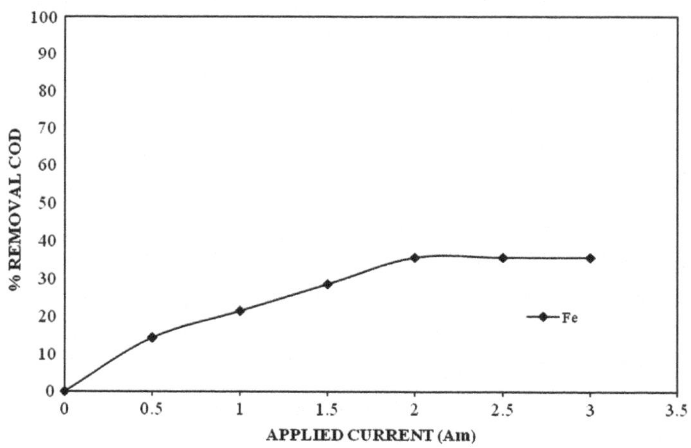

Fig. 12 Effect of applied current on % removal of COD for iron electrodes (No. of Electrodes = 6, Spacing = 1 cm, Voltage = 20 V, Time = 5 min, and Temp = 30 °C)

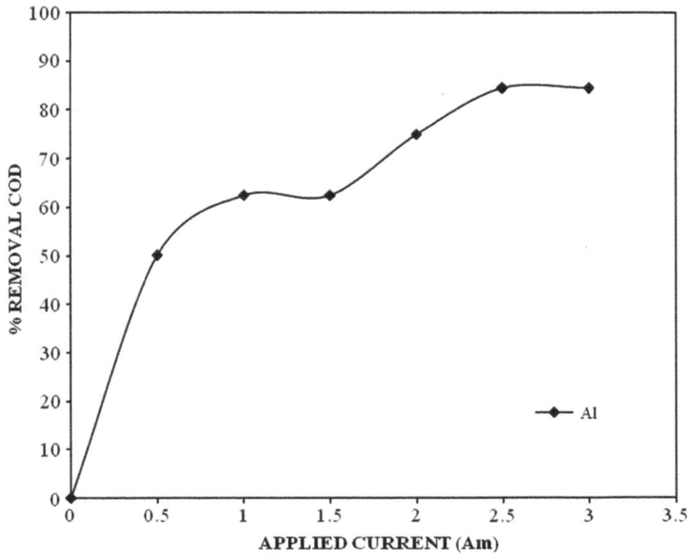

Fig. 13 Effect of applied current on % removal of COD for aluminum electrodes (No. of Electrodes = 6, Spacing = 1 cm, Voltage = 25 V, Time = 10 min, and Temp = 30 °C)

more white flocs. As for as the current is concerned, it can be compared that the COD removal of leachate was greater by aluminum than the iron electrode.

3.2.4 Effect of Number of Electrodes on Removal of COD

The number of electrodes is another important parameter which affects the COD removal efficiency during the ECT of leachate. The effect of number of electrodes on removal of COD from leachate was studied by varying number of electrode from 2 to 6 in the interval for 2 for both the electrodes. The results are shown in Figs. 14 and 15.

For Fe electrodes, the COD removal was found to be 14.61, 38.46, and 46.15% at 2, 4, and 6 electrodes, respectively. The color of leachate solution at the maximum removal of COD was found to be reddish yellow.

For Al electrodes, the COD removal was found to be 40.0, 67.0, 83.4% at 2, 4, and 6 electrodes, respectively. The color of leachate at maximum COD removal was found to be colorless.

For both Fe and Al electrodes, percentage removal of COD increases with increasing number of electrodes.

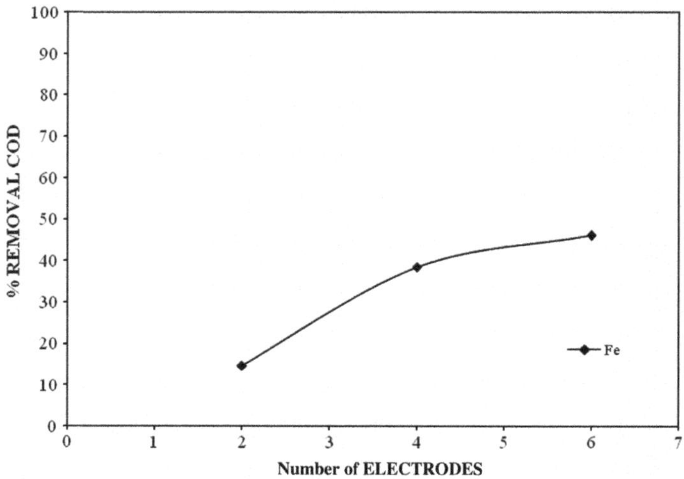

Fig. 14 Effect of number of electrodes on % removal of COD for iron electrode (Spacing = 1 cm, Voltage = 20 V, Time = 5 min, Applied Current = 2.0 A, and Temp = 30 °C)

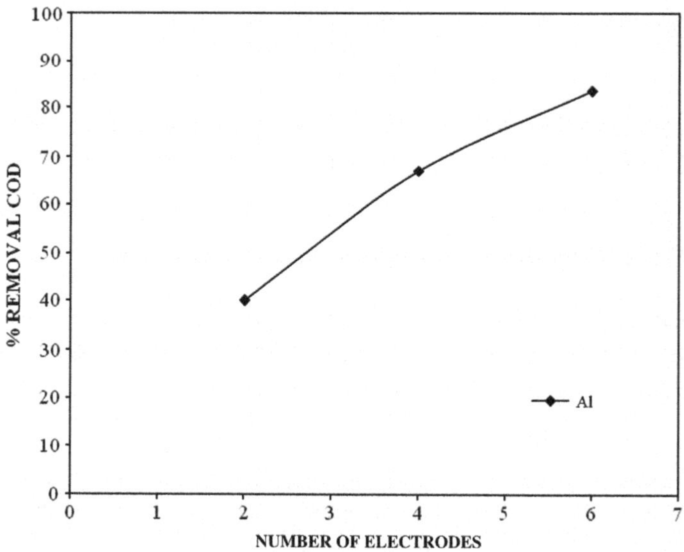

Fig. 15 Effect of number of electrodes on % removal of COD of leachate for aluminum electrode (Spacing = 1 cm, Voltage = 25 V, Time = 10 min, Applied Current = 2.5 A, and Temp = 30 °C)

3.2.5 Effect of Initial pH on Removal of COD

The effect of initial pH on the removal of COD from leachate by ECT was studied by using both Fe and Al electrodes. The pH of the leachate was varied from 2 to 12. The results are shown in Figs. 16 and 17.

For Fe, it was observed that the maximum COD get removed at the pH of 6 and 12. The COD removal at pH, i.e., 2, 4, 6, 8, 10, and 12 was found to be 11.11, 44.44, 55.55, 33.33, 44.44, and 55.55%, respectively. The minimum removal of COD is at pH 2. It was observed that at lower pH, the formation of bubbles takes place and the color of leachate solution becomes brownish black at the end of ECT process. As the pH increases, the color shifted toward reddish yellow and bubbling is also reduced up to pH 8, but after that there is no shifting of color, but the bubbling and the flocs formed in the solution during ECT have also increased.

For Al electrode, it was observed that the maximum COD gets removed at the pH of 6. The percent removal of COD was found to be 10.0, 45.6, 82.33, 80.0, 78.2, and 79.4% the pH of 2, 4, 6, 8, 10, and 12%, respectively, for 10 min ECT. It was observed that at lower initial pH, the size of bubbles formed during ECT was more and the top of leachate solution of ECR viscous bubble layer was formed with small quantity of flocs. As the initial pH increases, the size of bubbles becomes fine and the layer viscousness has demolished.

The variation of COD removal due to initial pH is not in a particular manner for Fe and Al electrode because of formations of various ionic and poly hydroxides species of Fe and Al into the ECR leachate solution during ECT process (Mouedhen et al. 2008; Moreno et al. 2007).

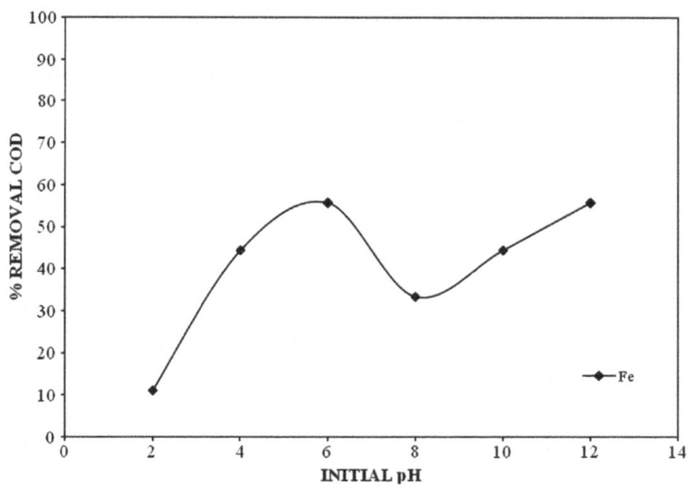

Fig. 16 Effect of initial pH on % removal of COD for iron electrode (No. of Electrodes = 6, Spacing = 1 cm, Voltage = 20 V, Time = 5 min, Applied Current = 2.0 A, and Temp = 30 °C)

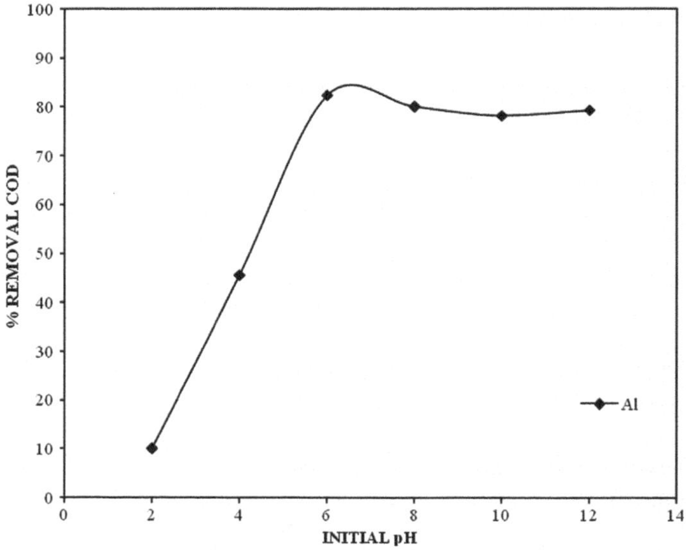

Fig. 17 Effect of initial pH on % removal of COD for aluminum electrode (No. of Electrodes = 6, Spacing = 1 cm, Voltage = 25 V, Time = 10 min, Applied Current = 2.5 A and Temp = 30 °C)

3.2.6 Effect of Electrode Spacing on Removal of COD

Electrode spacing is one of the effective parameters in the ECT process. The effect of spacing of electrodes on the performance of ECT was studied by varying the spacing from a cm to 2 cm in the present study for both the electrodes. The other parameters were kept constant (viz. time, temperature, pH, voltage, applied current, no. of electrodes). Figures 18 and 19 show the variation of COD percent removal with respect to the spacing between electrodes.

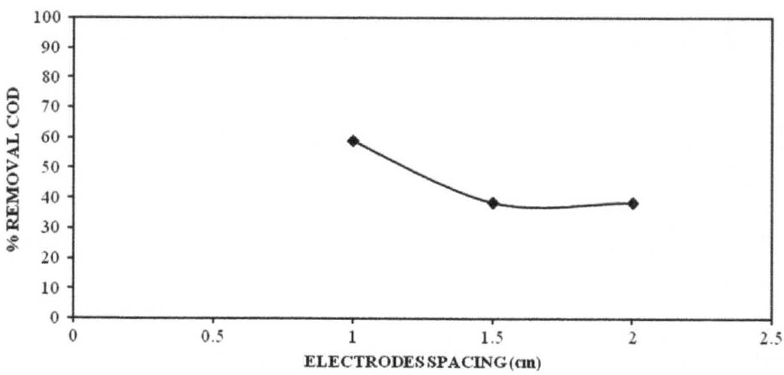

Fig. 18 Effect of electrodes spacing on % removal of COD for iron electrode (No. of Electrodes = 6, pH = 6, Voltage = 20 V, Time = 5 min, Applied Current = 2.0 A, and Temp = 30 °C)

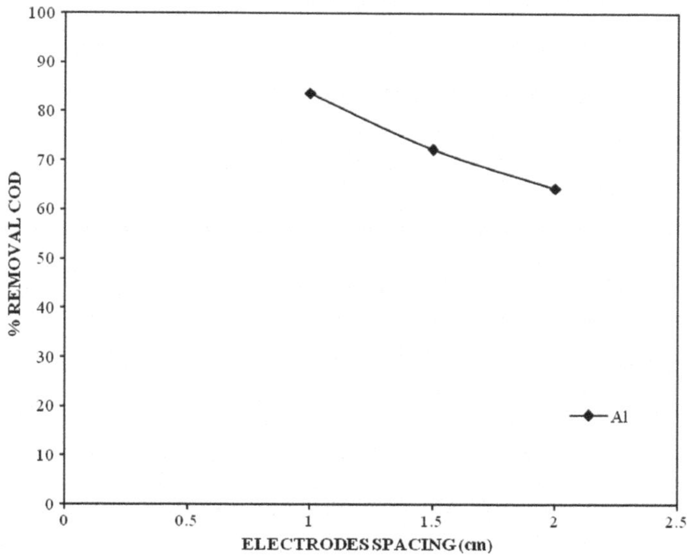

Fig. 19 Effect of electrodes spacing on % removal of COD for aluminum electrode (No. of Electrodes = 6, pH = 6, Voltage = 25 V, Time = 10 min, Applied Current = 2.5 A, and Temp = 30 °C)

For Fe electrodes, the percent removal of COD was found to be 58.88, 38.27, and 38.27% for the spacing of 1.0, 1.5, and 2.0 cm, respectively. It was observed that % removal of COD decreases with increasing the spacing of electrodes.

For Al electrodes, also the percent removal of COD was found to be decreasing with increasing spacing of the electrodes. The percent removal for COD by aluminum electrode was found to be 83.66, 72.26, and 64.34% at the spacing of 1.0, 1.5, and 2 cm, respectively, as shown in Fig. 19. It was observed that at 1 cm spacing, sufficient amount of bubbles were formed in the ECR leachate solution at the end of ECT and the color becomes somehow milky, but after very few time, it becomes colorless. Some parts of the flocs formed during the ECT rapidly settle at bottom, and another part floats at the top. As the spacing increased, the bubbles and flocs formed are reduced.

For both Fe and Al electrodes, percentage removal of COD increases with decreasing the spacing of electrode. The maximum percent removal of COD was found to be at the spacing of 1.0 cm; hence, further experiments were carried out at 1.0 cm for both the electrodes.

4 Conclusions

1. The solid waste collected from the VNIT SW dump site has the 1.427 kg of biodegradable and 0.573 kg of non-biodegradable solid waste mass.
2. From the characteristic analysis, it shows that the BOD-to-COD ratio for leachate is 0.8.
3. The COD removal efficiency of ECT for leachate by using aluminum electrode is found to be 80–85% at various conditions.
4. The COD removal efficiency of ECT for leachate by using iron electrode is found to be 40–45% at various conditions.
5. From the present study, it may be concluded that the aluminum electrode has higher efficiency than iron electrode.
6. The optimum time required to treat the leachate for iron electrode is 5 min, whereas for aluminum electrode, the optimum time required to treat is 10 min.
7. The optimum voltage to treat the leachate to remove maximum COD for iron electrode is 20 V and for aluminum electrode is 25 V.
8. The optimum applied current to treat the leachate to remove maximum COD for iron electrode is 2.0 A and for aluminum electrode is 2.5 A.
9. Percent removal of COD increases with number of electrodes in aluminum as well as iron electrodes. The maximum number of electrodes giving maximum efficiency was found by using 6 numbers of electrodes in both the cases of electrodes.
10. The maximum COD removal was found to be at the initial pH of 6 for both Fe and Al electrodes.
11. The COD removal by ECT was found to be decreasing with the increasing spacing between the electrodes in both the cases. The optimum spacing between the two consecutive electrodes is 1 cm for both Fe and Al electrodes. At the 1 cm spacing or gap, the removal of COD was found to be 58.8% by Fe electrode and 83.66% for that by Al electrode.
12. It was observed that the pH of the leachate after treatment is more than the pH before, so it reduces the further adjustment of pH and also reduced the chemical cost to increase the pH of leachate.

References

Amokrane, A., Comel, C., & Veron, J. (1997). Landfill leachate pretreatment by coagulation-flocculation. *Water Research, 31,* 2775–2782.

Bishop, P. L., & Kinner, N. E. (1986). Aerobic fixed-film processes. In W. Schonborn (Ed.), *Chapter III, Biotechnology*. H.-J. Rehm & G. Reed (Vol. Eds.), *Microbial degradation* (Vol. 8). Weinheim, New York: VCH.

Geenens, D., Bixio, B., & Thoeye, C. (2001). Combined ozone-activated sludge treatment of landfill leachate. *Water Science and Technology, 44*(2–3), 359–365.

Henry, G., & Heinke, G. W. (1996). *Environmental science and engineering* (2nd ed.). Englewood, NJ: Prentice-Hall.

Im, J.-H., Woo, H.-J., Choi, M.-W., Han, K. B., & Kim, C.-W. (2001). Simultaneous organic and nitrogen removal from municipal landfill leachate using an anaerobic-aerobic system. *Water Research, 35*(10), 2403–2410.

Loukidou, M. X., & Zouboulis, A. I. (2001). Comparison of two biological treatment processes using attached–growth biomass for sanitary landfill leachate treatment. *Environmental Pollution, 111*(2), 273–281.

Morawe, B., Ramteke, D. S., & Vogelpohl, A. (1995). Activated carbon column performance studies of biologically treated landfill leachate. *Chemical Engineering and Processing, 34,* 299–303.

Moreno, H., Cocke, D. L., & Jewel, A. G. (2007). Electrocoagulation mechanism for COD removal. *Seperation and Purification Technology, 56,* 204–211.

Mouedhen, G., Feki, M., Wery, M. D. P., & Ayedi, H. F. (2008). Behavior of aluminum electrodes in electrocoagulation process. *Journal of Hazardous Materials, 150,* 124–135.

Tatsi, A. A., Zouboulis, A. I., Matis, K. A., & Samaras, P. (2003). Coagulation–flocculation pre-treatment of sanitary landfill leachates. *Chemosphere, 53,* 737–744.

Ushikoshi, K., Kobayashi, T., Uemastu, K., Toji, A., Kojima, D., & Matsumoto, K. (2002). Leachate treatment by the reverse osmosis system. *Desalination, 150,* 121–129.

Yalmaz, G., & Oztürk, I. (2001). Biological ammonia removal from anaerobically pre-treated landfill leachate in sequencing batch reactors (SBR). *Water Science and Technology, 43*(3), 307–314.

Estimation of Greenhouse Gas Emissions from Matuail Landfill Site

Md. Maruful Hoque and M. Tauhid Ur Rahman

Abstract Landfill is the very common final solid waste disposal and treatment option in developing countries. The Matuail landfill site is a controlled dumping site grade 1 as per JICA study meaning the place for daily dumping is controlled without soil cover. So, greenhouse gases such as carbon dioxide (CO_2), methane (CH_4), and nitrous oxide (N_2O) that are emitted from the landfill site due to decomposition of solid wastes are contributing to local environmental pollution as well as global warming problem. The current study was focused on the quantification of greenhouses from the Matuail landfill site which is located at the south west of capital Dhaka, Bangladesh. The two popular models are used; the Landfill Gas Emissions model (LandGEM V3.02) provided by the United States Environmental Protection Agency (EPA) and the default methodology based on the intergovernmental panel on climate change (IPCC) 1996 revised guidelines. The models differ on their complexity and input data requirements and scientific approach for the estimation of greenhouse gases. The annual average CH_4 emission rates from Matuail landfill site for the period 2008–2016 are estimated. The LandGEM model was selected for the quantification of more representative estimation of landfill gas emission rates. Proper extraction and utilization of greenhouse gas emissions from Matuail landfill is highly suggested in order to reduce the harmful effects on the global warming potential and earning foreign currencies under clean development mechanism (CDM).

Keywords Greenhouse gas (GHG) · Landfill · LandGem · IPCC model · Solid waste

1 Introduction

For the local government of developing countries, municipal solid waste (MSW) is a continuous problem influenced by social, political, educational, and economic factors. Growth in economic, population, urbanization, and industrialization has led

---

Md. Maruful Hoque (✉) · M. Tauhid Ur Rahman
Climate Change Lab, Department of Civil Engineering, Military Institute of Science & Technology (MIST), Bangladesh, Mirpur Cantonment, 1216 Dhaka, Bangladesh
e-mail: marufpwd@gmail.com

S. K. Ghosh (ed.), *Urban Mining and Sustainable Waste Management*,
https://doi.org/10.1007/978-981-15-0532-4_15

to a large quantity of waste generation in developing countries, including Bangladesh is due to increase population, haphazard urbanization and recent foreign investment and unplanned industrialization. The Dhaka City of Bangladesh has been faced with highest damage and treats due to climate change and public health problems caused by the mismanagement of solid waste and its emissions of greenhouse gases (GHGs).

The greenhouse gases are methane (CH_4), carbon dioxide (CO_2), nitrous oxide (N_2O), and chlorofluorocarbons (CFCs). These greenhouse gases trap the heat of the sun in the atmosphere, causing the temperature rise to levels harmful to living beings and the environment. Figure 1 shows greenhouse gas emissions (CO_2e) by gas, World (Ritchie and Roser 2018). Municipal solid waste (MSW) treatment and disposal processes are considered worldwide as one of the main sources of anthropogenic emissions of greenhouse gases (GHGs) which contribute to climate change like global warming. The major waste management options that are contributed to the climate change are summarized in Table 1.

Landfill is the most economic waste treatment option used for 80% handling of municipal solid waste worldwide (Yang et al. 2015; Kumar and Sharma 2014). Where as in Europe, landfill sites are responsible 30% of anthropogenic sources of methane emissions (Georgaki et al. 2008). In the USA, landfill sites were known as the second major source of anthropogenic GHGs (Tian et al. 2013). The landfill gases (LFGs) are the sources of GHGs generated under favorable conditions by the anaerobic microbial decomposition of wastes in landfills. The LFG typically consists of methane (CH_4) from 45 to 60%, carbon dioxide (CO_2) from 40 to 60%

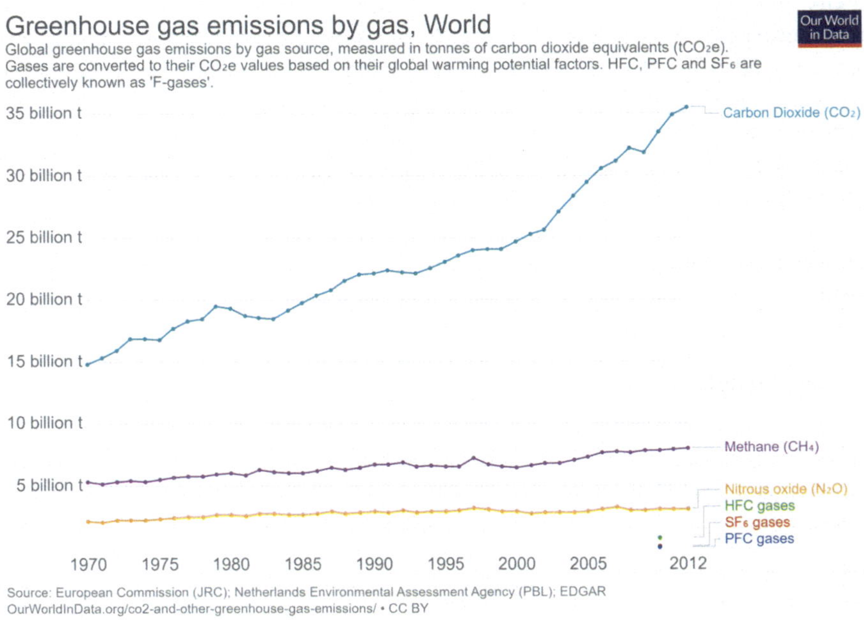

Fig. 1 Greenhouse emissions (CO_2e) by gas, world

Table 1 Contribution from solid waste to climate change

Serial	Waste management options	GHG emission contributing to global warming
1	Composting	CO_2 emission
2	Incineration	CO_2, N_2O, aerosol particle emission
3	Landfilling	CH_4 and methane precursor emissions
4	Recycling	emission

and trace gases from 0.01 to 0.5% like nitrogen (N_2), oxygen (O_2), ammonia (NH_3), hydrogen sulfide (H_2S), carbon monoxide (CO), sulfide (S_2), hydrogen (H_2), and non-methane organic compounds (NMOCs) such as trichloroethylene, benzene, and vinyl chloride. The methane (CH_4) is the most significant GHG gas produced from solid waste which has a global warming potential (GWP) more than 28 times that of CO_2. Additionally, high concentrations (5–15%) of methane can be explosive in nature when mixed with air. So, proper collection of methane is necessary (Kreith and Tchobanoglous 2002; Aydi 2012). To determine the quantity of CH_4, emissions from landfills is the most important step in managing and utilizing methane to reduce the adverse effects. Many one site investigation, filed testing, numerical model, and mathematical model have been introduced globally for estimating CH_4 emissions from landfills (Chiemchaisri and Visvanathan 2008; Di et al. 2011).

For example, to estimate methane emission from landfills, the Intergovernmental Panel on Climate Change (IPCC) has been developed the default methodology in its 1996 revised guidelines (Intergovernmental Panel on Climate Change 1996) and the United States Environmental Protection Agency (EPA) has been introduced the LandGEM model (Houghton 1997). Many researchers has been used the IPCC default methodology (Intergovernmental Panel on Climate Change 2000; Jensen and Pipatti 2002; Amini et al. 2013) where as other has been studied the LandGEM model (Tian et al. 2013; Waste Concern 2009; Hai and Ali 2005).

The objective of the study was to quantify methane gas production from the Matuail, the largest dumping site of Dhaka city. The both models have applied for CH_4 emission estimation from the Matuail landfill site. After literature review, it has been found that this area of research has not been covered in the Dhaka, Bangladesh; hence, it is expected that this study will bring a new window of knowledge to the stakeholders as well as researchers in Bangladesh. Basically, such kind of modeling is a predicting modeling technique for gas production in the landfill site based on waste disposal data in the past and future period in the site. The efficiency of the waste collection system can be evaluated by the output of the model. In addition, for developing a sanitary landfill project, the model is a key step to estimate the available recoverable amount of CH_4 as green energy source as well as foreign credit earning option under clean development mechanism (CDM). Besides, this study also aids in planning appropriate mitigation measures that aim at minimizing GHG emissions into the atmosphere.

2 Materials and Methods

Two different popular methods for estimating CH_4 quantities from the landfill were used in the study. Those methods are the IPCC default method and the LandGEM model. The both methods aimed at estimating the total methane (CH_4) quantities would be generated from Matuail, Dhaka landfill.

2.1 Matuail Landfill Site, Dhaka, Bangladesh

Dhaka (area about 306.38 km^2) is the capital of Bangladesh with about a population of 18.89 million people. Matuail landfill site is located about eight kilometers from *Gulistan* zero point in the south of Dhaka. It is used as a MSW crude landfill from the year 1993. In this study the second phase site, which is started to use from the year 2008 as a control MSW dumping site is considered. This site is also one of the two landfills serving Dhaka city. The site is used by the Dhaka South City Corporation (DSCC) area about 100 acres. The average waste height is about 60 feet (18 meters). There was few gas venting system which was part of adjacent semi-aerobic landfill but now it is recognized as a control dumping without any kind of sanitary measures like daily cover, layer by layer compaction, proper leachate collection system, gas collection and utilization system, etc.

2.2 Gas Emissions Estimations

2.2.1 IPCC 1996 Default Method (DM)

According to the Revised 1996 IPCC guideline for national greenhouse Gas (GHG) inventories, there are two methods to estimate methane (CH_4) emissions from solid waste disposal sites (SWDS):

1. First-order decay (FOD) method
2. Default method.

In this study, the IPCC default method has been used, and it is briefly described here. H. G. Bingemer and P. J. Crutzen in 1987 (Houghton 1997) at first developed the default method; then, IPCC has been adopted in 1996 as an accredited method for methane estimation from SWDS. The method is simple one and based on mass balance approach. The method assumes that the potential methane from the solid waste is released in the same year the waste is dumped in the SWDs (Jensen and Pipatti 2002). The parameters of the method are shown in Eq. (1) (Tables 2, 3, 4, 5, and 6):

Table 2 Contribution from solid waste to climate change

Parameter	Symbol	Value	Source
Annual total municipal solid waste	MSW_T, Gg/yr	Quantity from 2008–2016	Dhaka south city corporation
MSW disposed of in SWDS	[a]MSW_F	0.7	Unmanaged—deep (≥5 m waste)
Methane correction factor	MCF	0.6	IPCC (Houghton 1997) default value
Fraction of DOC that dissimilated	DOC_F	0.77	IPCC (Houghton 1997) default value
Volume fraction of CH_4 in the generated landfill gas	F	0.5	{50% of LFG} IPCC (Houghton 1997) default value
Recovered methane in a year	R	0	IPCC (Houghton 1997) default value
An oxidation factor in year	OX	0	IPCC (Houghton 1997) default value
Degradable organic carbon	DOC	0.156	IPCC (Houghton 1997) default value

[a]MSW_F has been taken equal to waste collection efficiency

Table 3 Physical composition of MSW in Dhaka

Waste type	Composition from all sources (%)
Paper and textiles	9.73
Garden and park waste and other organic putrescible	2.5
Food waste	67.65
Wood and straw waste	4.2
Metals	0.26
Glass and ceramic	1.13
Plastic, rubber, leather	5.1
Medicine/chemical	0.64
Rock, dirt, and misc	8.79

Source waste concern (2009)

$$CH_4 \text{ emissions}\left(\frac{\text{Gg}}{\text{yr}}\right) = \left(MSW_T \times MSW_F \times MCF \times DOC \times DOC_F \times F \times \frac{16}{12} - R\right) \times (1 - OX) \quad (1)$$

where

Table 4 Default DOC values for major waste streams

Waste stream		Percent DOC (by weight)
A	Paper and textiles	40
B	Garden and park waste and other organic putrescible	17
C	Food waste	15
D	Wood and straw[a]	30

[a]Excluding lignin *C*
Source IPCC guidelines (Houghton 1997)

Table 5 LandGEM model parameters (input) used for Matuail landfill methane estimation

Parameters	Symbol	Value	Unit
Landfill starting year		2008–2016	
Waste design capacity		5173328	Mg/year
Potential methane generation capacity	L_0	170	m3/Mg
Methane generation rate	k	0.2	Year-1
NMOC concentration		4000	ppmv as hexane

Table 6 Total methane (CH_4) volume generated from the Matuail landfill

Criterion	Unit	IPCC default method		USEPA LandGEM model
Total methane	m^3	3.756×10^7		3.647×10^7
Average	m^3		3.701×10^7	
difference	m^3		0.0554×10^7	
Relative standard deviation (RSD)	%		1.47	

MSW_T annual total municipal solid waste (MSW) generated from a community
MSW_F MSW disposed of in SWDS (fraction);
MCF methane correction factor
DOC degradable organic carbon (fraction)
DOC_F fraction of DOC that can be decomposed
F Volume fraction of CH_4 in the generated landfill gas;
$\frac{16}{12}$ molecular weight ratio of CH_4 to carbon
R recovered methane in a year T in Gg per year; and
OX an oxidation factor in year T (fraction). (Intergovernmental Panel on Climate Change 1996; Houghton 1997)

In the IPCC guidelines, a relation based on the weighted average of carbon content of various components of the municipal solid waste is provided to assist in degradable organic carbon (DOC) content calculation. In Table 4, a set of default DOC values for different waste categories are shown. The relation to calculate DOC is given as:

$$\text{DOC} = 0.4.\ A + 0.17.\ B + 0.15.\ C + 0.30.\ D \quad (2)$$

Waste types from Table 3 were categorized into the four classifications of Table 4. A value of DOC as 0.156 was obtained after applying certain approximations. In this case, default values for recovered CH_4 and oxidation factor were used in the absence of reliable values. All parameters for Eq. (1) along with the source information are already shown in Table 2. Sample methane emission for Matuail landfill MSW was calculated for the year 2016 as:

$$\begin{aligned}\text{Methane } (CH_4) \text{ emission} &= 747 \times 0.7 \times 0.156 \times 0.77 \times 0.5 \times 16/12 \\ &= 25.06\,\text{Gg/year}\end{aligned}$$

This default method gives methane emissions in terms of mass of methane generated in 1 year. The potential methane volume is 3.756×10^7 m^3/year. Although fairly good estimates using default method are found, it depends on the long-term uniformity of waste annual amount, waste composition, and final disposal. When precise results are required, a more sophisticated such as IPCC first-order decay (FOD) method or USEPA LandGEM can be used. This method reflects the time period of actual CH_4 emissions and considers the fact that CH_4 is emitted over a long period of time rather than rapidly.

2.2.2 Landfill Gas Emissions Model (LandGEM V3.02)

The LandGEM model CH_4 emission model was developed by the USEPA. It is one of the most extensively used models for estimating emission rates for total LFGs including CH_4, CO_2, NMOCs, and some particular air pollutants. LandGEM uses either default parameters if no site-specific data are available or site-specific data to estimate emissions. LandGEM is based on a first-order exponential decay rate (FOD) equation for quantifying CH_4 emissions from the MSW landfills waste decomposition for user-defined time period (default timing 80 years). In this study, LandGEM (Version 3.02-USEPA, 2005) model was used for estimating LFG emission from the landfill. The LandGEM CH_4 emission methodology can be described mathematically using the following equation.

$$Q_{CH_4} = \sum_{i=1}^{n} \sum_{j=0.1}^{1} \text{K}L_0 \left(\frac{M_i}{10} \right) e^{-kt_{ij}} \quad (3)$$

where Q_{CH4} is the annual CH_4 generation emissions in the year of the calculation (m^3/year); i, 1 year time increment and j, 0.1 year time increment; n is the duration of waste acceptance at landfill and equals the year of the calculation minus the initial year of waste acceptance; k is the methane (CH_4) generation rate constant(year^{-1});

L_0 is the potential methane (CH_4) generation capacity (m^3/Mg waste); M_i is the mass of waste accepted in the ith year (Mg); and t_{ij} is the age of the jth section of waste mass M_i accepted in the ith year (decimal years).

The crucial parameters that are used in the model are described as follows

(a) Landfill waste design capacity: The total amount of waste (default unit-Mg) that the landfill received during its full lifespan.
(b) Potential methane generation capacity (L_0): It depends only on the landfill waste type and composition. The higher the value of L_0 means the higher the cellulose content of the waste. Ideally, L_0 is to be found out experimentally. $L_0 =$ 170 m^3/Mg (Hai and Ali 2005) is used as calculated earlier for Dhaka MSW due to the absence of experimental data.
(c) Methane generation rate (k): The value of "k" is a function of waste pH, temperature, moisture content, and availability of the nutrients for methanogens. The higher the value of "k" indicates the faster the CH_4 generation rate increases and then decays over time. Five values of "k" were used in the software . $k = 0.2$/year (Hai and Ali 2005) is used as calculated earlier for Dhaka MSW in the absence of experimental data.
(d) Concentration of total non-methane organic compounds (NMOCs): The NMOC concentration (parts per million by volume {ppmv}) in landfill gas is depends on the waste types and products of waste anaerobic decomposition reactions. When NMOC emissions are being estimated, then it is used by LandGEM. The default value was retained for NMOC in the present study. A list of important parameters is shown in Table 5.

The LandGEM is able to estimate the total quantities of different gases along with variation in production of these gases with time as shown in Fig. 2. The model assumed that the resulted methane gas emission from waste decomposition process will remain up to more than hundreds of years beyond landfills closure. The peak potential methane amount is 24330 Mg/year or 24.33 Gg/year in the year 2017 just after the closure of the waste disposal at the landfill site.

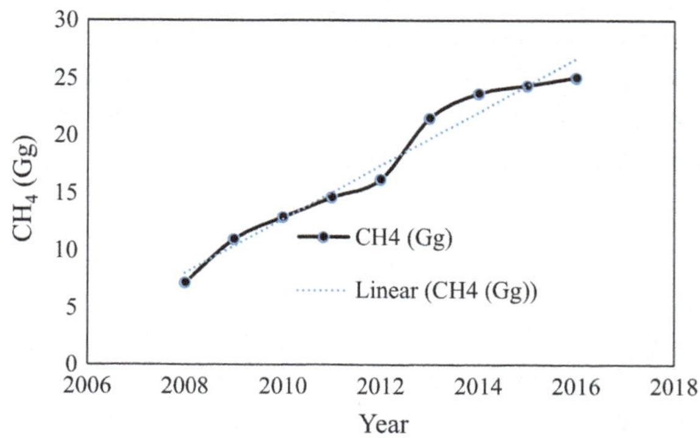

Fig. 2 Methane emission from solid waste landfills using the IPCC default method

3 Results and Discussions

The methane emissions were computed using the IPCC default method and the USEPA LandGEM models. Table 5 shows that the Matuail landfill is expected to have $3.701 \pm 0.0554 \times 10^7$ m^3 of methane (CH_4) emissions. The reliability of the results may be ensured because the amount obtained from the two methods are in reasonable agreement with the difference of 1.47%.

The IPCC default method showed that CH_4 emissions in the landfill will start in the first year of waste disposal and will remain up to the year 2016. On the contrary, the LandGEM model suggested that no methane emissions will be observed in landfill starting the year 2008; however, the emission will start in the year 2009. The LandGEM model also showed that potential CH_4 emissions are expected to continue even 132 years after the closure of the landfill that is up to the year 2148. In addition to CH_4 gas, landfill gases usually contain approximately 40–60% CO_2, NMOC which also showed in the graph. The pattern of respective gas emissions is same but the amount is different.

In both cases, it has been observed that the methane emissions are directly proportional to solid waste quantity disposed of in the landfill. In the case of IPCC method from Fig. 2, it is clear that the methane emission increases after the closing of the landfill in the year 2016. The methane production is linearly increasing with time. In Fig. 3, the time-series LFG, CH_4, CO_2 and NMOC gas production using LandGEM is presented. According to LandGEM results in Fig. 3, more than 95% of the LFG gas will be produced within 70 years (~2078) after starting of the landfill. The model also showed that the amount of gas produced within the last 78 year represents only 5% of total LFG emission including methane.

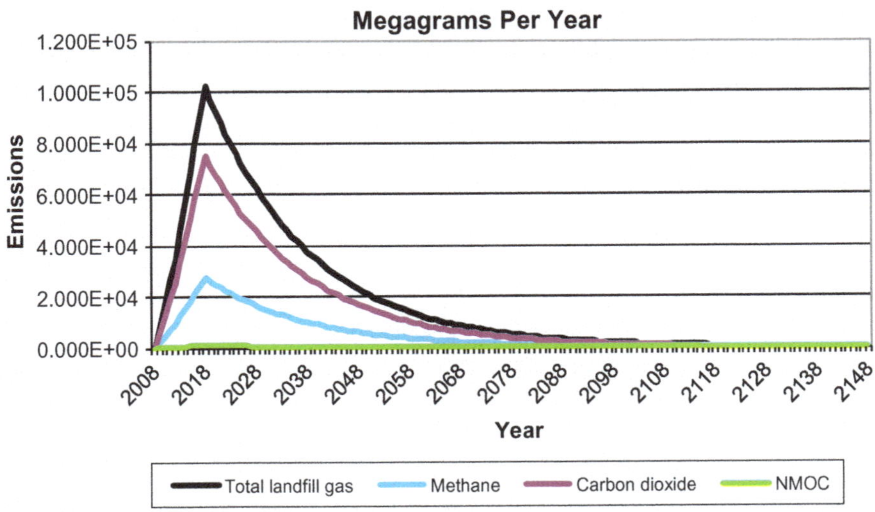

Fig. 3 LandGEM results of variation in LFG, CH_4, and CO_2 production

In IPCC method, the peak methane production has occurred in the year after 2016 that is after the closure of the landfill. In the case of LandGEM model, it is estimated that CH_4 peak production would occur just after the closure of the landfill, which is in agreement with the results obtained by the IPCC default method. A summary of results obtained by the IPCC default method and the LandGEM model are presented in Table 6.

Therefore, it can be concluded that the mathematical models diverge from each other with different assumptions, input parameters, outputs, time-bound, etc. However, in this study, although the methane emission results obtained from the two methods are very close, other researcher found quiet difference results for different mathematical models.

4 Conclusion

The present work deals with the comparability and of the two mathematical models (IPCC method and LandGEM) for GHGs especially methane (CH_4) emission estimation from the Matuail MSW landfill. In the study, both methods are estimated very close potential methane gas emission results. The study showed that the potential CH_4 emission from the landfill would be about 3.701×10^7 m^3 and gas generation would continue for more than 70 years after the closure of the landfill in case of LandGEM method. The LandGEM model predicted comparable results due to the fact that the model based on first-order decay (FOD) which recognized and verified in different literature with on-site experimentally obtained results. A number of uncertainties and factors are associated with results of both methods due to the lack of site-specific data like quantity of decomposable organic content, methane generation potential, placement conditions, etc. More extensive research is required in the future for Bangladesh condition to attain the factors or parameters used in the models. These amounts of GHG especially methane could be utilized properly for electricity generation by adopting improved gas collection system and converted control dumping into sanitary landfill and earn foreign currency through clean development mechanism (CDM).

References

Amini, H. R., Reinhart, D. R., & Niskanen, A. (2013). Comparison of first-order-decay modeled and actual field measured municipal solid waste landfill methane data. *Waste Management, 33*(12), 2720–2728. https://doi.org/10.1016/j.wasman.2013.07.025.

Aydi, A. (2012). Energy recovery from a municipal solid waste (MSW) landfill gas: A tunisian case study. *Hydrol Current Res, 3*(4), 1–3.

Chiemchaisri, C., & Visvanathan, C. (2008). Greenhouse gas emission potential of the municipal solid waste disposal sites in Thailand. *Journal of the Air and Waste Management Association, 58*(5), 629–635.

Di, B. G., Di, T. D., & Viviani, G. (2011). Evaluation of methane emissions from Palermo municipal landfill: Comparison between field measurements and models. *Waste Management, 31*(8), 1820–1826.

Georgaki, I., Soupios, P., Sakkas, N., Ververidis, F., Trantas, E., Vallianatos, F., et al. (2008). Evaluating the use of electrical resistivity imaging technique for improving CH_4 and CO_2 emission rate estimations in landfills. *Science of the Total Environment, 389*(2–3), 522–531.

Hai, F. Ibney, & Ali, M. (2005). A study on solid waste management system of Dhaka city corporation: Effect of composting and landfill location. *UAP Journal of Civil and Environmental Engineering, 1*(1), 18–26.

Houghton, J. T. (1997). *Revised 1996 IPCC guidelines for national greenhouse gas inventories.* Paris, France: Intergovernmental Panel on Climate Change (IPCC).

Intergovernmental Panel on Climate Change. (2000). *Good practice guidance and uncertainty management in national greenhouse gas inventories*. Paris, France: Intergovernmental Panel on Climate Change (IPCC).

Intergovernmental Panel on Climate Change, & Houghton, J. (1996). *Revised 1996 IPCC guidelines for national greenhouse gas inventories: Greenhouse gas inventory reporting instructions*. Paris, France: Organisation for Economic Cooperation and Development (OECD).

Jensen, J. E. F., Pipatti, R. (2002). CH_4 emissions from solid waste disposal, background paper. In *Good Practice Guidelines and Uncertainty Management in National Greenhouse Gas Inventories*, (pp. 419–439).

Kreith, F., & Tchobanoglous, G. (2002). *Handbook of solid waste management* (2nd ed.). New York, NY: McGraw-Hill Professional.

Kumar, A., & Sharma, M. P. (2014). Estimation of GHG emission and energy recovery potential from MSW landfill sites. *Sustainable Energy Technologies and Assessments, 5,* 50–61. https://doi.org/10.1016/j.seta.2013.11.004.

Ritchie, H., Roser, M. (2018). CO_2 and other greenhouse gas emissions. Published online at OurWorldInData. org. Retrieved from https://ourworldindata.org/co2-and-other-greenhouse-gas-emissions.

Tian, H., Gao, J., Hao, J., Lu, L., Zhu, C., & Qiu, P. (2013). Atmospheric pollution problems and control proposals associated with solid waste management in China: A review. *Journal of Hazardous Materials, 252–253,* 142–154.

Waste Concern. (2009). Waste data base of Bangladesh http://www.wasteconcern.org/documents/Waste%20Data%20Base_2009.pdf.

Yang, L., Chen, Z., Zhang, X., Liu, Y., & Xie, Y. (2015). Comparison study of landfill gas emissions from subtropical landfill with various phases: A case study in Wuhan, China. *Journal of the Air and Waste Management Association, 65*(8), 980–986. https://doi.org/10.1080/10962247.2015.1051605.

Stabilization of Contaminated Soil in a Landfill Site with Ground Granulated Blast Furnace Slag

Ramiz Raja and Supriya Pal

Abstract The aim of the present work is to examine the efficacy of ground granulated blast furnace slag (GGBFS) as an additive to improve the engineering properties of contaminated soft soil. The soil is contaminated with heavy metals such as zinc, lead and copper. In this study, emphasis was given on engineering property improvement of the stabilized soil. Various soil physical parameters were determined as per guidelines depicted in BIS 2720. The dry density, unconfined compressive strength (UCS) and specific gravity (*G*) values of the soil increased substantially with the increasing amount (5, 10 and 15%) of GGBFS addition in the contaminated soil. The specific gravity and UCS values of collected soil were improved from 2.31 to 2.65 and 95 to 1877 kN/m^2, respectively, when 15% of GGBFS was added to the untreated soil. The maximum dry density (MDD) of the untreated soil was improved from 1.32 to 1.90 g/cc for 15% addition of GGBFS. The concentrations of Zn, Pb and Cu in the contaminated soil were decreased by 70, 80 and 77%, respectively, after blending with GGBFS of 15% at a curing time of 28 days. The decrease in metal leachability and increase of strength properties of the contaminated soil clearly show that GGBFS material has a potential to be used as an additive for contaminated land stabilization and reclamation.

Keywords Contaminated soil · Heavy metals · Leaching · GGBFS · Shear strength · Stabilization

1 Introduction

The continuous increase of soil and water pollution due to the discharge of industrial and domestic waste in the landfill sites are of real concern to the people residing nearby the sites (Maheshwari et al. 2015). The toxic leachate from these sites migrates

R. Raja (✉) · S. Pal
Civil Engineering Department, National Institute of Technology Durgapur, Durgapur, West Bengal, India
e-mail: ramiz.nitdgp@gmail.com

S. Pal
e-mail: supriya_pal@rediffmail.com

S. K. Ghosh (ed.), *Urban Mining and Sustainable Waste Management*,
https://doi.org/10.1007/978-981-15-0532-4_16

through the sub-surface soil media causing serious depletion of surrounding soil and groundwater qualities which limits redevelopment of the area for future infrastructural activities (Xia et al. 2018). The lands near the industrial waste disposal sites, mine tailing ponds and abandoned mine sites in both developing and developed countries are heavily polluted with heavy metals (Zhang et al. 2015). Stabilization/solidification (S/S) technology is gaining popularity for in situ remediation of heavy metals-contaminated soils because of its higher efficiency, low cost and lesser time-consuming in comparison to the other available technologies (Du et al. 2014; Zhang et al. 2015). Xi et al. (2014) examined the potential of cement and lime, fly ash and their various proportions for solidification/stabilization of Pb-contaminated soil. They observed that proportional mixture of cement, fly ash and lime on mass basis of 2:1:1 is effective in remediating Pb-contaminated soil with a Pb^{2+} contamination level of 10,000 mg/kg. Rachman et al. (2018) applied S/S technology by using Portland cement additive for remediating Hg contaminated tailings from a gold mine in Indonesia. They observed more than 95% reduction of Hg contamination in the stabilized soil with an optimum composition of tailings: Portland cement of 90:10 and also a promising UCS value of 257 ton/m^2.

In the present study, an industrial waste product GGBFS was used as an additive to stabilize/solidify Zn-, Cu- and Pb-contaminated soil collected from a landfill site in Kolkata, West Bengal, India. The efficacy of GGBFS for reduction of heavy metal leachability and increase of strength properties of the stabilized soil were examined with an aim to reclaim the land in the study area for reusing for habitation and other industrial or infrastructural developments.

2 Material Used and Testing Procedure

The soil samples were collected by augur boring at a depth of 1 m below the ground level from a solid waste dumping site popularly known as Dhapa in Arupota (22° 32′ 33″ N, 88° 24′ 23″ E) near Eastern Metropolitan (EM) bypass in the eastern fringe of the metropolis of Kolkata, India as shown in Fig. 1. The soils were oven dried at a temperature of 100 ± 2 °C in a hot air oven and after then cooled down in room temperature. The physico-chemical properties of the collected soil samples were determined as per the guidelines depicted in BIS 2720 as shown in Table 1. Slag samples were collected from a slag pond of Durgapur Steel Plant, Durgapur, West Bengal, India, and pulverized in the laboratory using a wooden mallet for further testing. The chemical composition of the GGBFS was shown in Table 2.

The soil was then uniformly blended with 5%, 10% and 15% GGBFS, respectively. The specific gravity and compaction characteristics (MDD and OMC) of the blended soil were determined based on standard procedures. In order to perform the unconfined compressive strength (UCS) test, GGBFS blended soil sample of predetermined weight was kept in a cylindrical mould and water was added up to the OMC and mixed properly to prepare a homogeneous mixture. The mixture was then compacted in three layers under predefined load through static compaction until the

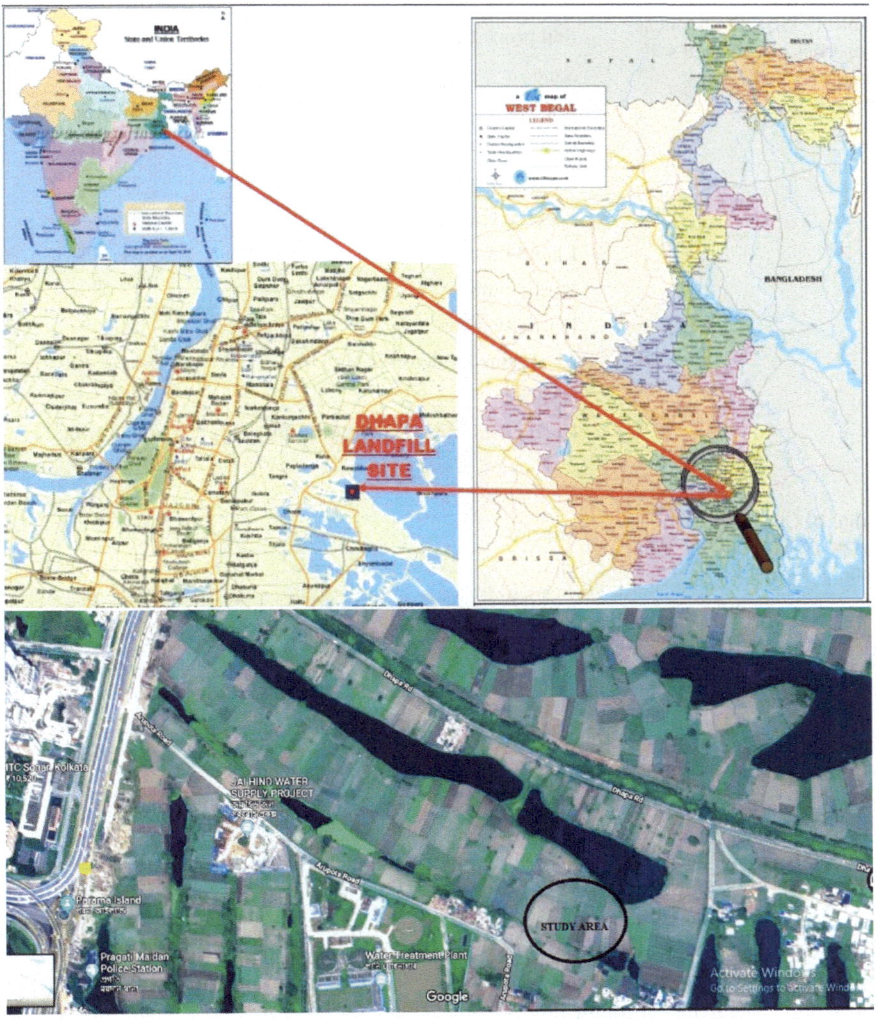

Fig. 1 Area of study. *Courtesy* Google map and West Bengal pollution control board

MDD was achieved. The cylindrical samples (diameter: 3.8 cm and length: 7.6 cm) was extracted through sampling tubes and cured for 28 days under water for UCS testing.

The leaching tests were performed as per USEPA (1992) method on the pulverized contaminated soil as well as stabilized soil-GGBFS mixtures. The soil samples were passed through 2.36 mm sieve. Deionized water was then added to each 10 g soil samples at a ratio of 1:20 (soil:water) and pH of the mixtures were maintained throughout the experiment near to 5.0 through addition of 0.5 N acetic acid. The soil water mixtures were then kept in a rotary shaker for 24 h at a speed of 40 rpm. The mixtures were then centrifuged at a speed of 7000 rpm for 10 min to separate

Table 1 Physico-chemical properties of collected soil sample

Soil properties	Value
Sand	35.40%
Silt	39.50%
Clay	25.10%
Field moisture content	8.16%
Specific gravity	2.06
Liquid limit	31.24%
Plastic limit	17.18%
Plasticity index	14.06%
Maximum dry density (MDD)	1.32 g/cc
OMC	24%
Unconfined compressive strength (UCS)	95 kN/m^2

Table 2 Chemical composition (%) of GGBFS sample

Minerals	Chemical composition (%)
SiO_2	18.25
Al_2O_3	12.50
CaO	26.85
Fe_2O_3	0.90
MgO	5.80
Na_2O	0.65
K_2O	0.85
TiO_2	0.90
SO_3	0.55

solid and liquid phase. The liquid phase was filtered through 0.22 μm membrane and the aliquot was then used for determination of metal concentration through atomic absorption spectrophotometer (AAS).

3 Results and Discussions

Table 3 shows the heavy metal concentrations in the collected soil. The soil was extremely polluted with Zn, Pb and Cu beyond the permissible limits as governed under the Canadian Soil Quality Guidelines (CCME 2007). The surface and groundwater in the area are also severely contaminated due to sub-surface migration of the contaminants from the study area as reported earlier elsewhere (Maiti et al. 2016). The soil quality of the study area is also very poor (Table 1) in terms of MDD (1.32 g/cc) and UCS (95 kN/m^2) values and this was happened due to long-term exposure of heavy metal contaminants in the natural soil (Estabragh et al. 2017).

Table 3 Heavy metal concentrations of the contaminated and GGBFS stabilized soil

Heavy metals	Metal concentrations (mg/kg) of untreated soil	Metal concentrations (mg/kg) of the stabilized soil after 5% GGBFS addition	Metal concentrations (mg/kg) of the stabilized soil after 10% GGBFS addition	Metal concentrations (mg/kg) of the stabilized soil after 15% GGBFS addition	Permissible values (mg/kg)
Zn	564.28	334.56	265.12	172.28	200
Pb	335.04	258.39	137.62	66.94	70
Cu	193.96	128.47	77.26	45.34	63

Table 3 also shows that concentrations of Zn, Pb and Cu in the contaminated soil were decreased by 70, 80 and 77%, respectively, after blending with GGBFS of 15% at a curing time of 28 days. The reduction of the heavy metals concentrations in the contaminated soil through addition of GGBFS may happen due to immobilization of contaminants and also due to change in the physical state of a contaminated material by formation of metal precipitates (Estabragh et al. 2017). The amount of Zn/Pb/Cu bearing insoluble and immobile metal precipitations also increased with the increment of the additive content and leads to a decrement of the leachable metals percentage (Wang et al. 2014). The reason behind this phenomenon was due to the fact that the stabilized soil possesses lower fraction exchangeable content and higher fraction of residual content than the untreated soil (Xia et al. 2018).

The variation of specific gravity (*G*) and unconfined compressive strength (UCS) of the stabilized soil with the increment of additive content are exhibited in Figs. 3 and 4. It is observed that G and UCS values are increased to 2.65 and 1877 kN/m^2, respectively, at a curing time of 28 days when GGBFS content increases to 15%. These was happened due to the formation of metal precipitates which increased the chemical bonding strength (Du et al. 2014) and thus enhanced higher compressive resistance (Xia et al. 2018).

The MDD values of the stabilized soil increased by 40% with respect to the untreated soil, when the GGBFS content increases to 15% as shown in Fig. 2. The formation of metal precipitates may fill the pore spaces in the soil with the increment of the additive content leading to the decrement of the total pore volume and facilitates denser soil structure (Horpibulsuk et al. 2009; Xia et al. 2018). The decrement of the pore volumes increased the inter-particle contacts in the stabilized soil and thus increased the dry density and UCS values of the treated soil with increasing additive contents (Ismail et al. 2002) (Figs. 3 and 4).

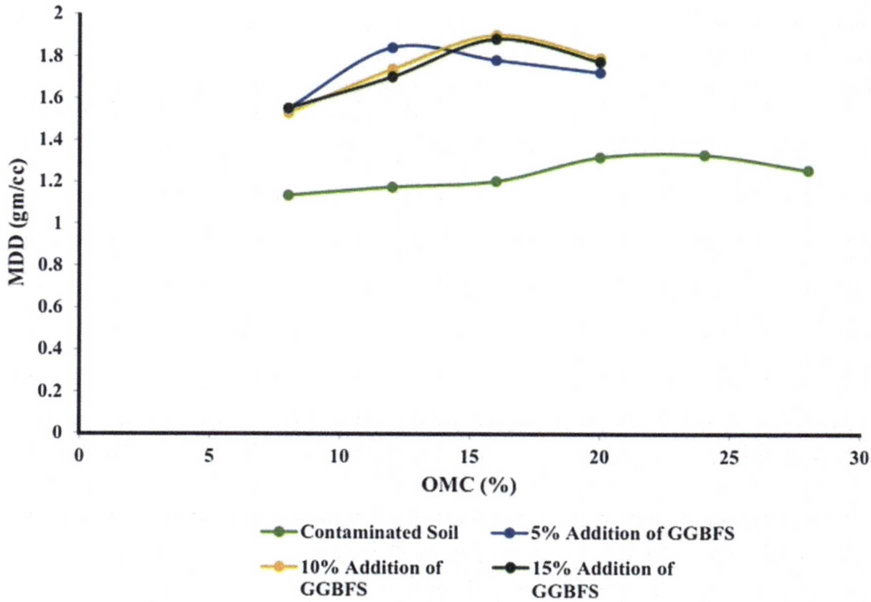

Fig. 2 Comparison of MDD versus OMC of stabilized soil for varying dosage of GGBFS

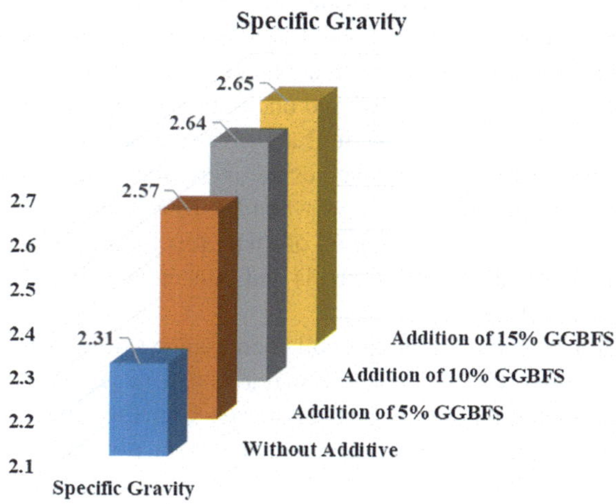

Fig. 3 Variation of specific gravity of the stabilized soil with the variation of GGBFS content

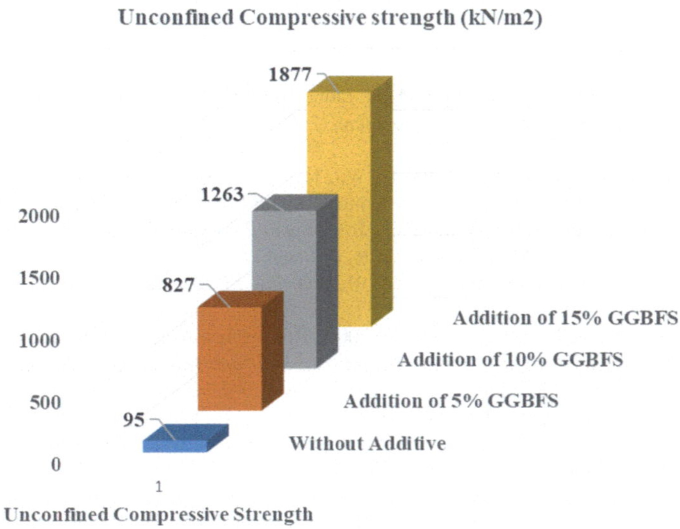

Fig. 4 Improvement of unconfined compressive strength of the stabilized soil with the increment of GGBFS percentage

4 Conclusions

After studying all the test data, following conclusions can be drawn:

- The maximum dry density and UCS values of the GGBFS stabilized soil increased with the increment of the additive content at a curing time of 28 days. This was happened due to the formation of metal precipitates and reduction of the pore volume of the treated soil under the application of GGBFS as a binder. The increased inter-particle contacts in the stabilized soil are prime responsible for the increased strength.
- The metal leachability of the stabilized soil reduced significantly due to the treatment with GGBFS additives. The addition of GGBFS as a binder lowered the exchangeable fraction of the metals and increased the residual fraction leads to the reduction of metal leachability from the contaminated soil.
- So, finally, it can be concluded that GGBFS can be used as an additive for improvement of engineering properties of contaminated soil. Thus, contaminated land can be reclaimed and constructional activities can be started without replacing the weak contaminated soil. Hence, the contaminated soil replacement cost can be minimized. However, further research studies are to be undertaken to increase the efficiency of GGBFS additives in reducing the metal leachability from contaminated soil through the addition of other co-stabilizers, viz. lime, cement, etc.

Acknowledgements The authors are thankful to the Director, National Institute of Technology, Durgapur-713209, West Bengal, India for providing necessary assistance for carrying out the present research.

References

BIS 2720. *Methods of test for soils*. New Delhi, India: BIS.

Canadian Council of Ministers of the Environment. (2007). *Canadian environmental quality guidelines*.

Du, Y. J., Wei, M. L., Reddy, K. R., Jin, F., Wu, H. L., & Liu, Z. B. (2014). New phosphate-based binder for stabilization of soils contaminated with heavy metals: Leaching, strength and microstructure characterization. *Journal of Environmental Management, 146,* 179–188.

Estabragh, A. R., Kholoosi, M. M., Ghaziani, F., & Javadi, A. A. (2017). Stabilization and solidification of a clay soil contaminated with MTBE. *Journal of Environmental Engineering, 143*(9), 04017054.

Horpibulsuk, S., Rachan, R., & Raksachon, Y. (2009). Role of fly ash on strength and microstructure development in blended cement stabilized silty clay. *Soils and Foundations, 49*(1), 85–98.

Ismail, M. A., Joer, H. A., Randolph, M. F., & Meritt, A. (2002). Cementation of porous materials using calcite. *Geotechnique, 52*(5), 313–324.

Maheshwari, R., Gupta, S., & Das, K. (2015). Impact of landfill waste on health: An overview. *IOSR Journal of Environmental Science, Toxicology and Food Technology, 01*(04), 17–23.

Maiti, S. K., De, S., Hazra, T., Debsarkar, A., & Dutta, A. (2016). Characterization of leachate and its impact on surface and groundwater quality of a closed dumpsite—A case study at Dhapa, Kolkata, India. *Procedia Environmental Sciences, 35*(2016), 391–399.

Rachman, R. M., Bahri, A. S., & Trihadiningrum, Y. (2018). Stabilization and solidification of tailings from a traditional gold mine using Portland cement. *Environmental Engineering Research, 23*(02), 189–1194.

USEPA (United States Environmental Protection Agency). (1992). *Toxicity characteristic leaching procedure (TCLP)*. Method 1311, Washington, DC.

Wang, J. R., Ma, B. G., & Li, X. G. (2014). The solidification and hydration products of magnesium phosphate cement with Pb^{2+}, Zn^{2+} and Cu^{2+}. *Journal of Functional Materials, 45*(5), 5060–5064.

Xi, Y., Wu, X., & Xiong, H. (2014). Solidification/stabilization of Pb-contaminated soils with cement and other additives. *Soil and Sediment Contamination: An International Journal, 23*(08), 887–898.

Xia, W. Y., Feng, Y. S., Du, Y. J., Reddy, K. R., & Wei, M. L. (2018). Solidification and stabilization of heavy metal contaminated industrial site soil using KMP binder. *Journal of Materials in Civil Engineering, 30*(6), 04018080.

Zhang, Z., et al. (2015). Screening and assessment of solidification/stabilization amendments suitable for soils of lead-acid battery contaminated site. *Journal of Hazardous Materials, 288,* 140–146.

Air Quality Survey of Some Major Dumpsites in Lagos State, Nigeria

Omowonuola Olubukola Sonibare, Adeniyi Saheed Aremu, Rafiu Olasunkanmi Yusuf and Jamiu Adetayo Adeniran

Abstract This study investigated the status of air quality at Olushosun, Abule Egba, and Solous dumpsite in Lagos, Nigeria. The parameters examined are total suspended particulates (TSP), CO, NO_2, SO_2, and odour. The Landfill Gas Emissions Model (LandGEM) was used to estimate the potential landfill gas emissions from the dumpsites. Sulphur dioxide (SO_2) was not detected at Olushosun and Abule Egba dumpsites. At Solous, the measured concentration is about 21-folds of the 1-h limit with 24-h concentration of 0.09 ppm which is about ninefolds of the FMENV's set limit. The measured concentrations of NO_2 are about 1.3-folds of the set limit at all the locations in Olushosun and Abule Egba but between 2.5 and 5.0-folds at Solous. In Olushosun, the measured 1-h CO concentrations range between 8.0 and 40 ppm, at Abule Egba and Solous, the range is 10.0–44.0 ppm and 1.0–179.0 ppm, respectively. At Olushosun, the 1-h TSP concentration is about 2.5-folds of the 1-h limit but 3.2-folds and 2.8-folds of the same limit at both Abule Egba and Solous, respectively. The odour levels from the three dumpsites are significantly high ranging between 4 and 5 on the "intensity by word" scale of 5. The measured air quality parameters concentrations signify a badly impaired airshed and can be classified as having poor air quality which calls for urgent attention. The methane potential using the LandGem tool is estimated to be between 4.04×10^2 and 4.97×10^4 ton/year which is equivalent to about 6.06×10^5 to 7.45×10^7 m^3/year. This can be used to generate about 0.4–49 MW of electricity if efficiently captured.

Keywords Dumpsite · Emission · Air quality · Lagos · Nigeria

O. O. Sonibare
Department of Family Medicine, General Outpatient Department (GOPD), Obafemi Awolowo University Teaching Hospitals Complex (OAUTHC), Ile-Ife, Nigeria

A. S. Aremu (✉)
Department of Water Resources and Environmental Engineering, University of Ilorin, Ilorin, Nigeria
e-mail: aremu_adeniyi@yahoo.com

R. O. Yusuf · J. A. Adeniran
Department of Chemical Engineering, Environmental Engineering Research Laboratory, University of Ilorin, Ilorin, Nigeria

S. K. Ghosh (ed.), *Urban Mining and Sustainable Waste Management*,
https://doi.org/10.1007/978-981-15-0532-4_17

1 Introduction

Landfilling is the oldest and the simplest form of waste disposal. Landfills are designed to dispose large masses of waste at economical costs with potentially less health and environmental effects (Powell and Chertow 2019; Zhang et al. 2010). Currently, the popular method to eliminate municipal solid wastes in most developing countries is through landfilling. However, improper landfill management may pose serious environmental and health risks. Wastes disposal in landfill sites where hazardous waste is dumped has increasingly caused concern about possible adverse health effects for populations living nearby in developing countries (Giusti 2009; Vrijheid 2000).

Waste deposited in landfills undergoes biological, chemical, and physical transformations that cause changes in its solid, liquid (leachate), and gas phases (Christensen et al. 2014; Schiopu and Gavrilescu 2010). Although municipal solid waste is not classified as a hazardous substance, the accumulation of solid waste in a location is unsafe for human health and the environment as harmful air pollutants are released into the environment due to high content of organic matters which are susceptible to biodegradation (Khalid et al. 2011; Salehpour et al. 2018).

Some of the greatest environmental impact of landfill is the emission of gaseous pollutants, like methane, carbon dioxide, nitrous oxide, and halogenated organic compounds which can contribute to both global warming and climate change by absorbing infrared radiation (Lim et al. 2016; Talaiekhozani et al. 2016); and odorous gases, like hydrogen sulphide, alkylated benzenes, esters, dimethyl sulphide (DMS), butyric acid, thiols (mercaptans), such as methanethiol and propanethiol, terpenes (*p*-cymene, limonene, α-pinene), carbon disulphide, and ammonia, gases which are mainly responsible for odour nuisance on a local scale (Liu et al. 2016, 2018; Wenjing et al. 2015). Other substances emitted from landfills which may pose serious health risks to living organisms, some of which are carcinogenic and mutagenic, include non-methane volatile organic compounds such as benzene, vinyl chloride, and tetrachloroethylene (de Morais et al. 2018; Petrovic et al. 2018).

Waste disposal and its possible effects on the health, environment, and the urban landscape is a burning national issue in Nigeria today. One of the focal Nigerian states currently battling with unlawful and haphazard dumping of refuse which is a clear violation of our environmental sanitation laws is Lagos state. Previous studies have attempted to evaluate the environmental impact associated with the major landfill sites in Lagos, Nigeria (Ife-Adediran and Isabota 2018; Ojo-Awo et al. 2018; Tongesayi et al. 2018). The most recent studies were majorly focused on the impact of the landfill sites on soil and groundwater quality (Aduojo et al. 2018; Asubiojo 2016; Ikem et al. 2002; Ojo-Awo et al. 2018). The study was undertaken to establish baseline levels information ambient air quality, at the three major dumpsites; identify their existing sources of air emissions; and to investigate the possibility of landfill gas capture at the Olushosun landfill.

2 Methodology

2.1 Sampling Locations

The study was carried out in three major landfill sites in Lagos, Nigeria. This includes Olushosun landfill site, Solous landfill site, and the Abule Egba landfill site. Olushosun is regarded as the largest government-owned landfill in Nigeria that covers about 42 ha of land (Ameloko and Ayolabi 2018). A total of sixteen sampling locations covering the upwind and downwind of the dumpsites were established for sampling activities. The locations of the landfill sites are shown in Fig. 1, while the coordinates of the selected sampling points in the landfill sites are as summarised in Table 1.

2.2 Sampling Strategy and Methodology

Air Sampling for TSP The TSP measurements were made by gravimetric air sampling method using Negretti air sampler. A metered volume of air was sampled through pre-weighed Whatman cellulose filter paper. Thereafter, the loaded filter was re-weighed and the concentration of the particulate sampled determined. Sample durations of both 1 h (to obtain 1-h TSP concentration) and 4 h (to obtain both the 8- and 24-h average concentrations on extrapolation) were adopted.

NO_X measurements NO_X concentrations were measured as NO_2 using an in situ single gas NO_X monitor (ToxiRAE Model PGM-1110). The monitor is a 9.3 cm × 4.9 cm × 2.2 cm measuring instrument weighing about 0.1 kg with an instantaneous direct readout displays through which current NO_2 concentrations can be continuously monitored in ppm (parts per million) with a detection range of 0–20 ppm and 0.1 ppm resolution.

It has facility for short-term exposure limit (STEL) from which the NO_2 concentration for the last 15 min can be determined; the time-weighted average (TWA) from which the accumulated reading of the gas concentration since the monitor was turned on is divided by 8 h; and the peak reading, which is the highest reading since the monitor was turned on.

The monitor was calibrated with Calibration and Test Certificate S/N 021-905130 from RAE Systems, 3775 North First Street, San Jose, California 95134, USA. For every field measurement, the "auto-zero at start-up" calibration is required and this was carried out during the study.

SO_2 Measurements To measure the SO_2 concentrations during the field study, an in situ single gas SO_2 monitor (ToxiRAE Model PGM-1130) was used. The monitor is a 9.3 cm × 4.9 cm × 2.2 cm measuring instrument weighing about 0.1 kg with an instantaneous direct readout displays through which current SO_2 concentrations

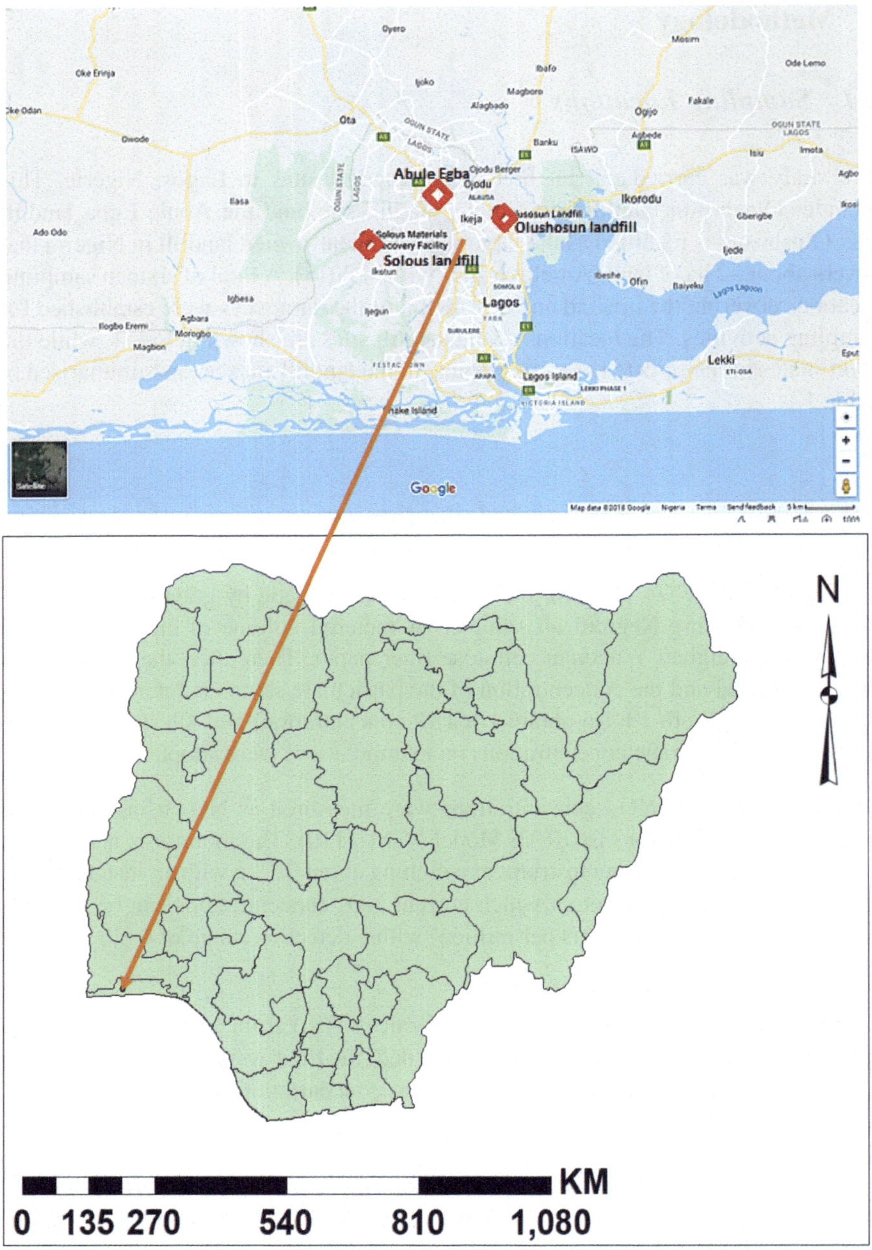

Fig. 1 Map of Nigeria showing the locations of the landfill sites

Table 1 Description of sampling locations in and around the dumpsites

S. no.	Station ID	Location	
		Name	GPS
Olushosun dumpsite			
1.	OAQ1	Centre of active side	Lat. N 06.59716°
			Long. E 003.37623°
2.	OAQ2	South-east flank	Lat. N 06.59790°
			Long. E 003.37710°
3.	OAQ3	Around a self-ignited source	Lat. N 06.59780°
			Long. E 003.37573°
4.	OAQ4	Non-active site	Lat. N 06.59535°
			Long. E 003.37811°
5.	OAQ5	Gate house by Ojota end	Lat. N 06.59717°
			Long. E 003.37932°
6.	OAQ6	Gate house by Oregun end	Lat. N 06.59280°
			Long. E 003.37571°
Abule Egba			
7.	AAQ1	Centre of active site	Lat. N 06.64061°
			Long. E 003.30171°
8.	AAQ2	First gate	Lat. N 06.64074°
			Long. E 003.30127°
9.	AAQ3	Around a self-ignited source	Lat. N 06.63960°
			Long. E 003.30171°
10.	AAQ4	Second gate	Lat. N 06.63834°
			Long. E 003.30023°
11.	AAQ5	South-west flank	Lat. N 06.63744°
			Long. E 003.30235°
12.	AAQ6	Non-active site	Lat. N 06.64058°
			Long. E 003.30292°
Solous dumpsite			
13.	SAQ1	Around a self-ignited source 1	Lat. N 06.57123°
			Long. E 003.25434°
14.	SAQ2	Around a self-ignited source 2	Lat. N 06.57120°
			Long. E 003.25431°
15.	SAQ3	Gate house by the site office	Lat. N 06.57204°
			Long. E 003.25447°
16.	SAQ4	Active site	Lat. N 06.57065°
			Long. E 003.25312°
17.	SAQ5	North-east flank	Lat. N 06.57124°
			Long. E 003.25486°

can be continuously monitored in ppm (parts per million) with a detection range of 0–20 ppm and 0.1 ppm resolution.

It has facility for short-term exposure limit (STEL) from which the SO_2 concentration for the last 15 min can be determined; the time-weighted average (TWA) from which the accumulated reading of the gas concentration since the monitor was turned on is divided by 8 h; and the peak reading, which is the highest reading since the monitor was turned on. The monitor was calibrated with Calibration and Test Certificate S/N 023-900636 from RAE Systems, 3775 North First Street, San Jose, California 95134, USA. For every field measurement, the "auto-zero at start-up" calibration is required and this was carried out during the study.

CO Measurements CO measurements were taken using an in situ non-integrated single gas carbon monoxide monitor (ToxiRAE Model PGM-1150). The monitor is a 9.3 cm × 4.9 cm × 2.2 cm measuring instrument weighing about 0.1 kg with an instantaneous direct readout displays through which current carbon monoxide concentrations can be continuously monitored in ppm (parts per million). It has facility for short-term exposure limit (STEL) from which the carbon monoxide concentration for the last 15 min can be determined; the time-weighted average (TWA) from which the accumulated reading of the gas concentration since the monitor was turned on is divided by 8 h; and the peak reading, which is the highest reading since the monitor was turned on. It has a detection range of 0–500 ppm with 1 ppm resolution.

The monitor was calibrated with Calibration and Test Certificate S/N O25-900527 from RAE Systems, 3775 North First Street, San Jose, California 95134, USA. However, for every field measurement, the "auto-zero at start-up" calibration is required and this was carried out during the measurements. In all the sapling locations, a sampling period of 1 h was adopted to obtain 1-h average concentrations and from which 8- and 24-h average-time concentrations were obtained on extrapolation.

Odour Measurements Standard practices for odour quantification at MSW landfills include characterisation by descriptors, intensity by word and butanol scales, threshold evaluation, and odour persistence (McGinley and McGinley 1998). Using simple word intensity scales or butanol intensity scales with standard odour descriptor nomenclature, direct field observation is a dependable practice for quantification of odours from MSW landfills. Odour intensity is a measure of the relative strength of an odour above the threshold and can be assigned a word descriptor or a number on a "5" or "10" scale. A common word scale is: 0 = No Odour; 1 = Very Faint; 2 = Faint; 3 = Noticeable; 4 = Strong; and 5 = Very Strong. In this study, the "intensity by word" approach with a number on a "5" scale was adopted.

3 Results and Discussion

3.1 Air Quality

The 1-h gaseous and TSP concentrations measured during the field study are as summarised in Table 2. Sulphur dioxide (SO_2) was not detected in any of the sampling locations at both Olushosun and Abule Egba dumpsites. At Solous dumpsite where it was detected at a location, the measured concentration is about 21-folds of the 1-h limit set for SO_2 by the Federal Ministry of Environment, Housing and Urban Development. Emission of heavy smoke was noted around this sampling location during the study and the elevated level of SO_2 in the environment might be attributed to the combustion of some sulphur-containing materials that might be partaking in the combustion activity. It should be noted that the burning was due to self-ignition, a common feature of the three studied dumpsites. Though the World Bank and Nigeria

Table 2 Measured concentrations of air pollutants in the study area

Locations	1-h gaseous concentrations (ppm)			TSP (μg/m^3)	
	SO_2	NO_2	CO	1-h	4-h
OAQ1	<0.1	0.1	9.0	1528	338
OAQ2	<0.1	0.1	8.0		
OAQ3	<0.1	<0.1	40.0		
OAQ4	<0.1	<0.1	<0.1		
OAQ5	<0.1	<0.1	<0.1		
OAQ6	<0.1	<0.1	<0.1		
AAQ1	<0.1	<0.1	<0.1	1944	334
AAQ2	<0.1	<0.1	10.0		
AAQ3	<0.1	<0.1	44.0		
AAQ4	<0.1	<0.1	<0.1		
AAQ5	<0.1	<0.1	<0.1		
AAQ6	<0.1	0.1	<0.1		
SAQ1	<0.1	0.2	1.0	1667	417
SAQ2	2.1	0.4	179		
SAQ3	<0.1	<0.1	<0.1		
SAQ4	<0.1	<0.1	<0.1		
SAQ5	<0.1	<0.1	<0.1		
FMENV limit[a]	0.1	–	–	600	–
World Bank limit[b]	–	0.08	26	–	–

[a]FEPA (1991)
[b]World Bank (2007)

Table 3 Extrapolated concentrations of air pollutants in the study area

Locations	8-h concentrations (ppm)			24-h concentrations (ppm)			TSP ($\mu g/m^3$)	
	SO_2	NO_2	CO	SO_2	NO_2	CO	8-h	24-h
OAQ1	–	0.01	1.10	–	0	0.38	169.00	56.33
OAQ2	–	0.01	1.00	–	–	0.33		
OAQ3	–	–	5.00	–	–	1.67		
AAQ2	–	–	1.25	–	–	0.42	167.00	55.67
AAQ3	–	–	5.50	–	–	1.83		
AAQ6	–	0.01	–	–	0.00	–		
SAQ1	–	0.03	0.13	–	0.01	0.04	208.50	69.50
SAQ2	0.26	0.05	22.38	0.09	0.02	7.46		
FMENV limit[a]	–	–	20	0.01	0.04–0.06	10	–	250
World Bank limit[b]	–	–	–	–	0.08	–	–	80

[a]FEPA (1991)
[b]World Bank (2007)

have no 8-h limit for SO_2, the extrapolated 24-h concentration of 0.09 ppm (Table 3) from the dumpsite is about ninefolds of the FMENV's set limit.

Nitrogen dioxide was detected at two locations each on the Olushosun and the Solous dumpsite but at a sampling location in Abule Egba. The 1-h measured concentrations at all these locations are above the World Bank set limit of 0.08 ppm. While it is about 1.3-folds of this limit at all the locations in Olushosun and Abule Egba, it is between 2.5 and 5.0-folds at Solous. However, the 24-h extrapolated concentrations of 0.01 and 0.02 ppm obtained at Solous are far below both the FMENVH&D and the World Bank limits.

Carbon monoxide is the commonest gaseous pollutant in all the three dumpsites and this might be attributed to incomplete combustion from the self-ignition characterising all the active sites. In Olushosun, the measured 1-h CO concentrations range between 8.0 and 40 ppm, at Abule Egba and Solous, the range are 10.0–44.0 ppm and 1.0–179.0 ppm, respectively. The World Bank limit for CO was exceeded in at least one sampling location from all the dumpsites. The elevated 1-h concentrations are about 1.5-folds, 1.7-folds, and 6.9-folds of this limit at Olushosun, Abule Egba, and Solous, respectively. Similarly, the extrapolated 8-h concentration at a location in Olushosun is about 1.1-folds of the FMENV, while the extrapolated 24-h concentrations are within the set limit in all the sampling locations of the three dumpsites.

In all the three dumpsites, the 1-h total suspended particulates (TSP) concentration is significantly higher than the FMENV's set limit. At Olushosun, it is about 2.5-folds of the 1-h limit but 3.2-folds and 2.8-folds of the same limit at both Abule Egba and Solous, respectively. In addition to carbon from incomplete combustion associated with self-ignition at the dumpsites, other identified sources of particulates include scavenging activities (Plate 1), mode of waste discharge from vehicles, and some

Plate 1 Scavengers at Olushosun dumpsite

open burning of used tyres for the recovery of some materials. On extrapolation, the 24-h TSP concentrations in all the three dumpsites are below both the FMENV and the World Bank limits.

3.2 *Odour*

As shown in Table 4, the odour level in all the sampling locations on the three dumpsites is significantly high ranging between 4 and 5 on the "intensity by word" scale of 5. In dumpsites or landfilling sites, odours can be either from fresh refuse and/or landfill gas. After several weeks, the character of the odour changes to a "sickly sweet" odour typical of landfill gas. However, the conversion from one type of odour to the other depends on the nature of the refuse and the amount of moisture available in the landfill.

3.3 *Landfill Gas Capture Possibility at Olushosun Landfill*

The terms of reference of this study does not include detailed gas pumping tests thus only a theoretical estimate of methane in place was carried out using the Landfill Gas Emissions Model, LandGem, a modelling tool from the United States Environmental Protection Agency. The Landfill Gas Emissions Model (LandGEM) is an automated estimation tool with a Microsoft Excel interface that can be used to estimate emission rates for total landfill gas, methane, carbon dioxide, non-methane organic compounds, and individual air pollutants from municipal solid waste landfills.

Table 4 Observed "word scale" odour intensity around the dumpsites

Locations	Intensity
OAQ1	5
OAQ2	5
OAQ3	5
OAQ4	4
OAQ5	4
OAQ6	4
AAQ1	5
AAQ2	5
AAQ3	5
AAQ4	4
AAQ5	5
AAQ6	5
SAQ1	5
SAQ2	5
SAQ3	4
SAQ4	5
SAQ5	5

During the study, information made available by the Lagos State Waste Management Authority (LAWMA) indicates that the average daily waste received by Olushosun, Solous, and Abule Egba is 1998 ton, 878 ton, and 1090 ton, respectively (Fig. 2). Though the dumpsite was opened in 1992, there is no information on the quantity of wastes received between then and now.

Assuming that just about 10% of the present-day waste was received at the site in the first year of operation with an annual increment of 10% until when stability is achieved in the year 2002; then, the total waste received between the year of commissioning to the year 2007 will be about 8,387,149 ton of waste. If the year

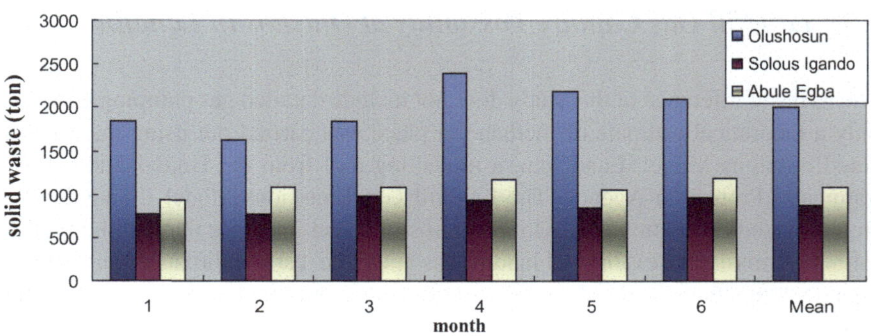

Fig. 2 Average daily wastes received in the last six months

of closure set at commissioning still remains at the year 2014, the present daily waste received and the assumed percentage of waste "in-place" at the initial year of operation will translate to a total of 13,492,371 ton of waste "in-place" at the year of closure (Fig. 3). This annual and total waste information is used in the LandGem to estimate the methane potential at Olushosun.

For the purpose of this investigation, it is further assumed that the waste design capacity of the Olushosun dumpsite is 16,044,981 ton; the methane generation rate is 0.05/year; the potential methane generation capacity is 170 m^3/ton, and that the methane component of the total landfill gas is 50% by volume. Within the limits of these assumptions, the methane potential from the LandGem tool will range between 4.04×10^2 and 4.97×10^4 ton/year which is equivalent to about 6.06×10^5 to 7.45×10^7 m^3/year (Fig. 3) that can be used to generate about 0.4–49 MW of electricity if efficiently captured (Fig. 4).

However, for methane to be captured at Olushosun, there will be a need to upgrade the facility to a proper landfill site with necessary accessories like gas well, collection

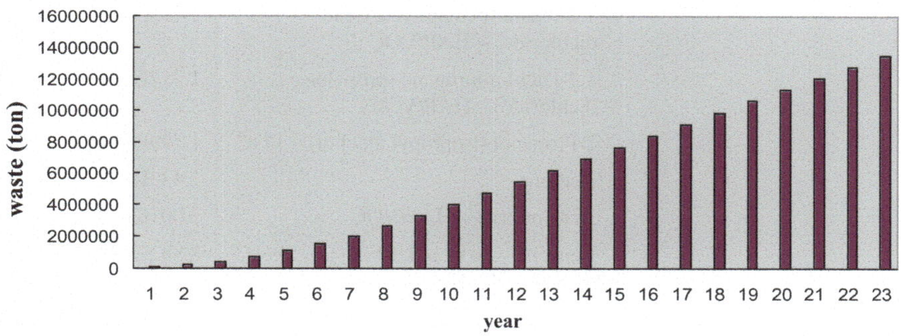

Fig. 3 Calculated annual wastes on-site at Olushosun (1992–2014)

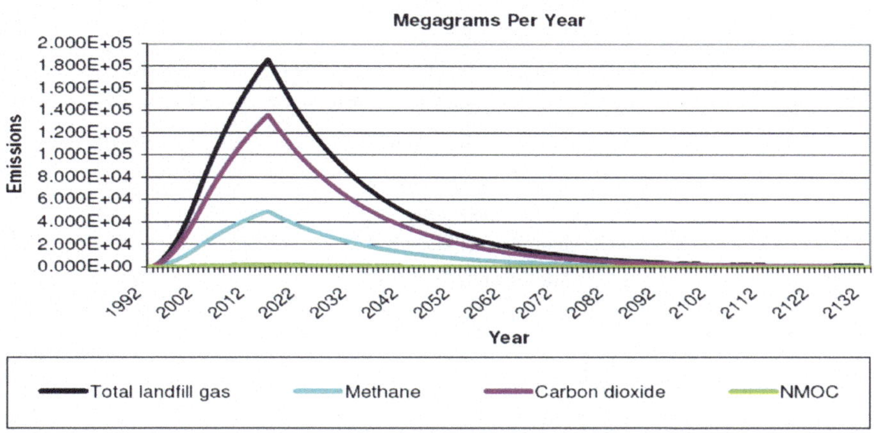

Fig. 4 Predicted landfill gas from Olushosun

piping, gas treatment systems, and gas flares. There is a need for comprehensive study to determine the present locations of landfill gas at the dumpsite but it is not unlikely that all the areas where heavy smoke are presently emitted are potential sources. Several other air pollutants that may be of concern and predicted by the LandGem software are summarised in Table 5.

Table 5 Estimated air pollutants from Olushosun dumpsite

Air pollutants	Emissions (ton/year)
Carbon dioxide	1.317E+05
NGOC	2.063E+03
1,1,1-Trichloroethane (methyl chloroform)—HAP	3.832E−01
1,1,2,2-Tetrachloroethane—HAP/VOC	1.105E+00
1,1-Dichloroethane (ethylidene dichloride)—HAP/VOC	1.422E+00
1,1-Dichloroethane (vinylidene chloride)—HAP/VOC	1.160E−01
1,2-Dichloroethane (ethylene dichloride)—HAP/VOC	2.428E−01
1,2-Dichloropropane (propylene dichloride)—HAP/VOC	1.217E−01
2-Propanol (isopropyl alcohol)—VOC	1.799E+01
Acetone	2.433E+00
Acrylonitrile—HAP/VOC	2.001E+00
Benzene—no or unknown co-disposal—HAP/VOC	8.882E−01
Benzene—co-disposal—HAP/VOC	5.142E+00
Bromodichloromethane—VOC	3.039E+00
Butane—VOC	1.739E+00
Carbon disulphide—HAP/VOC	2.643E−01
Carbon monoxide	2.347E+01
Carbon tetrachloride—HAP/VOC	3.683E−03
Carbonyl sulphide—HAP/VOC	1.762E−01
Chlorobenzene—HAP/VOC	1.684E−01
Chlorodifluoromethane	6.728E−01
Chloroethane (ethyl chloride)—HAP/VOC	5.020E−01
Chloroform—HAP/VOC	2.144E−02
Chloromethane—VOC	3.626E−01

(continued)

Table 5 (continued)

Air pollutants	Emissions (ton/year)
Dichlorobenzene—(HAP for para isomer/VOC)	1.847E−01
Dichlorodifluoromethane	1.158E+01
Dichlorofluoromethane—VOC	1.601E+00
Dichloromethane (methylene chloride)—HAP	7.117E+00
Dimethyl sulphide (methyl sulphide)—VOC	2.900E+00
Ethane	1.602E+02
Ethanol—VOC	7.446E+00
Ethyl mercaptan (ethanethiol)—VOC	8.552E−01
Ethylbenzene—HAP/VOC	2.923E+00
Ethylene dibromide—HAP/VOC	1.124E−03
Fluorotrichloromethane—VOC	6.249E−01
Hexane—HAP/VOC	3.404E+00
Hydrogen sulphide	7.343E+00
Mercury (total)—HAP	3.482E−04
Methyl ethyl ketone—HAP/VOC	3.064E+00
Methyl isobutyl ketone—HAP/VOC	1.139E+00
Methyl mercaptan—VOC	7.198E−01
Pentane—VOC	1.425E+00
Perchloroethylene (tetrachloroethylene)—HAP	3.672E+00
Propane—VOC	2.903E+00
t-1,2-Dichloroethane—VOC	1.624E+00
Toluene—no or unknown co-disposal—HAP/VOC	2.150E+01
Toluene—co-disposal—HAP/VOC	9.373E+01
Trichloroethylene (trichloroethene)—HAP/VOC	2.202E+00
Vinyl chloride—HAP/VOC	2.731E+00
Xylenes—HAP/VOC	7.624E+00

4 Conclusions

The measured air quality parameters concentrations signify a badly impaired airshed and can be classified as having poor air quality which calls for urgent attention by reconstructing all the sites to standard landfill locations. The average daily waste received by Olushosun, Solous, and Abule Egba was 1998 ton, 878 ton, and 1090 ton, respectively. Within the limits of assumptions, the total waste received by Olushosun between the year of commissioning to the year 2007 will be about 8,387,149 ton of waste and a total waste "in-place" of about 13,492,371 ton at the year of closure in 2014. From these, the methane potential using the LandGem tool is estimated to be between 4.04×10^2 and 4.97×10^4 ton/year which is equivalent to about 6.06×10^5 to 7.45×10^7 m^3/year. This can be used to generate about 0.4–49 MW of electricity if efficiently captured. Though some areas with significantly potential of landfill gas generation are identified, these need to be further investigated.

References

Aduojo, A. A., Ayolabi, E. A., & Adewale, A. (2018). Time dependent electrical resistivity tomography and seasonal variation assessment of groundwater around the Olusosun dumpsite Lagos, South-West, Nigeria. *Journal of African Earth Sciences, 147*, 243–253.

Ameloko, A. A., & Ayolabi, E. A. (2018). Geophysical assessment for vertical leachate migration profile and physicochemical study of groundwater around the Olusosun dumpsite Lagos, southwest Nigeria. *Applied Water Science, 8*(5), 142.

Asubiojo, O. (2016). Pollution sources in the Nigerian environment and their health implications. *Ife Journal of Science, 18*(4), 973–980.

Christensen, T. H., Cossu, R., & Stegmann, R. (2014). Landfill leachate: An introduction. In *Landfilling of waste* (pp. 17–30). CRC Press.

de Morais, C. R., Bonetti, A. M., Mota, A. A., Campos, C. F., Souto, H. N., Naves, M. P. C., et al. (2018). Evaluation of toxicity, mutagenicity and carcinogenicity of samples from domestic and industrial sewage. *Chemosphere, 201*, 342–350.

FEPA. (1991). *Guidelines to standards for environmental pollution control in Nigeria*. Lagos: Federal Environmental Protection Agency.

Giusti, L. (2009). A review of waste management practices and their impact on human health. *Waste Management, 29*(8), 2227–2239.

Ife-Adediran, O., & Isabota, O. (2018). Gamma dose profile and risk to scavengers and occupants near waste dumpsites in coastal Nigeria. *International Journal of Environmental Studies, 75*(5), 708–718.

Ikem, A., Osibanjo, O., Sridhar, M., & Sobande, A. (2002). Evaluation of groundwater quality characteristics near two waste sites in Ibadan and Lagos, Nigeria. *Water, Air, and Soil Pollution, 140*(1–4), 307–333.

Khalid, A., Arshad, M., Anjum, M., Mahmood, T., & Dawson, L. (2011). The anaerobic digestion of solid organic waste. *Waste Management, 31*(8), 1737–1744.

Lim, S. L., Lee, L. H., & Wu, T. Y. (2016). Sustainability of using composting and vermicomposting technologies for organic solid waste biotransformation: Recent overview, greenhouse gases emissions and economic analysis. *Journal of Cleaner Production, 111*, 262–278.

Liu, Y., Lu, W., Guo, H., Ming, Z., Wang, C., Xu, S., et al. (2016). Aromatic compound emissions from municipal solid waste landfill: Emission factors and their impact on air pollution. *Atmospheric Environment, 139*, 205–213.

Liu, Y., Lu, W., Wang, H., Huang, Q., & Gao, X. (2018). Odor impact assessment of trace sulfur compounds from working faces of landfills in Beijing, China. *Journal of Environmental Management, 220*, 136–141.

McGinley, C.M., & McGinley, M.A. (1998). Odor quantification methods and practices at msw landfills. *Proceedings of Air and Waste Management Association 91st Annual Meeting and Exhibition*. June 14-18, 1998. San Diego, CA.

Ojo-Awo, N. A., Agbabiaka, H. I., & Ilesanmi, A. O. (2018). Refuse dumpsite and its associated pollutants: Spatial variations of the impact of leachates on groundwater quality. *Management of Environmental Quality: An International Journal, 29*(3), 572–591.

Petrovic, M., Sremacki, M., Radonic, J., Mihajlovic, I., Obrovski, B., & Miloradov, M. V. (2018). Health risk assessment of PAHs, PCBs and OCPs in atmospheric air of municipal solid waste landfill in Novi Sad, Serbia. *Science of the Total Environment, 644,* 1201–1206.

Powell, J. T., & Chertow, M. R. (2019). Quantity, components, and value of waste materials landfilled in the United States. *Journal of Industrial Ecology, 23*(2), 466–479.

Salehpour, S., Jonoobi, M., Ahmadzadeh, M., Siracusa, V., Rafieian, F., & Oksman, K. (2018). Biodegradation and ecotoxicological impact of cellulose nanocomposites in municipal solid waste composting. *International Journal of Biological Macromolecules, 111,* 264–270.

Schiopu, A. M., & Gavrilescu, M. (2010). Options for the treatment and management of municipal landfill leachate: Common and specific issues. *CLEAN—Soil, Air, Water, 38*(12), 1101–1110.

Talaiekhozani, A., Masomi, B., & Hashemi, S. M. J. (2016). Evaluation of gaseous pollutants emission rate from Marvdasht landfills. *Journal of Advanced Medical Sciences and Applied Technologies, 2*(1), 162–175.

Tongesayi, T., Kugara, J., & Tongesayi, S. (2018). Waste dumpsites and public health: a case for lead exposure in Zimbabwe and potential global implications. *Environmental Geochemistry and Health, 40*(1), 375–381.

Vrijheid, M. (2000). Health effects of residence near hazardous waste landfill sites: A review of epidemiologic literature. *Environmental Health Perspectives, 108*(Suppl 1), 101.

Wenjing, L., Zhenhan, D., Dong, L., Jimenez, L. M. C., Yanjun, L., Hanwen, G., et al. (2015). Characterization of odor emission on the working face of landfill and establishing of odorous compounds index. *Waste Management, 42,* 74–81.

World Bank. (2007). *Pollution prevention and abatement handbook.* Retrieved from Washington DC.

Zhang, D. Q., Tan, S. K., & Gersberg, R. M. (2010). Municipal solid waste management in China: Status, problems and challenges. *Journal of Environmental Management, 91*(8), 1623–1633.

Remediation of bis(2-Ethylhexyl) Phthalate and Phenol, 4,4′-(1-Methylethylidene)bis—in Landfill Leachate Using Biopolymer

P. Agamuthu, A. Aziz, A. Hassan and S. H. Fauziah

Abstract Persistent organic pollutants (POPs) are global contaminants that have significant toxic effects and possess carcinogenicity, mutagenicity, neurotoxicity, immunotoxicity and endocrine disruption characteristics. This study aimed to investigate the effectiveness of Locust Bean Gum (LBG) in eliminating POPs such as phenol, 4,4′-(1-methylethylidene)bis-(bisphenol A) and bis(2-ethylhexyl) phthalate (DEHP) from landfill leachate and compared it with the removal efficiency by using alum. Optimization of the operating processes (flocculant dose, pH and stirring speed) was performed by using Box–Behnken design (BBD) to evaluate the removal efficiency of POPs. A partial cubic equation could model the bisphenol A and DEHP removal with R^2 values of 0.928 and 0.9732, respectively. Fourier transform infrared (FTIR) spectroscopy and scanning electron microscopy (SEM) were used to study the morphological structure and functional groups of the treated flocs. SEM micrographs revealed porous and rough cloudy surface in the LBG-treated flocs compared to alum. FTIR analysis indicated the presence of carboxyl, hydroxyl and amino groups as well as hydrogen bonding that are flocculating agents, which act as a bridging agent for the pollutant particles and encouraged aggregation to form flocs. At the optimal conditions of pH 7.5 and LBG dosage of 500 mg/L, removal efficiencies obtained were

P. Agamuthu · A. Aziz (✉) · A. Hassan · S. H. Fauziah
Faculty of Science, Institute of Biological Sciences, University of Malaya, Kuala Lumpur 50603, Malaysia
e-mail: aziz_barech@yahoo.com

S. H. Fauziah
e-mail: fauziahsh@um.edu.com

A. Aziz
Faculty of Marine Sciences, Lasbela University of Agriculture, Water and Marine Sciences, Uthal, Balochistan, Pakistan

P. Agamuthu · A. Aziz · A. Hassan · S. H. Fauziah
Faculty of Science, Center for Research in Waste Management, University of Malaya, Kuala Lumpur 50603, Malaysia

A. Hassan
Department of Biological Sciences, Faculty of Science, Federal University Kashere, Gombe State, Nigeria

S. K. Ghosh (ed.), *Urban Mining and Sustainable Waste Management*,
https://doi.org/10.1007/978-981-15-0532-4_19

76 and 69% for bisphenol A and DEHP, respectively. Results indicated that coagulation–flocculation using LBG can be highly effective in treating POPs from landfill leachate.

Keywords bis(2-ethylhexyl) phthalate · Phenol · 4,4′-(1-methylethylidene)bis · Landfill leachate · Locust bean gum · Flocs

1 Introduction

Persistent organic pollutants (POPs) are ubiquitous organic pollutants which have distinctive and very dangerous properties. POPs are highly recalcitrant in the environment, extremely resistant to biodegradation, travel long distance through sea or air, tendency to bioaccumulate in food chain, strong lipophilicity and elicit adverse effects on public health and ecological system (Hung et al. 2013). Moreover, toxic effects from POPs include neurotoxicity, reproductive and developmental problems, immunotoxicity, carcinogenicity, mutagenicity and endocrine disruption characteristics have been reported to be caused by human exposure to POPs (Kim et al. 2013). Therefore, removal of these pollutants from landfill leachate is crucial prior to its discharge in water bodies and causing deterioration and/or severe toxicity to the ecosystem and aquatic life.

Among the many treatment process, coagulation–flocculation (C–F) has been widely applied for the treatment of stabilized and old leachate prior to biological and physicochemical methods either pre or post treatment. It is an effective technique used for the treatment of landfill leachates due to its simple implementation, efficiency and cost effectiveness (Liu et al. 2012; Oladoja et al. 2017). C–F process can be affected by numerous parameters including type and dosage of flocculant, pH and mixing speed (Ayoub et al. 2014; Liu et al. 2012). Optimization of these parameters can result in excellent performance in POPs removal.

2 Materials and Methods

2.1 Site Description and Sample Collection

Landfill leachate was collected from Jeram Sanitary Landfill (JSL). Raw leachate was collected from an outlet of the leachate pipe before it went to the retention pond, and the leachate was filled into high density polyethylene bottles and stored in a cold room at 4 °C upon arrival at laboratory. Physicochemical parameters were measured immediately after all samples were collected. The characterization of landfill leachate was done in triplicates.

2.2 Coagulation–Flocculation (C–F) Test

C–F tests were carried out in jar test apparatus (Velp Scientifica JLT6 Flocculator tester) using LBG and alum. A fixed volume of 250 mL of leachate was placed in 1L of beaker and different concentration of the LBG that ranged between 100 and 600 mg/L was added. The pH of the leachate was adjusted using 1N H_2SO_4/NaOH solutions. Coagulants were added into leachate after 1 min of flash mixing stage (200 rpm for 5 min). The sample was subsequently subjected to two slow mixing at 70 and 50 rpm for 10 min each. The floc was allowed to settle for 45 min. The sample for analysis was taken 2 cm below the surface level for extraction and quantification of POPs.

2.3 Sample Extraction and Instrumental Analysis

Sample extraction was performed using liquid–liquid extraction (LLE) and analysed by GC–MS/MS. The method of extraction and quantification of POPs was done according to Aziz et al. (2018a) and Meng et al. (2011). For extraction and analysis of POPs, 100 mL of leachate sample was used. The quantification was carried out using gas chromatograph fixed with tandem mass spectrometry (Agilent technologies, USA).

2.4 Experimental Design and Data Analysis

Design-Expert software version of 8.0.6 was used for experimental design, modelling and data analysis. Response surface methodology fixed with Box–Behnken design (BBD) was performed for optimization of the key operating factors (concentration of coagulant, pH and agitation speed) with a central value of 1000 mg/L, pH 7.5 and mixing speed 200 rpm, respectively. Removal of bisphenol A and DEHP was analysed as the responses. An analysis of variance (ANOVA) was employed to investigate the statistical significance of the developed model.

2.5 FTIR and SEM Analysis

FTIR and SEM were performed on the flocs generated after treatment with LBG and alum. SEM coupled with SE1 detector JEOL JSM-7001F was carried out to study the morphology of the treated flocs. FTIR of the leachate and flocs was examined using ATR-FTIR (Perkin-Elmer 400 FT-IR/FT-FIR).

3 Results and Discussions

3.1 Removal Efficacy of Bisphenol A and DEHP from Leachate Using LBG

The concentration of bisphenol A and DEHP recorded in raw leachate was 37.05 and 13.93 mg/L. Figure 1 shows the percentage removal of POPs using LBG and alum at pH 7.5. The removal efficiency of bisphenol and DEHP was recorded 76 and 69% at the dosage of 500 mg/L, respectively. As the dosage increased from 100 to 500 mg/L, the removal efficiency of POPs also increased, and beyond this it was decreased.

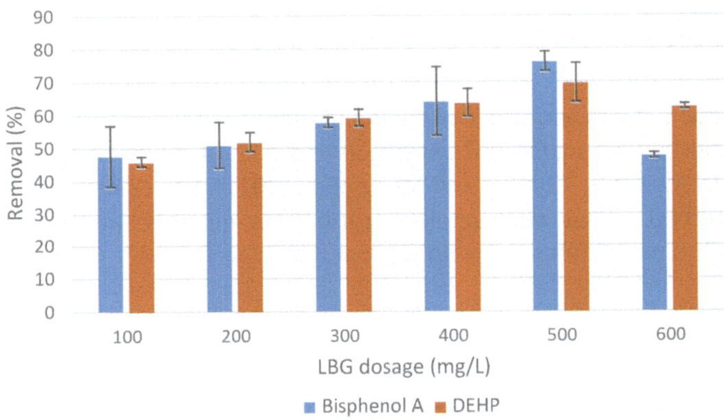

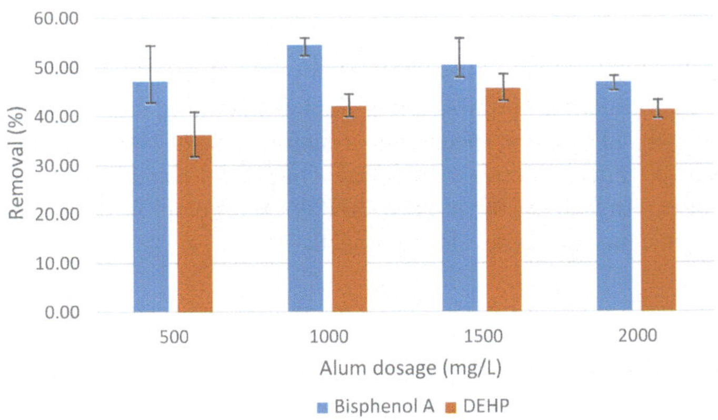

Fig. 1 Removal of bisphenol A and DEHP from landfill leachate using various dosages of **a** LBG and **b** alum at pH 7.5

3.2 Flocculation Studies

The experimental design of the parameters along with obtained removal of bisphenol A and DEHP is presented in Table 1. The equations were developed for DEHP and bisphenol A as follows:

DEHP removal (%) $= -676.9 + 21.5 \times \text{pH} + 1.6 \times \text{Dosage} + 2.5 \times \text{Speed} - 2.0\text{E-}003 \times \text{pH} \times \text{Dosage} - 0.01 \times \text{pH} \times \text{Speed} - 4.0\text{E-}005 \times \text{Dosage} \times \text{Speed} - 1.2 \times \text{pH}^2 - 1.6\text{E-}003 \times \text{Dosage}^2 - 6.1\text{E-}003 \times \text{Speed}^2$.

Bisphenol A removal (%) $= -603.6 + 11.6 \times \text{pH} + 1.6 \times \text{Dosage} + 2.1 \times \text{speed} - 6.4\text{E-}03 \times \text{pH} \times \text{Dosage} - 0.01 \times \text{pH} \times \text{speed} - 4.5\text{E-}04 \times \text{Dosage} \times \text{speed} - 0.3 \times \text{pH}^2 - 1.46\text{E-}0 \times \text{Dosage}^2 - 4.6\text{E-}03 \times \text{speed}^2$.

Table 2 shows the ANOVA for bisphenol A and DEHP removal efficiency. For bisphenol A and DEHP removal, the model F-values of 10.13 and 28.24, and P-values of 0.003 and 0.0001, respectively, imply that the model is significant (Ahmed et al. 2018). Whereas, high R^2 values of 0.92 and 0.97, respectively, imply the aptness of the model between the predicted and actual data (Table 3) (Huda et al. 2017). Less than 0.20 between predicted and adjusted R^2, indicated to be adequate model. In the case of bisphenol A and DEHP removal, the predicted R^2 values of 0.837 and 0.938 are in reasonable agreement with the adjusted R^2 values of 0.661 and 0.879, respectively. Adequate precision of 4 or more is desirable, in the case of bisphenol

Table 1 Experimental conditions and results of the Box–Behnken design

Run	Independent parameters			DEHP removal (%)	BPA removal (%)
	pH	Dosage (mg/L)	Mixing (rpm)		
1	7.50	600	150	34.18	42.97
2	7.50	400	150	33.99	33.11
3	11.00	500	250	29.67	46.87
4	7.50	500	200	66.17	64.54
5	7.50	600	250	32.24	30.66
6	7.50	500	200	62.27	57.86
7	7.50	500	200	68.13	56.40
8	7.50	500	200	59.57	55.33
9	4.00	500	150	36.33	39.14
10	4.00	500	250	34.05	41.37
11	7.50	400	250	32.28	29.92
12	11.00	400	200	37.28	49.22
13	7.50	500	200	70.08	77.76
14	4.00	400	200	33.98	33.28
15	11.00	500	150	39.00	51.65
16	4.00	600	200	31.55	38.635
17	11.00	600	200	31.92	45.60

Table 2 ANOVA output

Source	Sum of squares	Degree of freedom	Mean square	Rapport *F*	*p*-value prob > *F*	Remarks
For DEHP removal						
Model	3504.06	9	389.34	28.24	0.0001	Significant
A-pH	0.48	1	0.48	0.034	0.8580	
Residual	96.50	7	13.79			
Lack of fit	19.54	3	6.51	0.34	0.7999	Not significant
Pure error	76.96	4	19.24			
For bisphenol A removal						
Model	1996.22	9	221.8	10.13	0.003	Significant
A-pH	209.20	1	209.20	9.56	0.0175	
Residual	153.26	7	21.89			
Lack of fit	33.8	3	11.27	0.38	0.7754	Not significant
Pure error	119.46	4	29.87			

Table 3 ANOVA output for statistical parameters

	R^2	Adj R^2	Pred R^2	Std. dev.	C.V. %	Adeq precision	Press
DEHP	0.9732	0.938	0.879	3.71	8.61	12.26	432.84
Bisphenol A	0.928	0.837	0.661	4.68	9.14	8.137	727.44

A and DEHP, and the adequate precision values were 8.137 and 12.26, respectively, which imply an adequate precision.

3.3 Impact of Operating Factors

3.3.1 Impact of pH

The three-dimensional surface response 3D plots to assess the POPs removal over independent parameters pH and dosage are presented in Fig. 2. The contour plots indicated that as the pH of the leachate increased, the removal efficiency of both bisphenol A and DEHP increased and then reduced beyond optimum pH value. The maximum removal was found at pH 7.5 and 8.7, respectively, and showed better adsorption and bridging at higher pH.

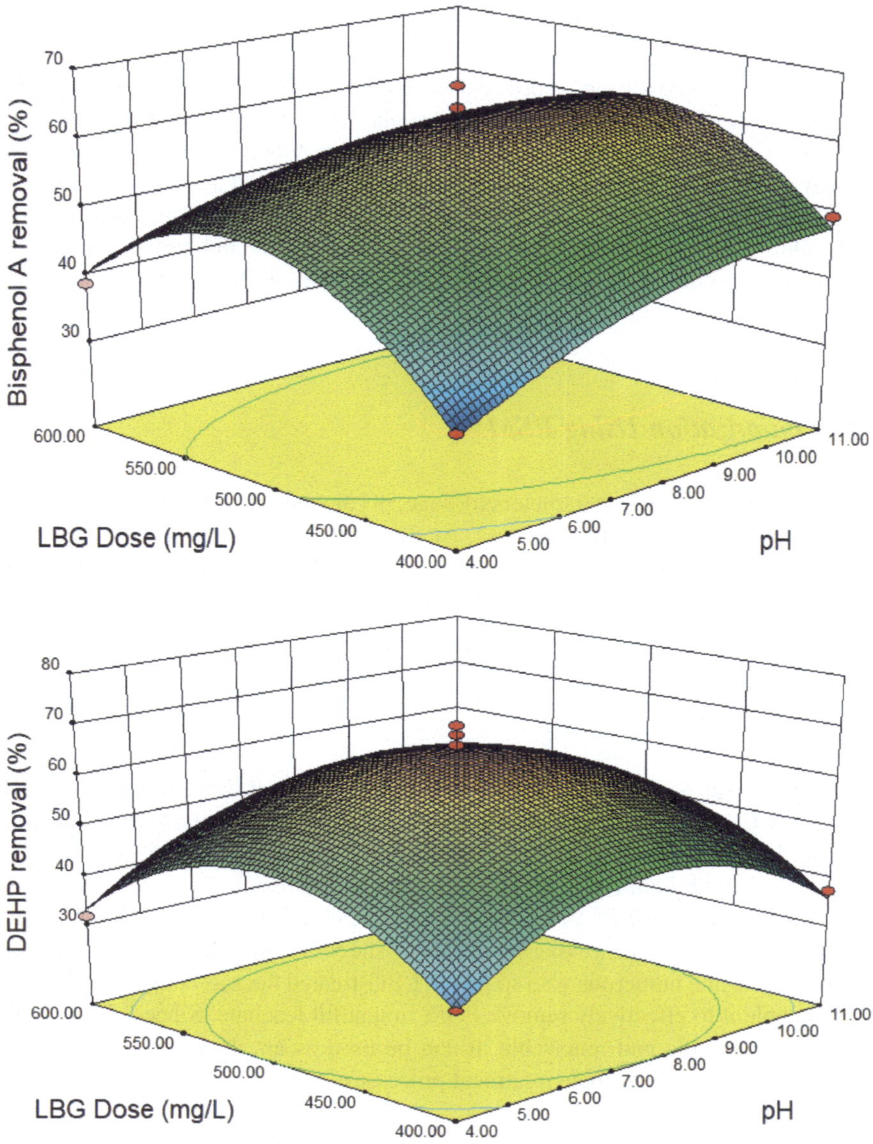

Fig. 2 3D response surface plots of **a** bisphenol A and **b** DEHP removal by LBG addition

3.3.2 Impact of Coagulant Dosage

Figure 2 indicates the response surface plots for removal of bisphenol A and DEHP. It can be observed from contour plot that as the LBG dosages slightly increased, it initially starts to increase the removal efficiency and then starts to reduce when reached at saturation point. The maximum removal was found at coagulant dosages of 502 and 496 mg/L, respectively. However, extra dosage addition results in low POPs removal due to the repulsive energy between polymer and particles at higher dosage, which inhibited the flocculation process (Aziz et al. 2018b; Mukherjee et al. 2018).

3.4 Optimization Using RSM

Optimization of operating parameters (dosage, pH and mixing speed) was optimized using numerical optimization process. At the optimum conditions, the predicted removal was 62.21 and 65.25% for bisphenol A and DEHP, respectively. This was achieved at pH 7.5, dosage of 497 mg/L and mixing speed of 197 rpm.

4 Conclusion

C–F process was applied for the removal of POPs in landfill leachate using LBG and alum. LBG was found to be an effective biopolymer for treating POPs in landfill leachate. LBG effectively removed about 76% of bisphenol A and 69% of DEHP from leachate at optimum dosage of 500 mg/L. ANOVA results for regression model equations indicated high R^2 values of 0.928 and 0.973, respectively, which indicate suitability of the model with experimental data. The SEM micrographs showed rough cloudy surface and numerous void spaces in LBG-treated flocs as compared to alum, which enable it to effectively remove POPs in landfill leachate. Since LBG is non-toxic, biodegradable and renewable, it can be used as an alternative to chemical coagulants for landfill leachate treatment processes.

References

Ayoub, G. M., BinAhmed, S. W., Al-Hindi, M., & Azizi, F. (2014). Coagulation of highly turbid suspensions using magnesium hydroxide: Effects of slow mixing conditions. *Environmental Science and Pollution Research, 21*(17), 10502–10513.

Aziz, A., Agamuthu, P., & Fauziah, S. (2018a). Removal of bisphenol A and 2,4-di-tert-butylphenol from landfill leachate using plant-based coagulant. *Waste Management & Research*. https://doi.org/10.1177/0734242X18790360.

Aziz, A., Agamuthu, P., & Fauziah, S. (2018b). Effective removal of p-tert-butylphenol and pyridine, 3-(1-methyl-2-pyrrolidinyl)-,(S)-from landfill leachate using locust bean gum. *Waste Management & Research*. https://doi.org/10.1177/0734242X18789062.

Huda, N., Raman, A., Bello, M., & Ramesh, S. (2017). Electrocoagulation treatment of raw landfill leachate using iron-based electrodes: Effects of process parameters and optimization. *Journal of Environmental Management, 204,* 75–81.

Hung, H., MacLeod, M., Guardans, R., Scheringer, M., Barra, R., Harner, T., et al. (2013). Toward the next generation of air quality monitoring: Persistent organic pollutants. *Atmospheric Environment, 80,* 591–598.

Kim, S., Park, J., Kim, H.-J., Lee, J. J., Choi, G., Choi, S., et al. (2013). Association between several persistent organic pollutants and thyroid hormone levels in serum among the pregnant women of Korea. *Environment International, 59,* 442–448.

Liu, X., Li, X.-M., Yang, Q., Yue, X., Shen, T.-T., Zheng, W., et al. (2012). Landfill leachate pretreatment by coagulation–flocculation process using iron-based coagulants: Optimization by response surface methodology. *Chemical Engineering Journal, 200,* 39–51.

Meng, C.-K., Szelewski, M., Zweigenbaum, J., Fürst, P., & Blanke, E. (2011). Non-targeted analyses for pesticides using deconvolution, accurate masses, and databases–screening and confirmation. In *Pesticides in the modern world–trends in pesticides analysis*, 391.

Mukherjee, S., Mukhopadhyay, S., Zafri, M. Z. B., Zhan, X., Hashim, M. A., & Gupta, B. S. (2018). Application of guar gum for the removal of dissolved lead from wastewater. *Industrial Crops and Products, 111,* 261–269.

Oladoja, N. A., Unuabonah, E. I., Amuda, O. S., & Kolawole, O. M. (2017). Mechanistic insight into the coagulation efficiency of polysaccharide-based Coagulants. In *Polysaccharides as a green and sustainable resources for water and wastewater treatment* (pp. 13–35). New York: Springer.

Biodegradation of Plastic Waste Using Marine Micro-Organisms

Rwiddhi Sarkhel, Shubhalakshmi Sengupta, Papita Das and Avijit Bhowal

Abstract Synthetic polymers are hazardous to the environment mainly in the marine environment as a lot of plastic wastes end up in the saline waters especially in the estuaries. Synthetic polymers obtained from plastic bottle wastes and polymethyl methacrylate (PMMA) are known to be non-biodegradable. In order to obtain biodegradability, polymers are reinforced with bio-fillers. Biodegradation is a convenient tool to degrade the synthetic polymers and bio-filler reinforcements increases the biodegradability of these polymers. Fungi are known to degrade organic materials and have developed adaptability to procure hard carbon sources. In this study, fungi isolated from the Sundarban areas (*three types of Aspergillus sp.*) have been used to biodegrade plastic bottle waste strips, PMMA and PMMA/cellulose composites. These polymer composite films were found to be degraded by the micro-organisms as ascertained by their weight loss analysis. The PMMA/cellulose composites lost 24% of its weight after 6 weeks by *Aspergillus sp*2, whereas the PMMA lost 16% of its weight by *Aspergillus sp3*. The plastic bottle waste lost about 20% of its weight after 6 weeks as a result of degradation by *Aspergillus* sp 1. Thus, these marine micro-organisms were found to degrade the polymer and polymer composites.

Keywords Synthetic plastics · Biodegradable polymers · Plastic bottle wastes · Fungi

R. Sarkhel · S. Sengupta · P. Das (✉) · A. Bhowal
Department of Chemical Engineering, Jadavpur University, Kolkata, India
e-mail: papitasaha@gmail.com

R. Sarkhel
e-mail: rwiddhisarkhel@gmail.com

S. Sengupta
e-mail: sengupta.shubha@gmail.com

A. Bhowal
e-mail: avijit_bh@yahoo.co.in

S. K. Ghosh (ed.), *Urban Mining and Sustainable Waste Management*,
https://doi.org/10.1007/978-981-15-0532-4_20

1 Introduction

Plastic contamination in marine environment have affects on marine litter and effects actions and policies taken worldwide. The amount of plastics has increased wildly in the marine environment due to the large number of plastics thrown into waterbodies everyday nowadays. Plastics in marine environment may not be easily biodegradable due to the long carbon chains present in it. Plastics have different types of advantages: having low-cost materials, lightweighted and unbreakable bonding. They also have high molecular mass and high yield of specific ability of deforming. Degradation by the breakdown of the compounds can occur by various ways like biotic or abiotic process as well as through biological methods like microbial degradation using different types of bacteria and fungi. Biological degradation is a phenomena which describes the breakdown or cut down of organic compounds by living organisms like bacteria, fungi. Recently, alertness of the environmental problems by plastic waste as a durable material has made biodegradable plastics an attractive alternative to conventional plastic materials. Zheng et al. (2005) reviewed biodegradable studies on plastic polymers that are mainly done by using microbial entities. Recently, experimental studies signify that awareness for different environmental problems can be harnessed by the plastics as studied by Zahra et al. (2012), and thus, the most effective way to harness the environmental problems is by the process of bioremediation which emphasizes on not accumulating the various hazardous chemicals to be replenished as environmental pollutants and to overcome the problems furnished by durable plastic wastes.

2 Literature Review

There are reports of biodegradation of plastic wastes by micro-organisms especially fungi:

It has been reported that the heterotrophic fungi identified as *Aspergillus sp.* were associated with the polymer degradation, and were isolated and identified by plating and growth method. Predominant fungal strains of *Aspergillus sp* were selected for polymer degradation by effective commercial pieces of LDPE polymers (plastic bottle wastes) under laboratory conditions. The degradation of the polymer film samples were studied for different time intervals. The rate of biodegradation of the samples were measured in terms of mean weight loss, which was nearly 14–30% after a period of 6 weeks explained by Raaman et al. (2012).

There are works reported on the biodegradability of polymers in marine environment using both natural and synthetic types of plastic. Many characterizations were explained for determining biodegradation of different types of plastic polymers using marine micro-organisms studied by Priyanka and Rachna (2011).

Plastic litter in marine and aquatic ecosystems mitigated various ecological and societal impacts explained by Ghosh et al. (2013). This work reflected on the view that only standard and international test methods are insufficient to describe realistically the predicted biodegradability of plastic litter in inland waters and marine ecosystems which has been reviewed in Shah et al. (2012).

Sain et al. (2014) studied that chemical modification of polymer matrix reinforced with cellulosic fibres by weight loss method is an effective way for making the polymers environment friendly and partly biodegradable.

3 Materials

The materials used for this research study is as follows:

Plastic bottle waste was procured from solid waste litter site, the commonly used polymers polymethyl methacrylate (PMMA) and polymethyl methacrylate/cellulose composites were synthesized in laboratory. Materials needed for the preparation process of Cellulose:

Sugarcane bagasse fibres were procured from local market. Sodium chlorite ($NaClO_3$), sodium hydrogen sulphite ($NaHSO_3$), sodium hydroxide (NaOH) and concentrated sulphuric acid (H_2SO_4) were obtained from Lobachemie, India.

For minimal media preparation, potassium dihydrogen phosphate (KH_2PO_4), sodium nitrate ($NaNO_3$), magnesium sulphate heptahydrate ($MgSO_4.7H_2O$), potassium chloride (KCl), iron (II) sulphate heptahydrate ($Fe_2SO_4.7H_2O$) and ammonium chloride (NH_4Cl) were procured from Lobachemie, India.

4 Methods

4.1 Preparation of Cellulose and Nanocellulose from Raw Sugarcane Bagasse

Cellulose yielding amounts of micro- as well as nano-fibres of about 20 g was treated with sugarcane bagasse fibres along with 0.7% sodium chlorite ($NaClO_2$) solution maintaining at pH 4 and kept for continuous stirring of 2 h. This procedure was performed twice and then filtered as the process explained in Sain et al. (2014) and Zheng et al. (2005). The obtained filtrate was then treated with 2% sodium hydrogen sulphite ($NaHSO_3$) for about 15 min, was filtered, dried and then again treated with 17.5% NaOH solution for 15 min, then again filtered, washed and thus, further acid hydrolysis was done by treating the above filtrate with 47% H_2SO_4 and kept for constant stirring of about 3 h at 50 °C. Hence, finally, we obtain powders of cellulose after the process of freeze-drying, also known as lyophilization at a temperature of about −30 °C by maintaining a constant pressure of about 10–15 kg/ms^2.

4.2 Synthesis of PMMA and PMMA/Cellulose Composites

PMMA and PMMA/cellulose composite were synthesized by ex situ dispersion method following the process given in Sain et al. (2014).

4.3 Biodegradation Studies: Degradation Through Marine Microbes

Biodegradation behaviour of the different polymer film samples was examined by the rate of degradation through marine microbes, namely different species of identified fungus: *Aspergillus* sp. 1, 2 and 3. The fungus was isolated from the marine waters near Sunderban area. The first film sample was plastic bottle waste obtained by cutting the Kinley mineral water bottle into square sized 1 cm diameter. The second film sample was Polymethyl methacrylate (PMMA) and third sample was PMMA/cellulose, obtained by the process discussed above and then was cut into small uniform pieces. The polymer film samples were taken out at 1, 2, 3, 4, 5 and 6 weeks for plastic bottle wastes, PMMA and PMMA cellulose composite respectively, washed with ethanol, were vacuum dried and thus, rate of biodegradation by these different polymer samples were speculated by examining the weight loss and was noticed that the degradation rate increases with time.

4.4 Weight Loss Study

The characteristics of these different polymer film samples can be speculated by the weight loss study measuring the rate of degradation at different time intervals. Plastic bottle wastes, PMMA, and PMMA cellulose composites were taken out from the minimal media after keeping it for 6 weeks at an incubator temperature of about 36–37 °C hence calculating the percentage of weight loss by the given equation,

$$\%\mathbf{Weight\ loss} = \{(\mathbf{S1} - \mathbf{S2})/\mathbf{S1}\} * \mathbf{100}$$

where $S1$ is the initial weight of the polymer sample (before putting it in media) and $S2$ is the final weight of the polymer sample (after taking out from media) at different time intervals (Fig. 1). Figure 1 represents the picture of a plastic bottle waste film and the growth of fungus around it.

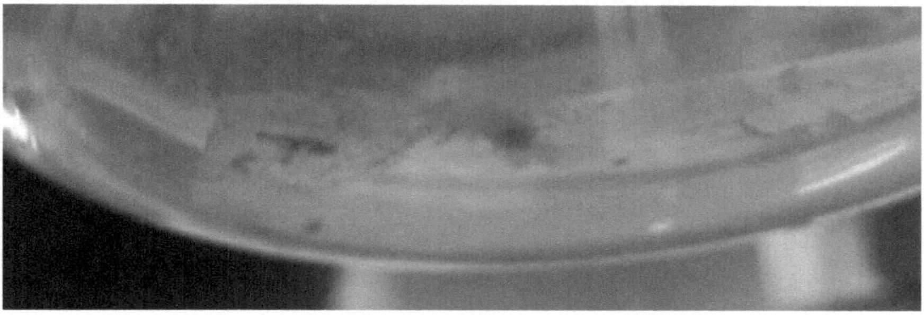

Fig. 1 Colonies of fungi growing encircling around the plastic bottle waste film

5 Results and Discussion

5.1 Evaluation of Weight Loss of Various Polymer Samples

The weight loss study of the films was evaluated by the graph as a function of rate of biodegradation with respect to time interval of 1 week upto 6 weeks obtained from the study of growth of the fungus on the different plastic polymer samples as shown in Fig. 2a–c. PMMA cellulose composites showed a higher % of weight loss as compared to plastic bottle waste and PMMA. This study reflects that PMMA cellulose composites can degrade the polymer samples at a higher rate with respect to the PMMA sample and plastic bottle waste. All the polymer samples showed a percentage of weight loss after keeping it for 6 weeks (plastic bottle waste around 20%, PMMA around 15% and PMMA cellulose composites around 24%).

6 Conclusion

In this present study, the influence of different synthetic polymers like plastic bottle waste sample and polymethyl methacrylate (PMMA) along with its composite PMMA cellulose was evaluated by microbial degradation in marine water by achieving the effective percentage of weight loss of all the plastic polymer samples. The results indicated that the degradation of PMMA composites was more since it showed a higher weight loss as compared to plastic bottle wastes and PMMA films with respect to 6 weeks data. Thus, the suitable method for microbial degradation by different polymers and its composites is thus also an efficient way for bioremediation as well as biodegradation.

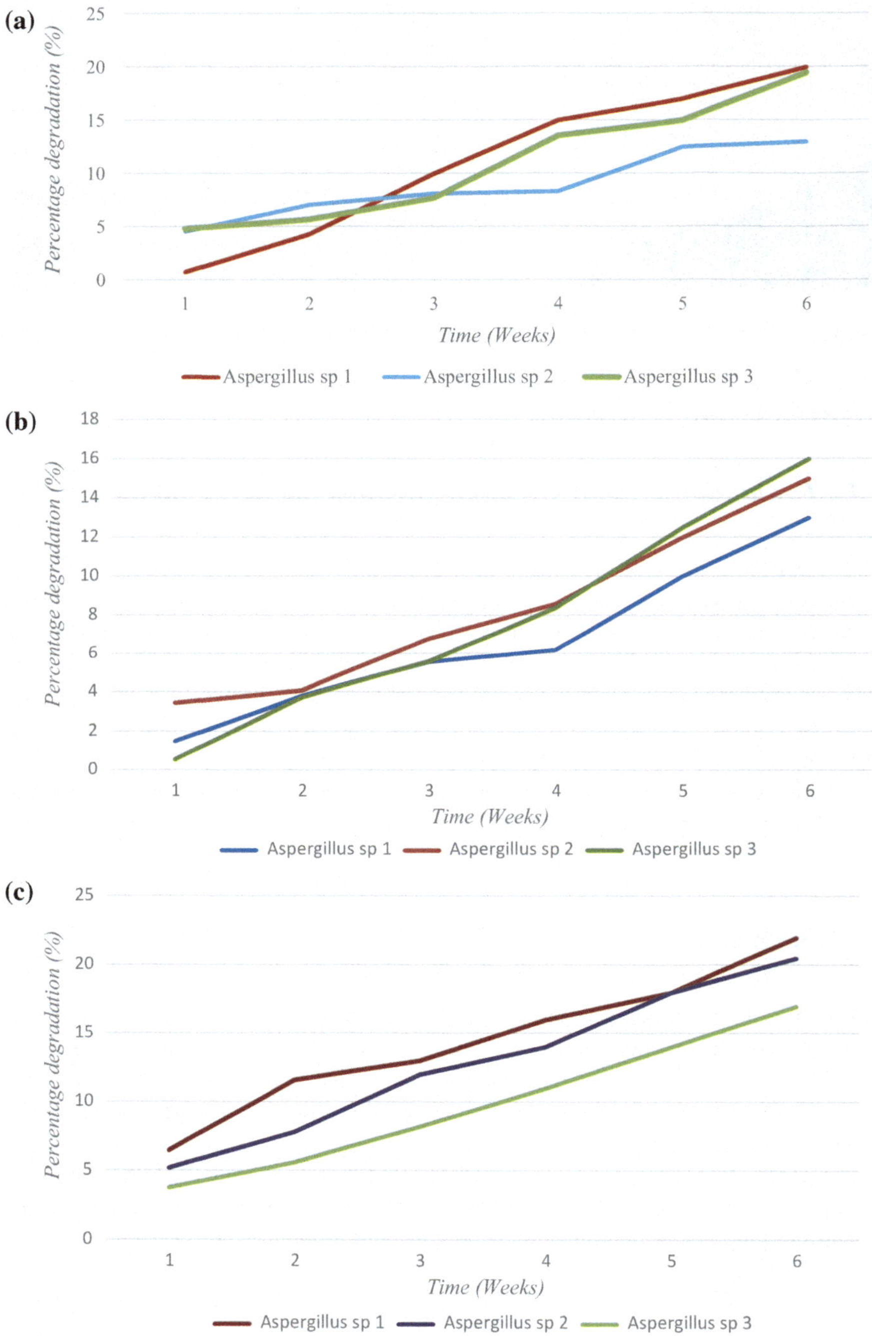

Fig. 2 **a** Rate of biodegradation by plastic bottle waste. **b** Rate of biodegradation by PMMA. **c** Rate of biodegradation by PMMA cellulose composite

Acknowledgements I am thankful to Jadavpur University, Department of Chemical Engineering for guiding me in a right way, especially to TEQUIP 3 and RUSA 2.0 for the financial support.

References

Ghosh, S. K., Pal, S., & Ray, S. (2013). Study of microbes having potentiability for biodegradation of plastics, 4339–4355.

Priyanka, N., & Rachna, T. (2011). Biodegradability of polythene and plastic with the help of microorganism, 2161-0525.

Raaman, N., Rajitha, N., Jayshree, A., & Jegadeesh, R. (2012). Biodegradation of plastic by Aspergillus spp. isolated from polluted sites around Chennai, 2278–5213.

Sain, S., Sengupta, S., Kar, A., Mukhopadhyay, A., Sengupta, S., Kar, T., Ray, D. (2014). Effect of modified cellulose fibres on the biodegradation behavior of in-situ formed PMMA/cellulose composites in soil environment: Isolation and identification of the composite degrading fungus, 156–165.

Shah, A. A., Hasan, F., Hameed, A., & Ahmed, S. (2012). Biological degradation of plastics: A comprehensive review, 246–265.

Zahra, S., Abbas, S. S., Mahsa, M. T., & Mohsin, N. (2012). Biodegradation of low-density polyethylene (LDPE) by isolated fungi in solid waste medium, 396–401.

Zheng, Y., Yanful, E. K., & Bassi, A. S. (2005). A review on plastic waste biodegradation, 243–250.

Current Scenario of Plastic Waste Management in India: Way Forward in Turning Vision to Reality

Tadinada Sri Sasi Jyothsna and Bandari Chakradhar

Abstract India pledges to combat plastic pollution as part of the World Environment Day, 2018. The theme of this paper is to comprehend the current scenario of Plastic Waste Management (PWM) in India, particularly in the context of achieving sustainable development goals 2030. It is estimated that out of the 8.3 billion metric tons of plastic produced globally as on date, 6.3 billion metric tons ends up as trash, whereas only 9% is recycled, 12% is incinerated and the rest 79% is accumulated in landfills. It is anticipated that global plastic waste totals to 4.9 billion tonnes and will reach 12 billion tonnes by 2050. India generates around 56 lakh tonnes of plastic waste annually and Delhi being the highest generator of plastic waste in India accounts for 9,600 mt per day among the top ten cities, followed by Chennai, Kolkata, Mumbai, Bangalore and few others. The paper presents a generic overview of plastic waste inventorization, composition and its effects on health and environment. It discusses about the legislations governing Plastic Waste Management and Handling Rules (2011, 2016 and 2018) in India, amendments, current status of implementation as well as opportunities and challenges in PWM. A small survey is carried out to thoroughly understand PWM in a community. Critical gaps between the legislations framed and their enforcement are identified. This paper also appraises the concept of 4R′s in PWM and their role in circular economy. Global initiatives on the ban of plastics, collection and recycling mechanisms of plastic waste are addressed and compared with the Indian scenario. Responsible production and consumption of plastics, emerging trends in PWM in relation to Integrated Waste Management (IWM) are highlighted. To conclude, the authors provide the best possible options and divergent solutions for tackling plastic wastes from Indian perspective towards turning vision to reality.

Keywords Plastic waste management · Inventorization · Integrated waste management · Circular economy

---

T. Sri Sasi Jyothsna · B. Chakradhar (✉)
Environmental Consultancy, Ramky Enviro Services Private Limited, 12th Floor, Ramky Grandiose, Gachibowli, Hyderabad 500032, India
e-mail: drchakradhar@ramky.com

S. K. Ghosh (ed.), *Urban Mining and Sustainable Waste Management*,
https://doi.org/10.1007/978-981-15-0532-4_21

1 Introduction

It is not a surprise to say that there is no day where plastics have not been used in our day to day activities. With the increasing demand of plastic products, the plastic industries have been showing their presence in every sector. The industry has grown at a compound annual growth rate (CAGR) of 10%, in terms of volume, from 8.33 MMTPA to 13.4 MMTPA in the past five years and is expected to grow at 10.5% for the next few years (Bhattacharya et al. 2016). Their convenient, lightweight, durable and affordable nature has led us over use plastics at such a rate that it is being littered uncontrollably without any concern for environment. Their rapid growth in production is surpassing other man-made materials (Geyer et al. 2017). It is estimated that out of the 8.3 billion metric tons of plastic produced globally as on date, 6.3 billion metric tons ends up as trash, whereas only 7% is recycled, 14% is incinerated and the rest 79% is accumulated in landfills. It is anticipated that global plastic waste totals to 4.9 billion tonnes and will reach 12 billion tonnes by 2050 (Parker 2017). Till date more than 300 million tonnes of plastic has been synthesized worldwide. Golam Kibria reported that China ranks first among the top 20 countries in the world which mismanage plastic wastes and India ranks 12th among them (Kibria 2017). Global consumption of plastic materials in 2015 is expected to be 45 kg/person/year. Highest consumption is expected to be in NAFTA region (North American Free Trade Agreement) which is 139 kg/person, followed by Western Europe, Japan, Central Europe, Asia, Middle East and Africa (Statista 2015).

To address the issue of Plastic Waste Management (PWM) in the country, the Plastic Waste (Management and Handling) Rules 2011, were introduced under the Environment Protection Act in 1986 by the Ministry of Environment and Forests, Climate Change (MoEF&CC), with further amendation in 2016 and 2018. (The Plastic Waste Rules 2016). Indian government has initiated programmes to make the country free from single-use plastics by 2022. The understanding about plastic hazards on human health and environment, due to usage and disposal is increasing in the public. Many countries have taken initiatives to completely ban plastics and substitute with safer materials. For a developing country like India, it is, therefore, necessary to reinforce the legislations framed and adopt good practices to attain the objectives of 4R′s and achieve Sustainable Development Goals (SDGs).

This paper discusses about the current situation of PWM in India. Reasons for lack of proper enforcement of these legislations have been highlighted. It appraises the concept of 4R′s in PWM and its role in circular economy. Emerging trends in PWM in relation to IWM are emphasized. The paper is concluded with best possible solutions for management of plastic waste in the country.

2 Results and Discussions

2.1 An Overview—Plastic Production, Consumption and Generation

The size of plastics industry in India is estimated to be about Rs. 110,000 Crores. There are over 30,000 units which produce plastic materials India. Approximately, 90% of these units are small- and medium-sized enterprises. 35% of plastic is consumed in packaging and 23% in building and construction.[1] Plastics consumption accounts to 13 MT per year. Malik (2013) reported that approximately 10,000 tonnes per day (TPD) of plastic waste is generated in India and the plastic waste content in municipal solid waste (MSW) on an average is estimated to be 1.5 kg/per person (Malik 2013). As on date, the country generates 25,940 tonnes of plastic waste daily, of which 40% is either left unattended or not treated properly. The average per capita consumption is reported to be about 11 kg. It is assumed that the annual per capita consumption in India would be 20 kg by 2022 (Goswami 2018).

It is anticipated that global plastic waste totals to 4.9 billion tonnes and will reach 12 billion tonnes by 2050. India generates around 56 lakh tonnes of plastic waste annually and Delhi being the highest generator of plastic waste in India accounts for 9,600 mt per day among the top ten cities, followed by Chennai, Kolkata, Mumbai, Bangalore and few others.[2] Figure 1 shows the typical composition of MSW in India. The share of plastics among the MSW is 9%. This shows how increasing

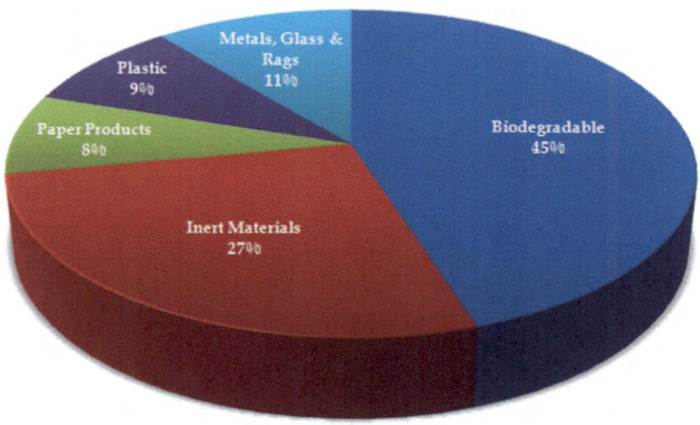

Fig. 1 Composition of MSW in India. *Source* http://yesinstitute.in/knowledgecontent/thought/statusofwte2016/

[1]Solid and Liquid Resource Management, Plastic Waste Management, file:///C:/Users/neelapriya.dvnl/Downloads/Resource%20book_Plastic%20Waste%20Management.pdf.

[2]Plastic Waste Crisis: What India Can Learn From Other Countries Around The World, https://www.ndtv.com/photos/news/plastic-waste-crisis-what-india-can-learn-from-other-countries-around-the-world-24272#photo-298497.

population, changing life styles and convenient usage of plastic products has significantly increased the per capita generation of plastic wastes. It is pathetic to know that much of this waste generation is rooted in urban areas. Further, it is observed that households are the biggest source of plastic waste (CIPET-CPCB 2015).

2.1.1 Impacts on Health and Environment

Plastic waste is a major concern to the environment. Accumulation of plastic products has left behind a huge trail of impact on the environment posing a serious threat to the health of humans and wildlife. The most possible plastics of concern are chlorinated plastics which contaminate the soil and water (Subba Reddy et al. 2014). As per recent studies, more than half a million tonnes of plastic wastes were dumped in oceans annually in India, which were found generally near the seas of Mumbai and other port cities. The aquatic creatures and their ecosystems were affected by this plastic pollution, attributed to entanglement and ingestion followed by injuries and death (Koushal et al. 2014; Li et al. 2016; Wagner et al. 2014).

2.2 Environmental Legislations

MoEF&CC has notified the Plastic Waste (Management and Handling) Rules in 2011 to address the issue of PWM in the country in a scientific manner, and further amended as PWM Rules, 2016 and 2018. Provisions made in the amended rules are discussed by comparing the PWM Rules 2011, 2016 and 2018 which are given in Table 1.

2.3 Opportunities and Challenges in PWM

Opportunities
Stringent legislations aimed at reducing pollutants discharged from the waste have triggered more investments in the plastic waste market. There is a huge demand for bio-based plastics or plastics produced from carbon dioxide which offer the same functionalities as traditional plastics with potentially lower environmental impacts.

Challenges in Management of Plastic Waste

2.3.1 Growth of Plastic Production and Consumption

Over the years, the role and importance of plastics in our economy have consistently grown. Global production of plastics has increased many fold crossing 300 million

Table 1 Comparison of PWM Rules 2011, 2016 and 2018

Title	Plastic waste (management and handling) rules, 2011	Plastic waste management rules, 2016	Plastic waste management (amendment) rules, 2018
Enforcement authority	• Manufacturing, recycling and disposal—SPCB/CPCB • Use, collection, segregation, transportation, disposal of post-consumer plastic waste—municipal authority	Not mentioned	–
Conditions	• Size of plastic sheet or multi-layered packaging shall not be less than 40 μ in thickness	• Size of plastic sheet or multi-layered packaging shall not be less than 50 μ in thickness • The manufacturer shall not sell or provide or arrange plastic to be used as raw material to a producer, not having valid registration from the concerned PCB committee • Provision of thickness shall not be applicable to carry bags made up of compostable plastic and conform to the Indian Standard: IS 17088:2008 • Plastic material, in any form including Vinyl acetate-maleic acid-vinyl chloride copolymer, shall not be used in any package for packaging gutkha, pan masala and tobacco in all forms	• The term 'non-recyclable multi-layered plastic' has been substituted by 'multi-layered plastic which is non-recyclable or non-energy recoverable or with no alternate use'

(continued)

Table 1 (continued)

Title	Plastic waste (management and handling) rules, 2011	Plastic waste management rules, 2016	Plastic waste management (amendment) rules, 2018
Plastic waste management	• Recycling, recovery or disposal of plastic be carried out as per rules, regulation and standards by Central government • Roles of municipal authority is to set-up, operationalize and coordinate safe collection, storage, segregation, transportation, processing and disposal of wastes	• Rules for plastic waste management is split to urban local bodies, local body gram panchayats, waste generator, producers, importers and brand owners as per their jurisdiction • Local bodies shall encourage the use of plastic waste for road construction as per IRC guidelines, energy recovery or waste to oil, etc., as per norms prescribed by specific authorities • Thermoset plastic shall be processed and disposed of as per CPCB guidelines • The responsibilities of municipal authority are now known as responsibilities of local body	–
Marking or labelling	• Details of name, the registration number of manufacturer and thickness of carry bag/multi-layered plastic to be given in English or any local language	• Details of name, registration number of manufacturer and thickness of carry bag/multi-layered plastic to be given in English only • Name and certificate number to be given in case of compostable plastic	–
Prescribed authority	–	• SPCB and pollution control committee are responsible for enforcement of these rules	–

(continued)

Table 1 (continued)

Title	Plastic waste (management and handling) rules, 2011	Plastic waste management rules, 2016	Plastic waste management (amendment) rules, 2018
Registration of producers, manufacturers and recyclers	• Registration was limited to only manufacturers and recyclers • Application for renewal of registration should be made 90 days before expiry of the validity of registration certificate	• Manufacturers need to fill Form-III for registration or renewal of registration for manufacturing of plastics • Application for renewal of registration should be made 120 days before expiry of the validity of registration certificate	• prescribed a central registration system by mandating brand owners and producers operating in more than two states to register with the CPCB
Responsibility of retailers and street vendors	–	Shopkeepers and retailers shall not sell plastic carry bags or multi-layered packaging or plastic sheet if they are not manufactured and labelled or marked as per the rules and if found are liable to pay fine if found	–
Explicit pricing of carry bags	• No carry bags shall be available free of cost, a minimum price is to be levied depending on quality, size of the material and waste management cost	• Registration under local body by retailers or street vendors to sell plastic bags by paying a fee of Rs. 48,000/- @ Rs. 4,000/-per month • This can be prescribed higher by local body based on sale capacity	• This rule has been omitted from here on

(continued)

Table 1 (continued)

Title	Plastic waste (management and handling) rules, 2011	Plastic waste management rules, 2016	Plastic waste management (amendment) rules, 2018
Annual reports	• SPCB/pollution control committee was responsible to submit annual report on implementation of rules to CPCB by 30th September each year • CPCB was responsible to prepare consolidated annual report on use of plastic waste and way forward along with recommendations to Central government before 30th December each year	• Every person engaged in recycling or processing of plastic waste shall prepare and submit an annual report in Form-IV to local body by 30th April each year • Every local body shall prepare and submit annual report in Form-V to concerned Secretary in-charge of urban development department by 30th June every year • Each SPCB/pollution control committee shall prepare and submit an annual report in Form-VI to CPCB on implementation of the rules by 30th July every year • CPCB shall prepare consolidated annual report on use and management of plastic waste and way forward along with recommendations to Central government before 30th August each year	–

tonnes. At the same time, plastic waste is a serious concern to the environment and public health. Plastic bags of all sizes and thickness are often found littered everywhere especially in urban areas. These appear to be aesthetically unpleasant causing visual pollution. Additionally, plastic bags tend to clog drains, gutters, thereby causing flooding of the areas for sparse rains. Further, they also pose a threat to the terrestrial animals. On a general note, 43% of the fabricated plastics are utilized in packing works and most of them are one-time use plastics. In most cases, plastic products like carry bags, coffee cups, water bottles, food packaging, etc., are thrown away after single use. This 'throw-away' culture without any concern for environment is causing a menace.

2.3.2 Laws and Their Enforcement

Despite the legislations framed, PWM has not received enough attention in India. CPCB has estimated that in 2014, out of 80.28% waste collected, only 28.4% was treated. The remaining quantities were either disposed of in landfills or open dumps. Under this circumstance, different studies have been done to examine the efficacy of the approaches and regulations confined in the nation. According to the reports, there are insufficient collection centres and needs a framework that takes backs plastics after its use. Although there is an increasing awareness about the detrimental effects of plastics on the environment, lack of stringent regulations has made the people forget their social responsibilities. Although there are provisions in legislations to penalize citizens for failing to comply with the rules framed, they are not being properly enforced.

2.3.3 Poor Infrastructure for Waste Collection and Disposal

In our country, PWM is an underestimated and improperly attended segment having poor waste collection and disposal mechanisms. The weak links in the system are inappropriate separation, collection and transportation. Many cities in India lack access to designed scientific landfills for waste disposal. Owing to these problems, dumping is a common practice to discard enormous amount of wastes generated from households and surroundings.

2.3.4 Informal Sector

The intermediation of informal sector is another major challenge. Majority of the plastic waste generated in India is managed by the unorganized/informal sector instead of going to the authorized recycling units. These scrap dealers buy the wastes and sell them to the big traders via the small trades. Ultimately, the waste is sold to the unauthorized recyclers/dismantlers. Because of these reasons, the waste does not reach to a proper recycling unit and further lacks treatment. But, since then under various economic scales, the informal sector still keeps on a noteworthy job in India for gathering and reusing of these wastes.

2.3.5 Recycling and Associated Problems

Technically, plastics can be divided into two categories—thermoplastics and thermosetting plastics. Across the country, 80% of post-consumer plastic waste is thermoplastics, which may contain any of these materials such as polyethylene, polyethylene terephthalate, polyvinyl chloride, polystyrene and polyesters. These thermoplastics need harsh thermal treatment methods to recycle them. A vast area in the research of plastic waste management has been untouched and needs to be explored. In comparison to glass, papers or metals, the life of plastics remains very low.

2.3.6 Quality of the Recycled Product

Sometimes, the recycling methods result in low efficiency, of the recycled product. The recycled plastics are more harmful to the environment than the virgin products due to the mixing of additives, colours, stabilizers, halogenated flame retardants and so on. Another major obstacle in the transformation of the plastic value chain is the poor demand for the recycled plastics.

2.4 A Case Study on the Existing Scenario of PWM

2.4.1 Need for the Study

Currently, the implementation of PWM in the country is inefficient, disposing the waste in an improper manner. Tons and tons of plastic debris lie in the environment as open dumps. Of the plastic products in use, carry bags are the major contributors of littered waste. With this background, a survey was undertaken with appropriate questionnaire in order to assess the awareness of vendors and consumers about the use of plastic bags and their adverse impacts on the environment up on their disposal.

2.4.2 Materials and Methods

The observational study was carried out for a period of one month. The survey was conducted in Hyderabad, Telangana, by selecting one locality randomly. Vegetable/fruit vendors, restaurants, kirana stores, super markets, medical shops, stationary and shopping malls which are the major sources of plastic bags generation were targeted and the questionnaire was distributed accordingly. The total sample size was 250. Further, questionnaires were also distributed to the common public including

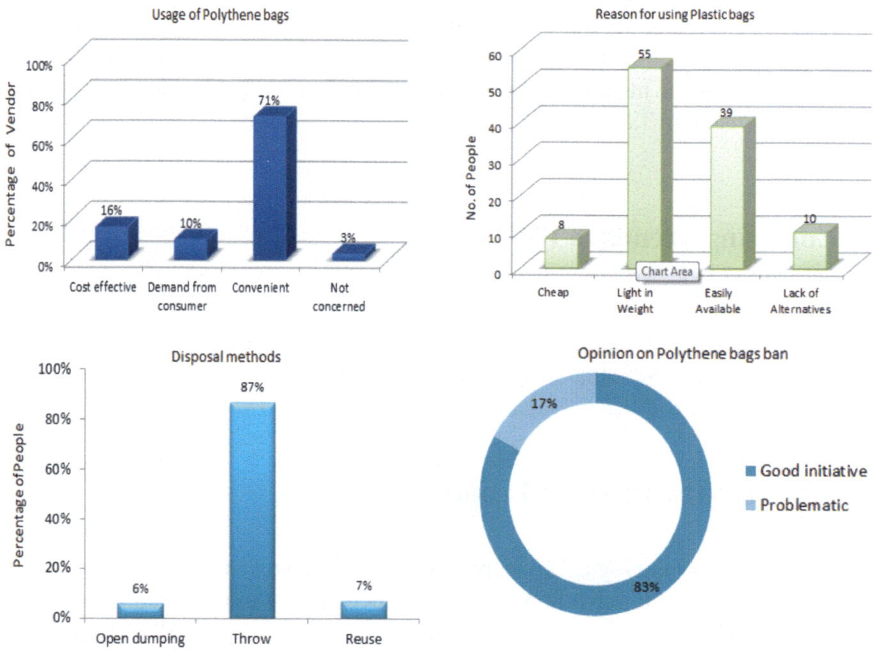

Fig. 2 Observations on PWM practices

households, school children and college students. In this instance, the total sample size was 120. Reasons for offering and using polythene bags, their mode of disposal, awareness on environmental impacts of plastic waste, percentage use of plastic bags, opinion on their ban, etc., are gathered. Multiple responses were obtained in some cases.

2.4.3 Observations and Discussions

The observations of the study are presented in Fig. 2. The graphs presented below are self-explanatory. 83% public felt that ban on plastic bags is a good initiative; whereas 17% felt that it is problematic.

3 Initiatives on Plastic Bags Ban

Worldwide, there is an increasing awareness on plastic waste and their associated impacts on the environment. Many countries are leading their way to end plastic pollution by implementing bans and imposing strict fines on single-use plastic bags. Some of these policies have been successful, reducing plastic bag consumption.

In the past few years, several policy initiatives have been taken by the state and central governments in India to ban specific plastic products. Many states have passed legislations to ban the manufacture, sale and use of plastic bags. Initiatives on plastic bags ban are listed in Table 2.

4 Emerging Trends in PWM

There are different conventional methods for PWM. Some of them are listed in Table 3. These methods involve plastic recycling and emerging technologies for use of discarded plastics as value-added resource.

5 4R′S and the Circular Economy

A circular economy is a logical concept in which we keep resources in use for as long as possible, extract the maximum value from them whilst in use, then recover and regenerate products and materials at the end of each service life. 4R′s (Reduction, Reuse, Recycling and Recovery) is a concept which must be sturdily regulated within a circular economy. The concept provides an ecologically sound and environmentally friendly approach to minimizing and managing wastes. The Integrated Solid Waste Management System Hierarchy under Swacch Bharath Mission, Ministry of Urban Development, has ranked recycling of materials as the second most preferred option of waste management. Processing non-biodegradable waste to recover commercially valuable materials is given top priority. Waste management strategies can be accomplished by recycling and recovery of materials. In the current economic system it is anticipated that someone assumes liability for the disposed materials. This has prompted for the formation of extended producer responsibility (EPR), where it is expected that producers should take the responsibility to collect their refused/used products from the market, recycle, process and again re-introduce. And nowadays, different NGOs and organizations are in process for recovering, buying and selling the discarded plastics. Therefore, it is a great responsibility for all those involved in collection and treatment of wastes to close the loop without leading to market distortions.

6 Turning Vision to Reality

The management of plastic waste needs innovative solutions to address critical challenges and for turning vision to reality. In view of the ever-increasing population and subsequent increase of plastic waste all around, imposing fines or penalties is not always the solution. The solutions should focus more on developing cleaner and

Table 2 Initiatives on the ban of plastic bags worldwide and in India

Country	PWM policies and initiatives
Sweden	• Presence of a number of recycling stations very near to the residential areas • Strict segregation at the household level into recyclable and biodegradable waste • Application of high environmental standards during energy recovery
France	• A strict ban on all disposable daily-use plastics including cutlery and bags • Phase-out these plastic goods by 2020, replacements products made of biologically sourced materials
Rwanda	• High fine structures for businesses that sell plastic bags or produce plastic waste • Citizens can be imprisoned for the use of plastic bags and imposed rigid zero tolerance policy • Conducting behaviour change campaigns to civilians
China	• Ban on the distribution of single-use plastic bags in grocery stores and shops around the country • Imposed a strict fine of 10,000 yuan, on the companies for illegal plastic bag distribution • Called for a ban on import of waste from EU nations and the US
Washington	• First to impose tax on plastic bags. The revenue collected from this tax goes to the Anacostia River Clean Up and Protection Fund
Chile	• Fines up to $300 USD are issued on businesses that continue to distribute plastic bags
Australia	• Banned plastic bags include all single-use polyethylene polymer bags that are less than 35 μ thick • Encourage citizens to bring reusable bags when shopping
Kenya	• Country-wide ban of plastic bags that falls on the distributors and producers of single-use bags • Imposing fine of Rs. 40,000/- or imprisonment for four years for the sale or use of plastic bags
India	
Sikkim	• The first state to ban plastic bags in 1998. Also the first to target single-use plastic bottles banned in 2016
Uttar Pradesh	• In December 2015, the Uttar Pradesh Cabinet approved a complete statewide ban on the distribution, manufacture and sale of polythene carry bags of 40 μ thickness and below • Recently, polythene below 50 μ has been banned and fine up to Rs. 1 lakh has been imposed
Karnataka	• Ban on manufacture, storage, distribution and use of plastics such as carry bags, banners, plastic plates, cups and spoons. All manufacturers in the state are banned from producing any single-use plastics
West Bengal	• Plastic bags less than 50 μ are banned
Maharashtra	• Recently, the state banned the manufacture, use, sale, transport, distribution and import of plastic bags with or without handle and disposable products made out of plastic and thermocol • The government enforced the ban with a fine of Rs. 5,000/-

(continued)

Table 2 (continued)

Country	PWM policies and initiatives
Kerala	• Petition filed by All Kerala River Protection Council seeking a ban on use of plastic carry bags in the state
Telangana	• Banned the use and sale of polythene bags which have a thickness of less than 50 μ

sustainable products for achieving Sustainable Development Goals (SDGs) 2030. Although reduce and reuse of a resource are the most potential solutions, some divergent solutions for PWM include:

- **Formalizing the informal sector**: Involving skilled manpower from the informal sector of the economy
- **Formalizing the plastic waste collection chain**: An appropriate framework including a proper collection network and linking GPS to points of segregation, tracking vehicles, etc.
- **Sustainable recycling**: Using sustainable technologies for producing value-added products and energy recovery. This should include cost-effective and environmental friendly methods.
- **Intervention of government**: The plastic manufacturers who come from unorganized and informal sectors should be strictly monitored. To ensure the effectiveness, regulations have to be implied strategically and cautiously.
- **EPR**: This can bring about some change in collection, recovery and recycling of plastic waste much efficiently. The producers of such waste need to be levied with the responsibility of addressing the disposal or recycling of plastic waste, which may prove to be a better option.
- **Product take-back mechanism**: To establish enough collection centres with respect to take back of plastic products after 'end of life' by the manufactures.
- **Alternatives to plastic bags**: Worldwide, research is going on to replace the synthetic plastics with plastics made from wood, food waste, compostable coatings, etc. Bioplastics made from natural materials such as corn starch are gaining importance these days.
- **Designing a better system**: Extending the life time of products and recycling them to be a part of circular economy
- **Awareness**: The easy-go-attitude of people and throw-away culture has made the people to litter plastic waste indiscriminately without any concern for environment. People should be made aware of the negative impacts of plastic waste on health and environment.
- **Shop friendly**: Carrying a paper or cloth bag when going for a shopping. Avoid bringing plastic bags to home and purchasing items with too much of packaging.

Table 3 Technological pathways for PWM

Type of treatment	Brief description	Disadvantages
Re-extrusion	Involves introduction of clean scrap of single types of plastics that can be re-entered into manufacturing processes to produce similar materials	Difficult because scrap plastic is not segregated by type and is not clean enough to re-enter manufacturing as it is
Mechanical recycling	It is primarily performed on simpler, single polymer plastics like polyethylene. Plastics are subjected to a variety of mechanical processes such as cutting/shredding, separation of contaminants, milling, washing and drying, extrusion and quenching with water. Road construction with plastic materials that have undergone mechanical processes	Difficult for requirement of homogenous plastic types that necessitate segregation by type
Chemical recycling	Uses advanced technical processes that convert plastic materials into smaller constituent molecules which can then be used as feedstock for the production of petrochemicals and plastics	Release of harmful chemicals as by-products
Thermal recycling (pyrolysis, gasification and hydrogenation)	Involves heating plastics under controlled temperatures with/without catalysts. Treatment with these methodologies yields gases with high calorific content or oils. These methods do not require segregation by plastic type, and are able to treat heterogeneous mixes. They produce useful and valuable feedstock to the petrochemical industry	By-products like chars can be particularly Harmful Process is energy intensive Economically not feasible
Energy recovery	Burn plastic to produce energy in the form of heat, steam and electricity. This process is commonly used to provide energy to cement kilns	Concerns due to the emission of carbon dioxide, and some carcinogenic pollutants

References

CIPET-CPCB. (2015). A study conducted by the CIPET-CPCB on the 'Assessment and Characterization of Plastic Waste in 60 Major Indian cities, 2015.

Geyer, R., Jambeck, J. R., & Law, K. L. (2017). Production, use, and fate of all plastics ever made. *Science Advances, 3*(e1700782), 1–5.

Goswami, U. A. (2018). *Environment: An 8.3-bn tonne problem: How to manage plastic waste—The Economic Times*. https://economictimes.indiatimes.com/industry/indl-goods/svs/paper-/-wood-/-glass/-plastic/-marbles/our-plastic-pollution-problem/articleshow/64420276.cms.

Kibria, G. (2017). *Plastic waste, plastic pollution- a threat to all nations*. Technical report.

Koushal, V., Sharma, R., Sharma, M., & Sharma, V. (2014). Plastics: Issues challenges and remediation. *International Journal of Waste Resources, 4,* 134.

Li, W. C., Tse, H. F., & Fok, L. (2016). Plastic waste in the marine environment: A review of sources, occurrence and effects. *Science of the Total Environment, 566–567,* 333–349.

Malik, P. (2013). Plastic waste and management: An environmental issue. *International Journal of Innovative Research and Development, 2*(13), 48–52.

Parker, L. (2017). A whopping 91% of plastic isn't recycled, you can't manage what you don't measure. https://news.nationalgeographic.com/2017/07/plastic-produced-recycling-waste-ocean-trash-debris-environment/.

Sailaja Bhattacharya, R. R. N., Chandrasekhar, K., Deepthi, M. V., Roy, P., & Khan, A. (2016). *Plastic waste management in india, challenges and opportunities*. Technical report by Tata Energy Research Institute, pp. 1–20.

Statista. (2015). *Per capita consumption of plastic materials worldwide in 2015 by region (in kilograms)*. https://www.statista.com/statistics/270312/consumption-of-plastic-materials-per-capita-since-1980/.

Subba Reddy, M., Srinivasulu Reddy, P., Venkata Subbaiah, G., & Venkata Subbaiah, H. (2014). Effect of plastic pollution on environment. *Journal of Chemical and Pharmaceutical Sciences*, pp. 28–29.

The Plastic Waste (Management and Handling) Rules. (2016). http://www.moef.gov.in/sites/default/files/PWM%20Rules%2C%202016.pdf.

Wagner, M., Scherer, C., Alvarez-Muñoz, D., Brennholt, N., Bourrain, X., Buchinger, S., et al. (2014). Microplastics in freshwater ecosystems: What we know and what we need to know. *Environmental Science Europe, 26,* 12.

Modelling of Post-consumer Plastic Flow in Municipal Solid Waste Stream: A Case Study in Few Major Local Authorities of Sri Lanka

Yushani Alahakoon and Anurudda Karunarathna

Abstract The quantitative assessment of material flow dynamics has a great influence on decision-making in waste management planning, which is still not a common practice in developing countries. Thus, the post-consumer plastic waste (PCPW) flow was analysed in order to examine the flow patterns and the components of the PCPW management system in Sri Lanka. PCPW flow of nine local authorities; Kandy Municipal Council (KMC), Akkaraipattu Municipal Council (AMC) and seven local authorities in Western Province (WP) were numerically analysed to view the plastic recovery and recycling processes as a network of flows, stocks and processes. Material flow analysis (MFA) is the tool used in quantitative assessment, where the principle of mass conservation was the underlying concept. All five major processes of MFA, consumption, collection, processing, manufacturing and dumpsite disposal, were identified in well-established KMC and WP, though only three major processes of consumption, collection and dumpsite disposal were available in AMC. Numerical analysis of collected data helped in illustrating Material flow as Sankey diagrams by using software STAN® for each local authority. It was estimated that the total post-consumer plastic waste generation in KMC, WP and AMC were 12.0, 54.4 and 4.8 tonnes/day, and out of that PCPW recycling rates were 33%, 6.6% and 0%, respectively. The direct accumulation of the majority of PCPW in the open dumpsites was estimated as 7.23, 43.13 and 1.35 tonnes/day in KMC, WP and AMC, respectively. Thereby the obtained results illustrated a clear picture on the current situation of the PCPW flow through the waste management system, which emphasized the burden of PCPW accumulation in the environment through open dumping, as well as the inefficient resource recovery practices in Sri Lanka.

Keywords Material flow analysis · Post-consumer plastics · Plastic recycling

Y. Alahakoon
Board of Study in Agricultural Engineering, Postgraduate Institute of Agriculture, University of Peradeniya, Kandy, Sri Lanka
e-mail: yushaniamyw@gmail.com

A. Karunarathna (✉)
Department of Agricultural Engineering, Faculty of Agriculture, University of Peradeniya, Kandy, Sri Lanka
e-mail: anujica@yahoo.com

S. K. Ghosh (ed.), *Urban Mining and Sustainable Waste Management*,
https://doi.org/10.1007/978-981-15-0532-4_22

1 Introduction

Increasing production, consumption and discharge of non-degradable solid wastes and different toxic materials in massive quantities to the environment are one of the major challenges we face in the recent history. High amount of non-reported wastes, lack of solid waste management (SWM) policies, lack of waste quality and composition data enhance the egregiousness of this issue (Tchobanoglous and Kreith 2002). Together with that, low waste collection services, lack of suitable treatment facilities, unsatisfactory recycling services and lack of public and political commitment increase the burden of this issue especially in the urban and suburban areas around the world (Majumder 2012; Vidanaarachchi et al. 2006; JICA 2003). Thereby sustainable coexistence between nature and humankind has fallen into danger, which alerts the upcoming massive environmental issues in the world. Finding solutions through minimizing energy and natural resource use, minimizing waste and gas emissions to the environment and enhancing 3R concept were found as key solutions to the above-mentioned issues (Binder et al. 2009).

As described by Bandara and Hettiarachchi (2010) and (Liyanage et al. 2015), municipal solid waste (MSW) management and post-consumer plastic waste (PCPW) management in Sri Lanka are unsystematic and selling of cleaned PCPW to private sector or export to overseas is the popular PCPW management method. The issue with the MSW is the major portion of the commercially valuable waste materials are collected by the private sector prior to entering the waste management system, and the government SWM sector is left with MSW which majorly comprised of wastes with no commercial value (Kuruppuge and Karunarathna 2014). Together with that, the implementation of new plastic recycling projects is done using conventional waste generation and composition data where most of these recycling facilities are informal. In industrial ecology, the most important aspect regarding plastic recycling is the extent of the contribution of recycled plastics to reduce plastic primary production. To address this aspect, it is essential to see the links between different processes of plastic waste management system (Kuczenski and Geyer 2010). Thus, there is a need to study the "dynamic flow" of PCPW within the waste management system for the implementation of a sound PCPW recycling policy framework. That will enable us to identify the most critical components of the PCPW management system and act as required. Material flow analysis (MFA) was used as the method of analysing processes and flows of the PCPW management system under this study.

2 Methodology

2.1 Approach

MFA is a popular method used in different industries for obtaining quantitative solutions for environmental problem solving and decision-making (Chang et al. 2007).

MFA can be defined as the analysis of the throughput of process chains comprising extraction or harvest, chemical transformation, manufacturing, consumption, recycling and disposal of material (Ayres and Ayres 2002). In simple, MFA is a systematic assessment of flows, stocks of material (goods and substances) within an arbitrarily system defined in space and time (Cencic and Rechberger 2008). The importance of MFA is increasing due to its vital use in industrial ecology due to the relevance to the policy making (Graedel and Allenby 2003). The basic methodological principle used in this analysis is the principle of mass balance (Cencic and Rechberger 2008). The net change of a particular material stock is given by the total inflows (production and imports) minus the total outflows (consumption and exports) from the stock, using the principle of conservation of mass (Kuczenski and Geyer 2010; Geyer et al. 2007).

$$\Delta \text{Stock}_{i,k} = \text{Production}_{i,k} - \text{Consumption}_{i,k} + \text{Import}_{i,k} - \text{Exports}_{i,k} \quad (1)$$

MFA was widely used to study many material flows during previous researches. Muthaa et al. (2006) have done a MFA for plastic in India, and they were able to develop a MFA diagram for plastics. In Taiwan, another MFA study was done to identify pollution sources of cadmium in soil (Chang et al. 2007). Bureecam et al. (2018) also done a MFA to model plastic waste flow in Thailand.

This study was done with the help of software STAN® (short name for substance flow analysis) which was developed for the purpose of MFA (Cencic and Rechberger 2008). In this software, results of the calculations are presented in the form of Sankey diagrams. In Sankey diagrams, mass flows of goods and substances are displayed as Sankey arrows and width of a Sankey arrow is proportionate to the mass flow value (Schmidt 2008). The material flow analysis is presented here by considering entire plastic waste as the "material", independent from their Resin type.

2.2 Study Areas

It is important to define the geographical and temporal system boundaries for the MFA. Local authorities named Kandy Municipal Council (KMC), Akkaraipattu Municipal Council (AMC) and seven local authorities from Western Province (WP) in Sri Lanka were used in this research to study and compare the PCPW management systems operating in those local authorities. The model was developed for a period of one year. An overall understanding of the Sri Lankan PCPW management system was also obtained with the final research outcomes. The locations of three areas are illustrated in Fig. 1.

KMC is the capital city of Central Hills of Sri Lanka, and it is the second largest secondary city in the country. Due to the cultural and religious importance in the Kandy city, there is a significant daily population flow to the Kandy City which is about 400,000. The daily waste generation is 176 tonnes/day, and there is about 130–150 tonnes/day of daily waste collection (Dissanayake 2013).

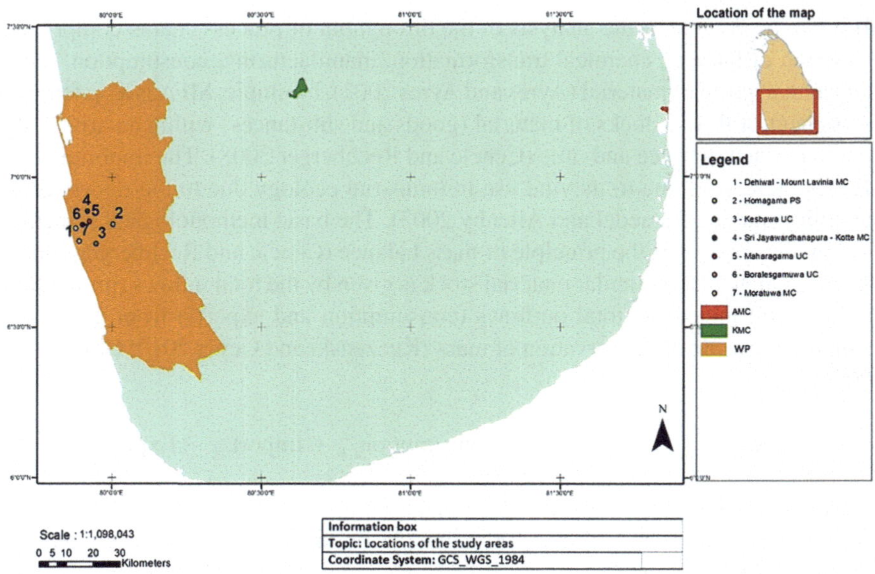

Fig. 1 Locations of the selected local authorities

Akkaraipattu Municipal Council is located in Eastern coastal area of Sri Lanka, and the average daily waste generation is about 50 tonnes/day. Seven major local authorities were considered from Western Province as mentioned in Table 1. All these seven local authorities used one common open dumpsite named *Karadiyana Dumpsite* as their final disposal point. The modelling was done for WP by considering that dumpsite as the basis.

2.3 Data Collection

The methodology adopted for the study is briefly illustrated in Fig. 2. Four main stakeholder groups were identified namely waste generators, waste collectors, waste buyers, waste recyclers and other relevant authorities. An initial secondary data collection survey was done at several relevant government authorities related to plastic waste management.

A basic material flow diagram was developed with available secondary data by using software STAN incorporating all possible processes, flows and stocks for the system (Fig. 3). The required primary data to complete the material flow diagram was identified, and samples for data collection were determined.

All required data to develop the material flow diagram for Western Province were collected through secondary data. Updated data on plastic collection and recycling in KMC was not available, and therefore, a primary data collection was done for KMC.

Table 1 Summary of the basic information of the selected local authorities

Local authority		Location	Area (km^2)	Population	Waste generation (T/day)
Kandy Municipal Council (KMC)		7.29057° N 80.63372° E	28.53	102,500	176
Akkaraipattu Municipal Council (AMC)		7.2194° N 82.8498° E	6.5	35,000	50
Western Province (WP)	Dehiwala-Mount Lavinia Municipal Council	6.8301° N 79.8801° E	21.1	182,996	180
	Homagama Pradeshiya Sabha	6.8433° N 80.0032° E	134.75	250,845	45
	Kesbewa Urban Council	6.7787° N 79.9473° E	41.5	184,360	55
	Sri Jayawardenepura Kotte Municipal Council	6.8868° N 79.9187° E	17.04	121,832	110
	Maharagama Urban Council	6.8522° E 79.9247° N	39.4	200,170	75
	Boralesgamuwa Urban Council	6.8412° N 79.9025° E	13.5	68,788	40
	Moratuwa Municipal Council	6.7881° E 79.8913° N	23.34	166,857	164

A questionnaire survey was conducted to collect information regarding waste collection and recycling in KMC. Sampling was done by "snowball sampling method" (Biernacki and Waldorf 1981) due to the lack of information on plastic waste recyclers and collectors within KMC area. In AMC, only a composition analysis of dumpsite waste was conducted to calculate plastic waste flow to the dumpsite as the other necessary data was available with AMC.

3 Results and Discussion

According to the collected data, the following major processes were identified in the post-consumer plastic flow in considered local authorities.

- **Plastic consumption and waste generation**: The general inflow to the process is commercial plastic products, and the outflow of the process is PCPW once discharge to waste stream.

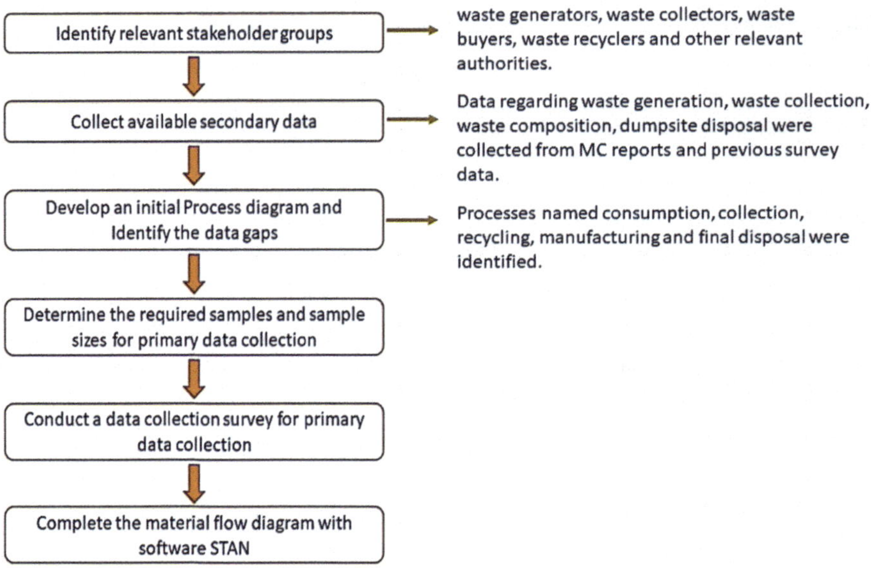

Fig. 2 Methodology used to conduct the material flow analysis

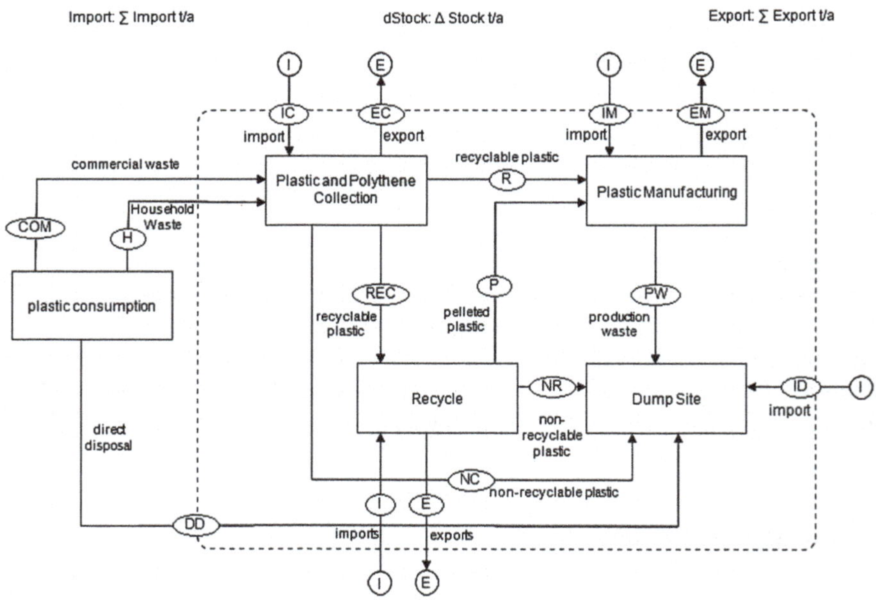

Fig. 3 Developed initial material flow diagram

- **Collection**: Plastic wastes are collected and transported either to recycling factory or to disposal site. In this process, some collected wastes are transported beyond the physical boundary of PS or LA (exports).
- **Final disposal**: Most of the collected municipal wastes in Sri Lanka are disposed at open dumpsites, so plastic wastes are aggregating permanently in dumpsites if not recover through recycling or energy generation.
- **Recycling (Processing/Crushing)**: Collected plastic wastes are processed in these recycling facilities by crushing into pellets. Wastes collected from outside areas are also used in the recycling facilities (imports).
- **Manufacturing**: The plastic products are generated from plastic industries by using virgin and recycled plastic pellets (masterbatch) as raw materials.
- **On-site Disposal**: The non-collected wastes were disposed at the waste generation sources. This was commonly happened with households where there is no municipal waste collection mechanism in action. Open burning and burying of wastes are the common practices used.

3.1 *Material Flow Diagrams*

As shown in Fig. 4, the total annual plastic waste generation from the consumption process in KMC is nearly 4000 tonnes and 60% (2400 tonnes) from the generated waste is disposed directly to the "Gohagoda dumpsite" in Kandy. A portion of PCPW of about 6.25% is not collected, and this portion includes household accumulation, burning and trans-boundary flow. Plastic waste collectors in Kandy area collect approximately 1300 tonnes/annum of plastic for recycling, and it is equal to 33% of

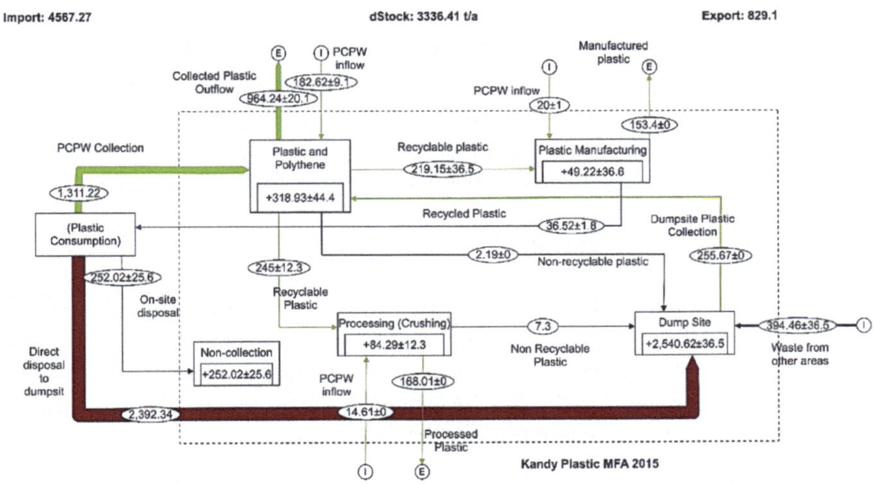

Fig. 4 Material flow diagram for the Kandy Municipal Council

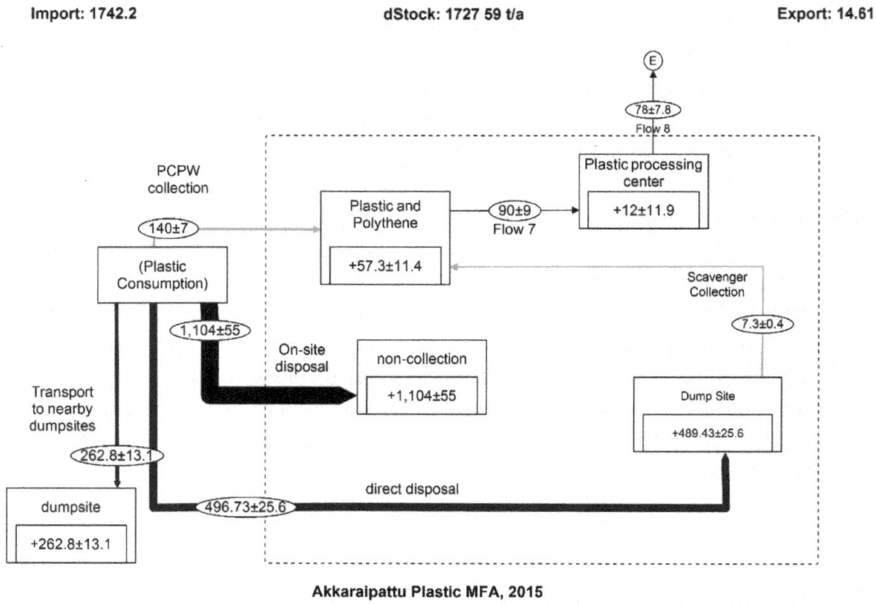

Fig. 5 Material flow diagram for the Akkaraipattu Municipal Council

total plastic waste generation. Moreover, the major proportion of the plastic waste is still disposed to the dumpsite with a stock generation of about 2500 tonnes/annum. A scavenger collection of 255 tonnes/annum is the one and only PCPW recovering method from the dumpsite. Thus, there is a potential to increase the recycling rate in KMC.

The total annual plastic waste generation from the consumption process is nearly 2,000 tonnes (Fig. 5) in AMC; however, only 25% (500 tonnes/annum) of generated plastic waste is disposed to the dumpsite. Another 61% of plastic waste is not collected and disposed on-site by generators. Plastic waste collection for recycling in Akkaraipattu area is only 0.9% which equals to 18 tonnes/annum. Therefore, the plastic recycling rate in Akkaraipattu Municipal council area is at a very low rate compared to Kandy Municipal Council, primarily accounted to inefficient waste collection system.

As shown in Fig. 6, out of the total 19,600 tonnes/annum post-consumer plastic waste generation within the seven local authorities in Western Province, 15,500 tonnes/annum is transported to the Karadiyana dump site and the rest 3,720 tonnes/annum disposes illegally or disposes on-site. Moreover, 1,800 tonnes/annum of plastics waste are collected in recyclable collection centres, and 1,300 tonnes/annum will be recycled for manufacturing of goods which account an overall plastic recycling rate of 6.6% in Western Province.

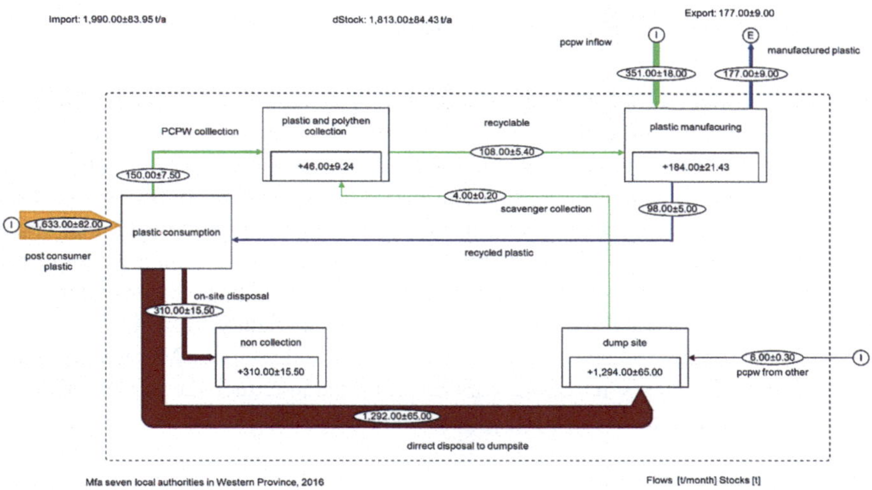

Fig. 6 Material flow diagram for seven local authorities in Western Province

3.2 Plastic Waste Classification Based on Physical Appearance

PCPW can be categorized basically to two categories as soft and hard plastics based on their physical appearance. Rigid plastic items were classified into hard plastics, and film plastic items were classified as soft plastics. The summary of the plastic classification for the three local authorities is illustrated in Fig. 7.

As shown in Fig. 7, plastic waste composition shows similar behaviour in both KMC and WP where both local authorities are economically important. Consumption of film plastics in packaging and rapping purposes is high in commercial areas, and the use of rigid plastic content is high in AMC due to insufficient plastic waste recovery for recycling.

Fig. 7 Summary of the PCPW classification

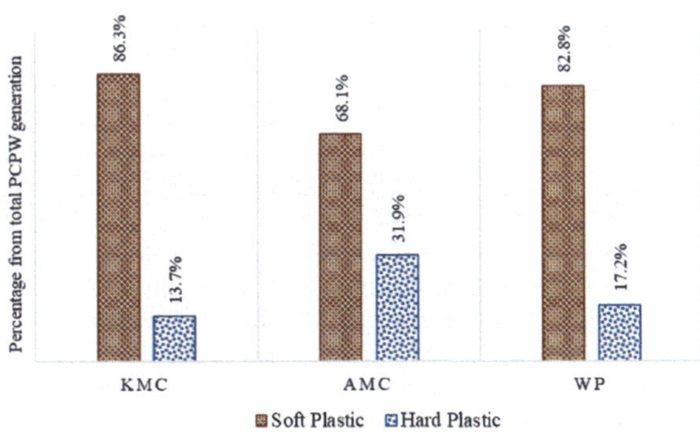

4 Conclusion and Recommendation

The study showed that the material flow analysis is a vital tool that can be successfully used for the quantification of PCPW flow in a local authority or province. The STAN® software can be effectively used for material flow analysis with adequate data on material flows. The study revealed that major post-consumer plastic waste flow paths for Sri Lanka has consisted of consumption, collection, processing, manufacturing and dumpsite disposal. Admit many recycling promotion attempts, dumpsite disposal still remains the major final destination for PCPW. The post-consumer plastic recycling rates of the three study areas were 33% in Kandy Municipal Council, 6.6% in Western Province and nearly zero recycling in Akkaraipattu Municipal Council. In perspective, plastic recycling shall be promoted to reduce the burden on open dumpsites in Sri Lanka.

Acknowledgements All procedures performed in studies involving human participants were in accordance with the ethical standards of the institutional research committee, and informed consent was obtained from all individual participants in the study. Authors appreciate Kandy and Akkaraipattu municipalities and also Waste Management Authority of Western Province for generous support towards the successful completion of the study.

References

Ayres, R. U., & Ayres, L. W. (Eds.). (2002). *A handbook of industrial ecology*. Cornwell: Edward Elgar Publishing. ISBN 1-84064-506-7.

Bandara, N. J. G. J., & Hettiarachchi, J. P. A. (2010). Environmental impacts with waste disposal practices in a suburban municipality in Sri Lanka. *International Journal of Environment and Waste Management, 6*(1–2), 107–116.

Biernacki, P., & Waldorf, D. (1981). Snowball sampling: Problems and techniques of chain referral sampling. *Sociological Methods and Research, 10*(2), 141–163.

Binder, R. C., Van Der Voet, E., & Rosselot, K. S. (2009). Implementing the results of material flow analysis: Progress and challenges. *Journal of Industrial Ecology, 13*(5), 643–649.

Bureecam, C., Chaisomphob, T., & Sungsomboon, P. Y. (2018). Material flows analysis of plastic in Thailand. *Thermal Science, 22*(6A), 2379–2388.

Cencic, O., & Rechberger, H. (2008). Material flow analysis with software STAN. *Environmental Management, 18*(1), 3–7.

Chang, I. C., Hsiao, T. Y., Yu, Y. H., & Ma, H. W. (2007). Identification of pollution source of cadmium in soil. Application of material flow analysis and a case study in Taiwan. *Environmental science and Science and Pollution Research International, 14*(1), 49–59.

Dissanayake, N. D. (2013). *Master plan for Kandy Municipal Council, Solid Waste Management Division, KMC (2014–2024)*. Sri Lanka: Kandy Municipal Council.

Geyer, R., Davis, J., Ley, J., He, J., Clift, R., & Kwan, A., et al. (2007). Time-dependent material flow analysis of iron and steel in the UK: Part 1: Production and consumption trends 1970–2000. *Resources, Conservation and Recycling, 51*(1), 101–117.

Graedel, T. E., & Allenby, B. R. (2003). Industrial ecology (vol. 7632, pp. 83–187). Prentice Hall: Englewood Cliffs, New Jursey (1995).

JICA. (2003). *Solid waste management plan for Kandy—Study on the solid waste management for secondary cities in Sri Lanka*. (Final report) Colombo, Sri Lanka.

Kuczenski, B., & Geyer, R. (2010). Material flow analysis of polyethylene terephthalate in the US, 1996–2007. *Resources, Conservation and Recycling, 54*(12), 1161–1169.

Kuruppuge, R. H., & Karunarathna, A. K. (2014). *Issues in Management of Municipal Solid Waste: Institutional Capacity of Local Authorities in Sri Lanka*. https://doi.org/10.13140/2.1.4102.7841.

Liyanage, B. C., Gurusinghe, R., Herath, S., & Tateda, M. (2015). Case study: Finding better solutions for municipal solid waste management in a semi local authority in Sri Lanka. *Open Journal of Civil Engineering, 5*(1), 63–73.

Majumder, S. C. (2012). Urban solid waste management: A study on Comilla City Corporation. *Journal of Economics and Sustainable Development, 3*(6), 53–61.

Muthaa, N. H., Patel, M., & Premnath, V. (2006). Plastics materials flow analysis for India. *Resources Conservation and Recycling, 47*(3), 222–244.

Schmidt, M. (2008). The Sankey diagram in energy and material flow management Part 1: History. *Journal of Industrial Ecology, 12*(1), 82–94.

Tchobanoglous, G., & Kreith, F. (2002). *Handbook of solid waste management* (2nd ed.). USA: The McGraw-Hill Companies. ISBN 978-0-07-150034-0.

Vidanaarachchi, C. K., Yuen, S. T. S., & Pilapitiya, S. (2006). Municipal solid waste in the Southern Province of Sri Lanka, issues and challenges. *Waste Management, 26*(8), 920–930.

Youth Engagement for E-waste Management by Urban Local Bodies in India

Sai Dinesh Shetty

Abstract The National Youth Policy, 2014 (NYP-2014) aims to empower the youth of the country to achieve their full potential, and through them enable India to find its rightful place in the community of nations. As per the NYP 2014, there is a trend of increasing urbanization and anonymity that characterizes urban life and hence it is all the more important for the government to act as an intermediary and create channels and processes by which young Indians can engage with urban decision makers and contribute to urban governance. To achieve the objectives of the NYP - 2014, one of the strategies is Youth engagement. Youth engagement is the result when young people are involved in responsible, challenging actions to create positive social change. Urban governance, in general, has many dimensions and waste management is one of the major problems in urban areas. In particular, E-waste management is emerging as a major problem in India. In this context, the Urban Local Bodies (ULBs) have a crucial role in E-waste management. This paper is an attempt to connect the objectives of Youth engagement of the National Youth Policy, 2014 for addressing the issue of E-wastes in India. To achieve this, the paper proposes a model for youth engagement in E-waste management for urban local bodies in India. The main focus of this model is to channelize E-wastes from the informal sector to registered E-waste Recycler/Dismantler. To aid this channelization, organizations like NGOs, Educational Institutes can act as an intermediary. For the working of these organizations in a fruitful manner, the model aims at working towards creating a systematic channel for engagement between the government and young citizens.

Keywords Youth engagement · E-waste management · Participation · Urban governance · National youth policy, 2014

S. D. Shetty (✉)
Department of Public Administration and Policy Studies, Central University of Kerala, Kasargod, Kerala, India
e-mail: saidshetty007@gmail.com

S. K. Ghosh (ed.), *Urban Mining and Sustainable Waste Management*,
https://doi.org/10.1007/978-981-15-0532-4_23

1 Introduction

India lies on the cusp of a demographic transition; however, in order to capture this demographic dividend, it is essential that India's economy has the ability to support the increase in the labour force which can be achieved by ensuring that India's youth has the appropriate education, skills, health awareness along with other enablers to productively contribute to the nation's economy (National Youth Policy 2014, p. 3).

The (National Youth Policy 2014) describes youth in the age group of 15–29 years, and they comprise 27.5% of the country's population and about 34% of India's Gross National Income (GNI) (p. 3). It also states that "there exists a huge potential to increase the contribution of this class of the nation's citizenry by increasing their labour force participation and their productivity" (p. 3). The Policy aims, "to empower the youth of the country to achieve their full potential and through them enable India to find its rightful place in the community of nations" (p. 5).

The importance of youth is that this age group of 15–29 years comprise 27.5% of the population in a situation where India is expected to become the 4th largest economy by 2025, contributing about 5.5–6% to the world GDP (p. 10). The Policy mentions that

> While most of these countries face the risk of an ageing workforce, India is expected to have a very favourable demographic profile. The working population of India is expected to increase to 592 million by 2020, next only to China (776 million), pointing to the fact that youth will make a significant contribution to the economic development of the country. This "demographic dividend" offers a great opportunity to India. (p 10)

To achieve the objectives of the NYP-2014, one of the strategies suggested is Youth engagement. Youth engagement is the result when young people are involved in responsible, challenging actions to create positive social change (What is Youth Engagement, Really? 2018).

The NYP-2014 described the trend in urbanisation process and its links with urban governance as follows:

> There is a trend of increasing urbanisation and anonymity that characterises urban life, and hence, it is more important for the government to act as an intermediary and create channels and processes by which young Indians can engage with urban decision-makers and contribute to urban governance. (p. 61)

Accordingly, one of the priorities of NYP-2014 is Participation in politics which includes "Engage youth outside of the political system, Create governance mechanisms that youth can leverage, Promote youth engagement in urban governance and governance and Youth Engagement" (p. 6) and also has another priority as youth engagement which includes "Measure and monitor effectiveness of youth development schemes, Create a platform for engagement with youth" (p. 6).

In India, urban governance, in general, has many dimensions among which E-waste management has emerged as a major problem in urban areas, and hence, the urban local bodies (ULBs) have a crucial role in E-waste management.

The (E-waste (Management) Rules 2016) have made the urban local bodies (Municipal Committee or Council or Corporation) as one of the major authorities and have laid down the following corresponding duties:

(i) To ensure that E-waste if found to be mixed with municipal solid waste is properly segregated, collected and is channelised to authorised dismantler or recycler.

(ii) To ensure that E-waste pertaining to orphan products is collected and channelised to authorised dismantler or recycler. (pp. 23–24)

This paper is an attempt to connect the objectives of youth engagement of the NYP-2014 for addressing the issue of E-wastes in India. To achieve this, the paper proposes a model for youth engagement in E-waste management for ULBs in India. The purpose of this research paper is to find innovative solutions to implement the NYP-2014 and E-waste Management Rules, 2016 (Rules 2016). The methodology used is to collect primary data from government publications and reports and create a new innovative model by connecting and linking various policy and legal provisions in Rules, 2016 and the NYP-2014.

2 Proposed Model to Youth Engagement in E-waste Management for Urban Local Bodies in India

2.1 Stakeholders Involved in This Model as Follows

1. Urban Local Body (ULB)
2. Youth Connect Platform
3. Collection centres
4. Registered Recycler/Dismantler
5. Youth partnered NGOs and educational institutions.

2.2 Roles and Responsibilities of Various Stakeholders

The various roles and responsibilities as per the existing law on E-waste Management and National Youth Policy 2014 for each stakeholder are as follows:

1. **Urban Local Body (ULB)**

The legal role of ULBs is to ensure E-waste is segregated, collected and is channelised to authorised dismantler or recycler which is registered with the Central Pollution Control Board (CPCB). The (National Youth Policy 2014) tells that the government should "act as an intermediary and create channels and processes by which young Indians can engage with urban decision-makers and contribute to urban governance" (p. 61).

2. **Youth Connect Platform (YCP)**

The Policy has created YCP as a platform which is given the responsibility for "engaging with the youth and by ensuring youth develop leadership and other interpersonal skills, the Government of India will help create a generation of individuals that are committed to civic, social and political progress" (p. 62). The policy also suggests that "the government should work towards leveraging the vast number of stakeholders that are already working to support youth development and participation, and expanding its own reach and access to the youth through the networks of these organisations" (p. 85).

3. **Registered Recycler/Dismantler**

The (E-waste (Management) Rules 2016) clearly state that the Registered Recycler/Dismantler should "ensure that the facility and recycling processes are in accordance with the standards or guidelines prescribed by the Central Pollution Control Board from time to time" (p. 9).

4. **Collection centres**

The E-waste Rules, 2016 also make it mandatory that the Collection centres must "collect E-waste on behalf of producer or dismantler or recycler or refurbisher including those arising from orphaned products" (p. 7).

5. **Youth partnered NGOs and educational institutions**

The (National Youth Policy 2014) emphasises on the need to "institutionalise community engagement and to design and streamline schemes such that they cater to the non-homogenous youth population" (p. 55). The Policy has given the observation that "There are some unstructured interactions between policy makers and young Indians in forums such as educational institutions. However, there are no systematic channels for engagement between the government and young citizens and no mechanisms for youth to provide inputs to government" (p. 63). Thus, for its effective implementation, the NYP-2014 recommends that "The government can leverage partner organisations like NYKS and NSS, NGOs and educational institutions to connect with the youth and use technology and social media in order to achieve the objective of creating a platform for engagement with youth" (p. 64). For practical purposes, it suggests that "the government can identify representative educational institutions, youth groups and other partners to create a channel to engage with the youth" (p. 64).

2.3 Functioning of Various Stakeholders in the Model

The main focus of this model is to channelise E-wastes from informal sector to registered E-waste Recycler/Dismantler. To aid this channelisation, organisations like

NGOs, Educational Institutes can act as intermediary. For working of these organisations in fruitful manner, the model aims at working towards creating a systematic channel for engagement between the government and young citizens. This channel will ensure to implement the legal mandate of ULBs of ensuring that E-wastes are segregated, collected and are channelised to authorised dismantler or recycler. This will also aid in better implementation of E-waste Management Rules, 2016 where there is legal mandate of Registered Recycler/Dismantler to ensure that the facility and recycling processes are in accordance with the standards or guidelines prescribed by the Central Pollution Control Board from time to time. In this process, the legal responsibility of collection centres to collect E-waste on behalf of producer or dismantler or recycler or refurbisher including those arising from orphaned products is fulfilled by being key points of interaction between recyclers/dismantlers and Youth Organisations working for proper environmentally sound management of E-wastes.

2.4 Working of the Model

Urban Local Body (ULB): It will set up a formal way to get NGOs and various youth organisations to aid and assist the ULB in fulfilling its mandate of E-waste management in the city. Getting formal registration or some kind of association with the ULB will render authenticity and recognition to the NGO/Educational Institution which will participate as an intermediary.

Youth Connect Platform (YCP): This platform is simply the basis on which all stakeholders come together. This can be in the form of a special initiative from the part of the ULB or the NGO/CSO participating together. This is not a physical entity but a virtual entity made on paper for fulfilling the requirements of all stakeholders coming together on this platform. This platform will act as foundation for the working of various stakeholders connected with the model.

Youth partnered NGOs and educational institutions: These agencies will work in co-ordination with ULB and set up E-waste Collection centres in different parts of the city. They will serve interior locations of the city for better penetration of collection services. They are the ones who will also involve in promoting the YCP set up especially for the purpose of effective E-waste management in the city.

Collection centres: These will be run by youth partnered NGOs, Educational Institutes and volunteering agencies. They will collect E-waste from people directly, from companies and producers, from housing societies, schools, colleges, etc. These are actual point of contact between public at large and the YCP created for working of this model. These collection centres should be spread at length and breadth of the city with maximum penetration to maximum people for effective collection of E-wastes. This helps to sustain the model as collection of E-wastes is starting point where later processing and actual recycling will follow.

Registered Recycler/Dismantler: It will collect the bulk E-wastes from collection centres and transport it to place of recycling and dismantling. Since E-waste also has monetary value and important components, the recycler pays the collection centres by weight of E-waste collected. All these are to be documented in relevant forms in the E-waste Management Rules, 2016. This is the last stage of the model where the mandate of the Rules, 2016 will be implemented effectively where large amount of E-waste which was hitherto collected by informal agents will now go to these formal recyclers route. This is the desired outcome in the form of channelisation of E-wastes is done in proper manner.

3 Ensuring Smooth Working of the Model and Its Sustainability in Long Run for Effective E-waste Management in the City

The main incentive to sustain this model is the value contained in E-wastes. Valuable materials and spare parts from E-wastes collected have value in its own and cannot be categorised as something to discard completely. The model will sustain as long as Recyclers and Dismantlers are getting enough quantity of E-wastes for their business sustainability. People in general and IT growth and high usage of consumer electronics and electrical goods will be providing more E-wastes as the recycling capacity of all recyclers in India is below 8% (Business Standard 2018, para 7).

ULBs will require assistance of a body whose responsibility is to set up Collection Centres for E-waste management in the city. This model proposes a reputed, registered NGO or Civil Society Organisation (CSO) having experience in environmental and waste management issues in city as key agencies which will work in association with the ULB. Youth is focussed to run and manage the working of NGOs/CSOs which can make good use of latent energy and time of young students and volunteers. Since ULBs are engaging Youth in waste management and including them as a part of Civic Engagement, in many ways the mandate of National Youth Policy, 2014 and E-waste Management Rules, 2016 are implemented in better manner.

For success of such initiatives in Indian context, strong political will of the elected municipal representatives and strong support of municipal officials and bureaucrats can make this model a grand success. This model can be easily modified to suit local requirements and resources availability. For pilot implementation, large metro cities can use this model with modifications to suit their local requirements and resources by which they can start managing E-wastes in their cities. This model provides a blueprint of how civic engagement can work in a better manner by associating partner organisations from private sector and engaging citizens in active urban governance. This engagement will not only enhance sense of civic responsibility among people but will work for improving the relationship between government and people. This model can also act as positive confidence-building measure where citizen co-operation can be garnered for similar other work or initiative for good urban governance.

This paper concludes by giving recommendation to ULBs in India to implement this workable model. It has been made in Indian context looking into local resources and local constraints in the various cities of India. This model is flexible and can be modified to suit local needs for its effective implementation. The model itself is the part of recommendations to be presented on behalf of author to the various governmental agencies and also to NGOs/CSOs for giving their practical wisdom on working of this model. This working model can also be improvised and given further refining by implementing agencies and through the various stakeholders involved in it. The effective co-operation, inter-agency co-ordination and strong commitment for achievement of desired goals will help to manage E-wastes in India efficiently and effectively.

Note This paper is part of the Ph.D. Programme and Thesis undertaken by the author. The author is an independent Ph.D. Scholar.

Acknowledgements Dr. Ashalekshmi B. S., Assistant Professor, Department of Public Administration and Policy Studies, Central University of Kerala.
Shri. Vikas Patil, Chairman of the Environment Conservation Association (ECA), Pune.
Dr. Rajesh Manerikar, CEO, Poornam Ecovision Foundation, Pune.

References

Business Standard. (2018). *India among the top five countries in e-waste generation: ASSOCHAM-NEC study*. Retrieved from https://www.business-standard.com/article/news-cm/india-among-the-top-five-countries-in-e-waste-generation-assocham-nec-study-118060500207_1.html.

Ministry of Environment, Forest and Climate Change, Government of India. E-waste (Management) Rules, 2016. (2016). New Delhi.

National Youth Policy. (2014). Retrieved from https://yas.nic.in/sites/default/files/National-Youth-Policy-Document.pdf.

Salient Features of E-waste (Management) Rules, 2016 and its likely implication. (2016). Retrieved from http://cpcb.nic.in/cpcbold/E_waste_rules_2016.PDF.

What is Youth Engagement, Really?. (2018). Retrieved from http://actforyouth.net/youth_development/engagement/.

Improvement in Engineering Behaviour of Expansive Soil Reinforced with Randomly Distributed Waste Plastic Strips

Rachita Panda and Sudhira Rath

Abstract With rapid urbanisation in developing countries like India, there is an increase in the generation of plastic wastes. Disposal of these plastic wastes creates not only environmental issues but also consumes a large area of lands as landfills. Furthermore, due to no availability of sufficient recycling facility in India, most of the waste plastic could not be reused. Utilisation of these plastic wastes for construction purposes can not only solve the waste disposal problem but also create a sustainable construction methodology. The performance of pavement construction over expansive soil subgrade is greatly affected by moisture content in the soil. During monsoon/rainy season, expansive soil properties change drastically due to increase in moisture content. Due to migration of moisture content into the expansive soil, the volume of soil increases but strength reduces significantly. This variation in volume and strength may lead to catastrophic failure of the pavement. Therefore, it has become very much vital to adopt innovative construction approach to counteract the damaging effect of expansive soil subgrade. In this study, a new approach was attempted to study the effectiveness of polyethylene terephthalate (PETE) strips obtained from waste plastic glass as a reinforcing agent to improve the engineering properties of expansive soil. California Bearing Ratio (CBR) tests were performed to study the effect of plastic strip content and size of the plastic strip on swelling and strength behaviour of plastic reinforced expansive soil samples in series. The plastic strip length (10–30 mm) and quantity (0–4% by weight of soil) were varied to study the effect of the plastic reinforcement on the properties of the expansive soil. After the completion of series of experiments, it was observed that there was a significant improvement in the strength and swelling characteristics of the soil with polymer reinforcement to the expansive soil.

Keywords Plastic waste · PETE strips · CBR · Expansive soil

R. Panda (✉)
K.I.I.T, Bhubaneswar, India
e-mail: rachitapanda184@gmail.com

S. Rath
V.S.S.U.T, Burla, India
e-mail: sudhira.rath1@gmail.com

S. K. Ghosh (ed.), *Urban Mining and Sustainable Waste Management*,
https://doi.org/10.1007/978-981-15-0532-4_24

1 Introduction

With rapid population growth and urbanisation in developing countries like India in the last decade, the per capita production of plastic has increased by 890% (Banerjee and Srivastava 2012). The per capita consumption of plastic in India is going to be 30 kg by the year 2022 (Bhattacharya et al. 2018). Presently in India, only 60% of plastic waste generated is being recycled (Mathiyazhagan et al. 2016). The remaining wastes not only take huge land resource for disposal but also create various environmental hazards. Bulk utilisation of these wastes can be achieved by its civil engineering applications (Choudhary et al. 2010; Jha et al. 2014; Siddique et al. 2008).

Furthermore, in developing countries like India, there is a shortage of good quality soil to use as subgrade and sub-base materials for highway pavement construction due to unavailability of land resources. Expansive soil is one of the problematic soils, which shows swell–shrink behaviour with change in moisture content and has very less strength at higher water content. Due to these characteristics, it is difficult to build highway pavements over it. Expansive soil covers 20% land area of India (Ranjan and Rao 2000). Hence, in developing countries like India, alternate arrangements needed to be made to use these expansive soils by modifying and stabilising its engineering properties. In recent past, various researches have shown that adding various industrial wastes like fly ash (Sabat and Pradhan 2014), WRP (Choudhary et al. 2017), red mud (Kalkan 2006) to expansive soil will not only stabilize the soil but also provide a cost-effective approach to utilise the stabilised soil as a construction material. Likewise, various researchers have used LDPE (Rao and Dutta 2004), HDPE (Choudhary et al. 2010), PET (Botero et al. 2015) plastic strips as reinforcing agents in soil for improvement in its engineering behaviour.

In this present approach, the feasibility of utilising PETE plastic strips obtained from waste plastic glass to reinforce expansive soil is evaluated from the experimental study. A series of CBR tests were carried out both on untreated and treated soils. The effect of strip content and strip aspect ratio on CBR characteristics and expansion ratio has been studied by varying the strip content and strip aspect ratio in reinforced sample. The test results show that increase in strength and decrease in swelling can be obtained by PETE plastic strip reinforcement in expansive soil.

2 Experimental Study

2.1 Materials Used

2.1.1 Soil

The expansive black cotton soil was collected from Attabira, Bargarh District of Odisha. To avoid contamination by organic matters, the soil sample was taken from

Table 1 Properties of expansive soil

Property	Value
Specific gravity	2.4
Differential free swell (%)	60
Liquid limit (%)	69
Plastic limit (%)	28
Plasticity index (%)	41
Classification (as per IS:1498-1970)	CH (high plastic clay)
Optimum moisture content (%)	16
Maximum dry density (kN/m^3)	17.62
CBR (%)	1.46

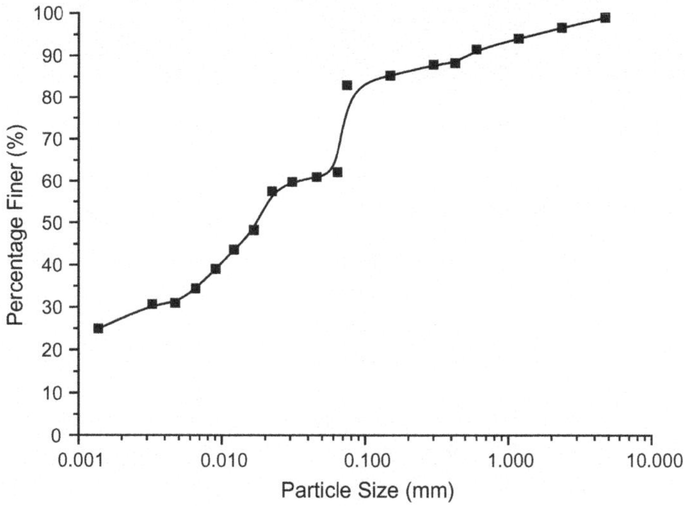

Fig. 1 Grain size distribution of soil

50 cm below the ground level. The various engineering properties of the soil are given in Table 1. The grain size distribution curve of the soil is presented in Fig. 1.

2.1.2 Waste Plastic

Waste plastic taken for the study was collected from used plastic glass disposed of in the nearby disposal sites of Sambalpur and cut into strips of length 10 mm, 20 mm and 30 mm, while keeping the constant strip width of 10 mm, thus making strips of aspect ratio (AR) of 1, 2 and 3, respectively. The strips should be cut in such a manner that the mould diameter will remain at least four times higher so that there will be sufficient room for the strips to deform freely and independently inside the

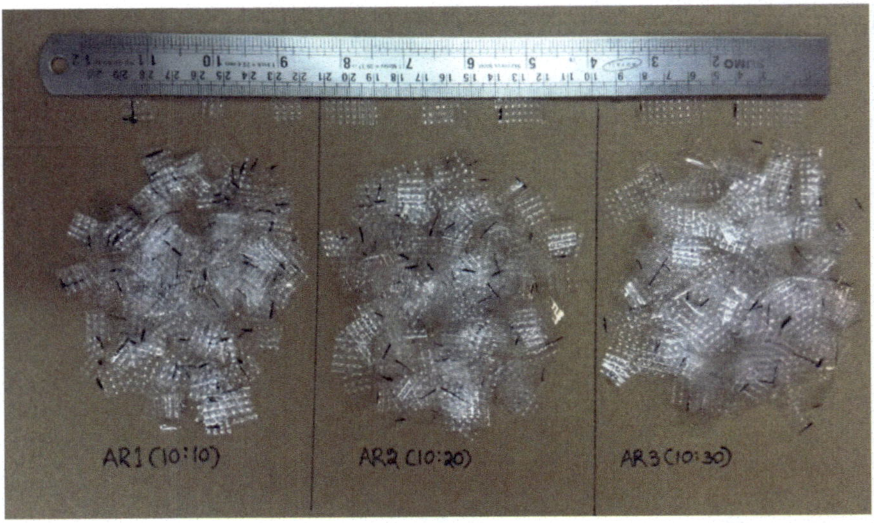

Fig. 2 Waste plastic strips taken for study

Table 2 Properties of waste plastic strips

Property	Value
Thickness (mm)	0.5
Tensile strength (kN)	0.35
Failure strain (%)	25

mould (Choudhary et al. 2010). Figure 2 shows the picture of sample plastic strips of different aspect ratio taken for present study. The basic properties of plastic strips are tabulated in Table 2.

2.1.3 Reinforced Soil Sample Preparation and Testing

PETE plastic reinforced soil samples were prepared at the maximum dry density and optimum moisture content of unreinforced soil. The plastic strip content in the reinforced soil samples was taken as percentage weight of dry soil (Ranjan et al. 1994; Jha et al. 2014). In the present study, reinforced soil sample having strip content of 0.25, 0.50, 1.00, 2.00 and 4.00% were studied for aspect ratio of 1, 2 and 3. The waste plastic strips to be added to the soil were considered a part of the solid fraction in the void solid matrix of the soil. For preparation of plastic strip reinforced soil, the plastic strips were mixed thoroughly to the dry pulverised soil by hand to get a uniform mixture.

In the present study, a series of laboratory CBR tests were conducted on unreinforced as well as reinforced soil specimens.

For reinforced samples, the plastic strips were mixed thoroughly to the oven-dried pulverised soil sample. Required amount of water corresponding to OMC was added and mixed properly to get a homogeneous mixture. The mix was kept in the desiccator for 3 h to achieve a homogeneous moisture regime. The mixture was then transferred to the CBR mould. Now, the mix is compacted by dynamic compaction so as to achieve the maximum dry density. The CBR mould-soil system was submerged in water for 96 h. A surcharge weight of 25 N was placed over it, and a dial gauge was attached to measure the expansion due to moisture migration into the soil.

The CBR tests were performed in accordance with IS 2720-Part XVI (Standard 1987) for both unreinforced and reinforced soil samples at varying strip content and aspect ratio. The load was applied on uniform rate of 1.25 mm/min through the loading plunger. The loads were recorded up to the penetration of 12.5 mm. The load–penetration curves were plotted for each case. In the present investigation, no corrections to load–penetration curves are needed. Since in the present investigation, CBR value corresponding to 2.50 mm was greater than that of 5.00 mm; hence, the CBR value corresponding to 2.50 was reported as CBR value.

3 Results and Discussion

CBR test is a penetration resistance test for soil, and it is the most widely used test for evaluating the suitability of subgrade, sub-base and base materials for pavement construction and design. The load–penetration curves of the composite samples prepared by mixing plastic strips of aspect ratios 1, 2 and 3 with different strip contents are shown in Figs. 4, 5 and 6. From Figs. 3, 4 and 5, it can be observed that for aspect ratio 1, 2 and 3, respectively, with increase in plastic strip content, the penetration resistance increases for same penetration depth. The penetration resistance of the reinforced soil for any strip content increases with increase in the aspect ratio.

In the present case as the CBR value corresponding to 2.5 mm penetration is greater than 5.0 mm penetration in all cases, the CBR value corresponding to 2.5 mm is considered for the study. The effect of aspect ratio and strip content on CBR value is shown in Fig. 6. The unreinforced soil has a CBR value of 1.46%, and the greatest increase (six times) in CBR value occurs at a strip content of 4% for aspect ratio 3.

The quantitative improvement in penetration resistance with reinforcement is expressed by two-dimensionless parameters, namely California Bearing Ratio Index (CBRI) and Peak Piston Load Resistance (PPLR) (Jha et al. 2014). CBRI is defined as the ratio of CBR value of treated soil to that of untreated soil, and PPLR is defined as the ratio of load resistance corresponding to 12.5 mm penetration of reinforced soil to that of unreinforced soil. Figure 7 shows the effect of strip content on both CBRI and PPLR values for strip aspect ratio of 3. It can be observed that there is a noticeable increase in both CBRI and PPLR values with increase in strip content, but improvement in CBRI value is much more than PPLR value with increase in strip content.

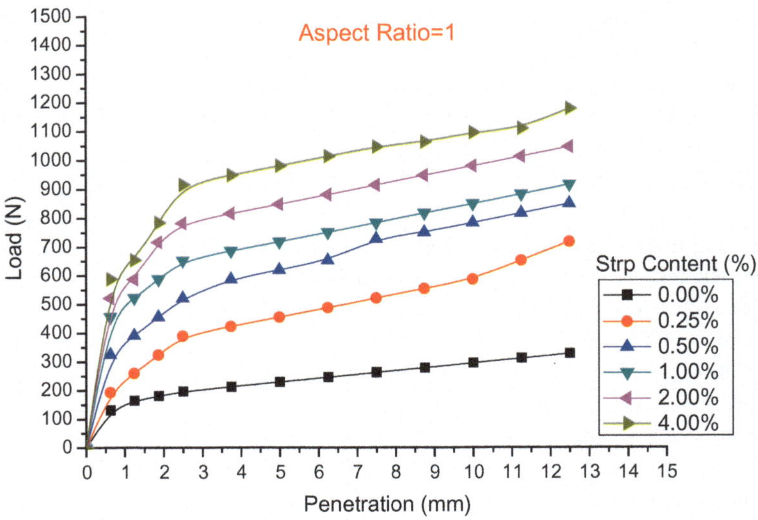

Fig. 3 Effect of plastic strip content on load–penetration curve at AR = 1

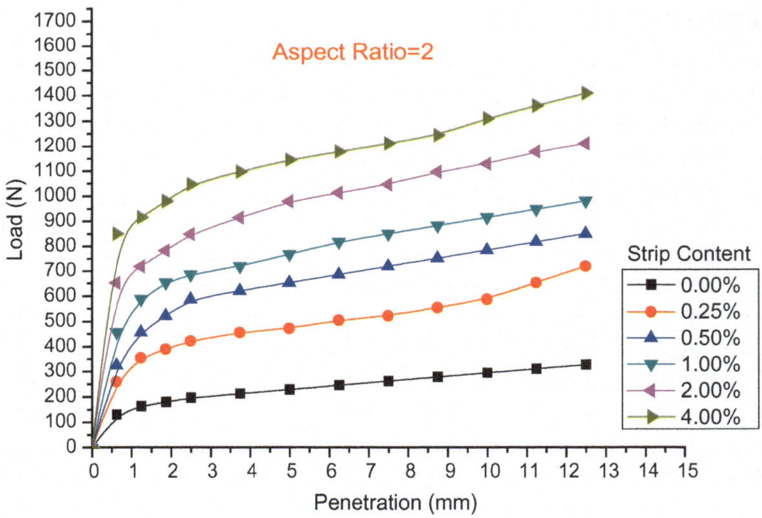

Fig. 4 Effect of plastic strip content on load–penetration curve at AR = 2

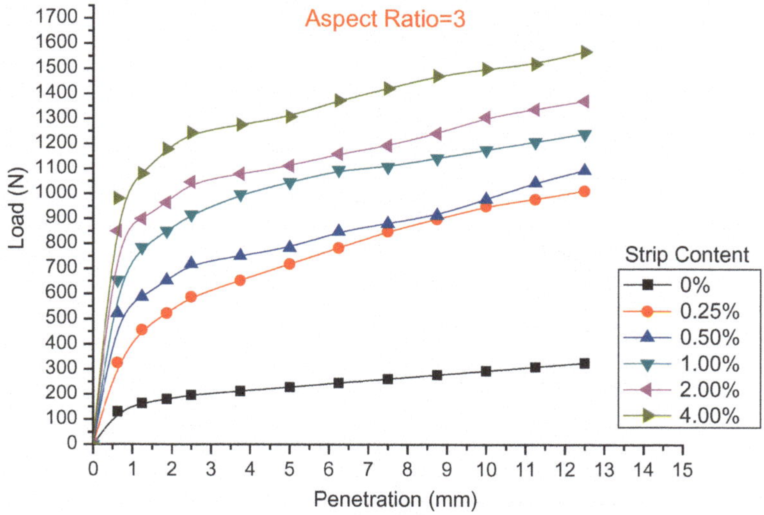

Fig. 5 Effect of plastic strip content on load–penetration curve at AR = 3

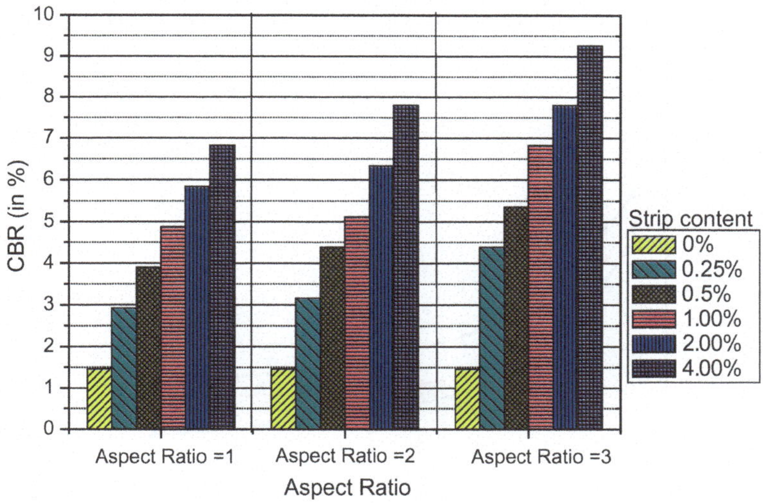

Fig. 6 Effect of plastic strip content and aspect ratio on CBR

Effect of aspect ratio and strip content on expansive ratio is presented in Fig. 8. It can be seen that, with increase in strip content, the expansion ratio decreases. Lowest expansion ratio (0.56%) is achieved at strip content of 4% for aspect ratio 3. This decrease in expansion ratio with increase in strip content is partially due to replacement of expansive soil by non-swelling plastic strips and due to generation of frictional resistance between soil and strip surface.

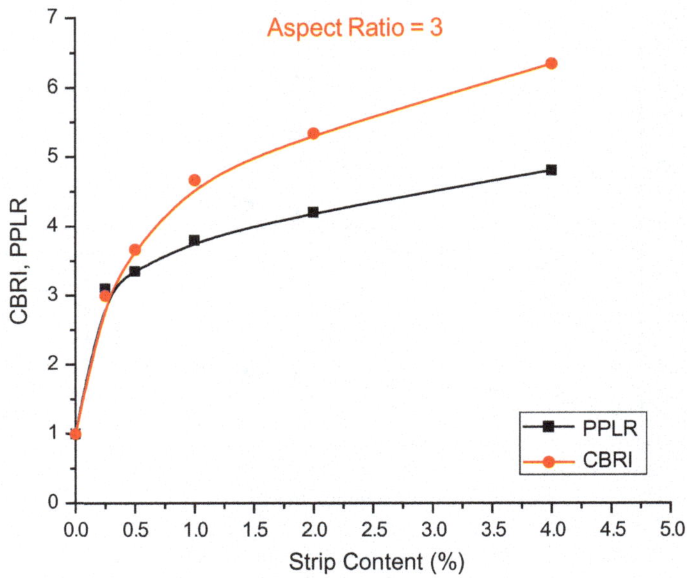

Fig. 7 Effect of plastic strip content on CBRI and PPLR for aspect ratio of 3

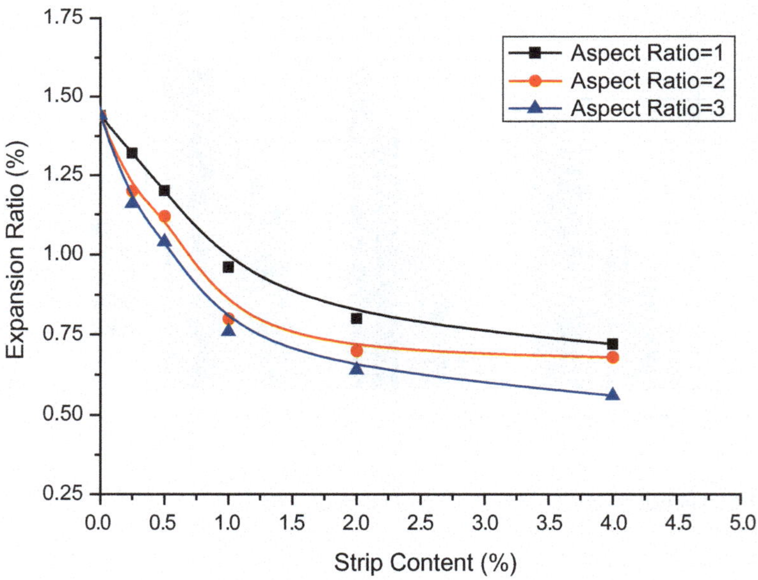

Fig. 8 Effect of plastic strip content and aspect ratio on expansion ratio

After completion of the test, the fibres were removed from the samples and studied. It is observed that the strips were stretched, but none of them were torn. This may be due to the shear deformation caused by loading which is far less to cause tensile failure of the strips (Falorca and Pinto 2011). When load is applied on the reinforced soil, the soil starts deforming, and due to the deformation in soil matrix, the load is transferred to the strips by active and passive resistances (Shukla 2017). The strips resist the loading by mobilising tensile stress. As the plastic strips have high tensile strength, it can resist large deformation without tearing. This results in the increase in penetration resistance and CBR value. The increase in penetration resistance is greater in the initial stage of penetration but decreases with increase in deformation.

4 Conclusion

In the present study, the feasibility of using reclaimed PETE plastic strips as a reinforcing agent in expansive soil was investigated by conducting CBR tests on both unreinforced soil and soil reinforced with varying percentage of strip content for different aspect ratios. The following conclusions are drawn from the experimental results.

1. With increase in plastic strip content and aspect ratio, the penetration resistance increases.
2. Addition of 4.00% PETE plastic strip having aspect ratio of 3 to expansive soil increases its CBR by six times.
3. The maximum improvement in PPLR value is obtained at strip content of 4% for aspect ratio 3.
4. The inclusion of plastic strips as a reinforcement in expansive soil resulted in decrease in expansion ratio, and lowest expansion ratio is achieved at strip content of 4% for aspect ratio 3.

The results of this current approach suggest that reclaimed PETE plastic strips cut from waste plastic glass can be utilised as soil reinforcement in expansive soils and can be used in highway subgrade and sub-base materials.

References

Banerjee, T., & Srivastava, R. K. (2012). Plastics waste management and resource recovery in India. *International Journal of Environment and Waste Management, 10*(1), 90–111.

Bhattacharya, R. R. N., Chandrasekhar, K., Roy, P., & Khan, A. (2018). *Challenges and opportunities: Plastic waste management in India.*

Botero, E., Ossa, A., Sherwell, G., & Ovando-Shelley, E. (2015). Stress–strain behavior of a silty soil reinforced with polyethylene terephthalate (PET). *Geotextiles and Geomembranes, 43*(4), 363–369.

Choudhary, A. K., Jha, J. N., & Gill, K. S. (2010). A study on CBR behavior of waste plastic strip reinforced soil. *Emirates Journal for Engineering Research, 15,* 51–57.

Choudhary, A. K., Sahoo, T. K., Jha, J. N., & Shukla, S. K. (2017, December). Swelling, shrinkage and compaction characteristics of expansive soil treated with waste recycled product. In *Indian Geotechnical Conference, GeoNEst.*

Falorca, I. M. C. F. G., & Pinto, M. I. M. (2011). Effect of short, randomly distributed polypropylene microfibres on shear strength behaviour of soils. *Geosynthetics International, 18*(1), 2–11.

Jha, J. N., Choudhary, A. K., Gill, K. S., & Shukla, S. K. (2014). Behaviour of plastic waste fibre-reinforced industrial wastes in pavement applications. *I*nternational Journal *of* Geotechnical Engineering, *8*(3), 277–286.

Kalkan, E. (2006). Utilization of red mud as a stabilization material for the preparation of clay liners. *Engineering Geology, 87*(3), 220–229.

Mathiyazhagan, K., Haq, A. N., & Baxi, V. (2016). Analysing the barriers for the adoption of green supply chain management-the Indian plastic industry perspective. *International Journal of Business Performance and Supply Chain Modelling, 8*(1), 46–65.

Ranjan, G., & Rao, A. S. R. (2000). *Basic and applied soil mechanics* (2nd ed.). New age International (P) Limited.

Ranjan, G., Vasan, R. M., & Charan, H. D. (1994). Behaviour of plastic-fibre-reinforced sand. *Geotextiles and Geomembranes, 13,* 555–565.

Rao, G. V., & Dutta, R. K. (2004). Ground improvement with plastic waste. In *Proceeding, 5th International Conference on Ground Improvement Technique* (pp. 321–328), Kaula Lumpur, Malaysia.

Sabat, A. K., & Pradhan, A. (2014). Fiber reinforced-fly ash stabilized expansive soil mixes as subgrade material in flexible pavement. *EJGE, 19,* 5757–5770.

Shukla, S. K. (2017). *Fundamentals of fibre-reinforced soil engineering*. Springer.

Siddique, R., Khatib, J., & Kaur, I. (2008). Use of recycled plastic in concrete: A review. *Waste Management, 28*(10), 1835–1852.

Standard, I. (1987). Laboratory determination of CBR. *IS 2720 (Part XVI).*

Willingness of Students and Academicians to Participate in E-waste Management Programmes—A Case Study of Bangalore

Khushbu K. Birawat, Biswajit Debnath, Shushmitha. L. Gowda and Sadhan Kumar Ghosh

Abstract The world is marching forward to embrace sustainability along with development, and the concept of circular economy has become relevant today more than ever. Electronic items have become an inseparable part of today's lifestyle. From mining of raw materials to manufacturing of the electronic items to its sales and then to its disposal, the burden on the nature of these electronics is humongous. By reusing and recycling, we can reduce that burden to some extent. But motivation precedes action, and hence, it is essential to understand the level of awareness, understanding and willingness among the common citizens regarding e-waste, its hazards and safe disposal. Previous studies have identified that lack of awareness has been a major issue which restricts the sustainable management of e-waste in India. The purpose of this study is to understand the awareness about e-waste management in the educated mass, as well as their willingness to participate in formal e-waste management programmes. A structured questionnaire was circulated in the academic circles (random sampling) in Bangalore. The data thus obtained was analysed using qualitative software. The results thus obtained can be used by various stakeholders to design and implement plans for heightened success of the e-waste management in these urban centres.

Keywords E-waste management · Circular economy · SPSS · Academicians · Willingness to participate · International Society of Waste Management, Air and Water · Sustainable development

---

K. K. Birawat (✉) · Shushmitha. L. Gowda
Department of Civil Engineering, Global Academy of Technology, Bangalore, India
e-mail: khushbu.kb@gat.ac.in

B. Debnath
Department of Chemical Engineering, Jadavpur University, Kolkata, India

S. K. Ghosh
Department of Mechanical Engineering, Jadavpur University, Kolkata, India

S. K. Ghosh (ed.), *Urban Mining and Sustainable Waste Management*,
https://doi.org/10.1007/978-981-15-0532-4_25

1 Introduction

The electronic industry has been one of the most prospering and profitable industries in a developing country like India (Mehta and Rajan 2017). For almost a decade now, there has been a revolution in the old industries as well as small scale budding startups in order to get this kind of electronic environment. Since the demand for 'e-goods' has increased in this span of time, the industries have also come up with more affordable products in order to reach all kinds of people. In this process, the companies started manufacturing products of lower quality and less durability. This was the way the companies managed to design electronic goods of lesser price. This led to the rapid growth of these industries, thereby increasing the rate of requirement, so did the increase in rate of manufacturing came to demand (Debnath et al. 2018). The electronic industry is expected to reach \$400 billion in 2022 from \$69.6 billion in 2012 (Corporate Catalyst India 2015). This growth will lead towards more piles of e-waste in the anthropogenosphere to handle. E-waste is the fastest growing toxic waste in the world (Basel Convention 2014). The world is marching forward to embrace sustainability along with development, and the concept of circular economy has become relevant today more than ever. Electronic items have become an inseparable part of today's lifestyle. From mining of raw materials to manufacturing of the electronic items to its sales and then to its disposal, the burden on the nature of these electronics is humongous. Global e-waste generation has increased from 41.8 million metric tonnes in 2014 to 44.7 million metric tonnes in 2016 (Balde et al. 2015; Balde et al. 2017). This is equivalent to 4500 Eiffel towers (Balde et al. 2017). In 2017, expected WEEE generation will exceed 46 million metric ton and 52.2 MMT by 2021. BRICS nations generated nearly 20% of the total WEEE generated in 2016. China generated 7.2 million metric ton of WEEE in the year 2016 which is both the highest in Asia as well as globally. India generated 2 million ton of WEEE in the year 2016 (Balde et al. 2017).

Reduce, Reuse, Recycle, Refurbish and Recover—these 5Rs are the best solution to deal with this problem of e-waste. Motivation precedes action, and hence, it is essential to understand the level of awareness, understanding and willingness among the common citizens regarding e-waste, its hazards and safe disposal. Limited studies have been found in this area, specifically in the Indian context. Bhatt et al. (2017) has studied the consumption pattern of mobile phones and general awareness on waste mobile disposal in the Delhi-NCR region. Borthakur and Govind (2017) have presented an overview of e-waste disposal behaviour and awareness in India. Dwivedy and Mittal (2013) established a path towards understanding the willingness of residents for e-waste disposal in India. Reported literature focuses on either specific electronic item or the country as a whole. Studies on city-specific awareness level are scant. The main objective of the present study is to evaluate the willingness of the educated mass to participate in holistic e-waste management and their awareness and attitude towards e-waste disposal. The management of e-waste is to be promoted at household level (at source of generation), and thus, this kind of data will provide input for the government bodies/NGOs to focus on the target group effectively. This data will especially serve as a basis for future studies in the study area—Bangalore and also pave way for comparison with other metropolitan cities.

2 Methodology

This study was divided into two sub groups: Literature review and survey involving academicians, students—participants are mainly linked to academics. Literature review helped in improving our understanding of framing the survey questions and analysing them.

To conduct the present study, a questionnaire was made by researchers to measure the willingness of respondents to participate in e-waste management programme. The designated institutional board's consent was taken to conduct this study with interested participants.

The data obtained was then analysed using the SPSS Statistical Software Version 20.

The respondents were asked 29 questions in 06 categories. Their demographic data, their awareness about e-waste, their attitude towards e-waste management, their willingness to participate in e-waste recycling programmes and what are the factors that drive the attitude and their understanding and inputs about what government and individuals can do about proper management of e-waste. Answers for the questions varied from short answers, multiple choice—one answer, open-ended, scaled ratings and multiple choice-multiple answers. The survey was conducted using a researcher made online survey using Google platform. Totally 57 responses were obtained from 77 requests. The survey was open from September 25, 2018 to October 25th, 2018.

3 Results and Discussions

Survey results were analysed by using the overall response to each question, 1-1 variable co-relation and 1-many variable co-relation. SPSS ver 20 was used for qualitative analysis. The quantitative and qualitative data obtained gave a baseline understanding of willingness of the people in academics to participate in e-waste management and their level of understanding around e-waste.

When asked to define e-waste, various responses were recorded. Thus, based on the underlying meaning, we could classify the responses into five definitions according to the respondents. The qualification of the respondent and their understanding of the e-waste is shown in Fig. 1. Majority of the respondents understand e-waste as an electronic appliance, big or small, which is no more functioning, or in simple words, it means that the waste generated by electronic and electrical goods. Also from the questions 'Are you aware of health hazard of e-waste?' and 'how do you dispose e-waste?' we noticed that the fact that e-waste is toxic to the environment is

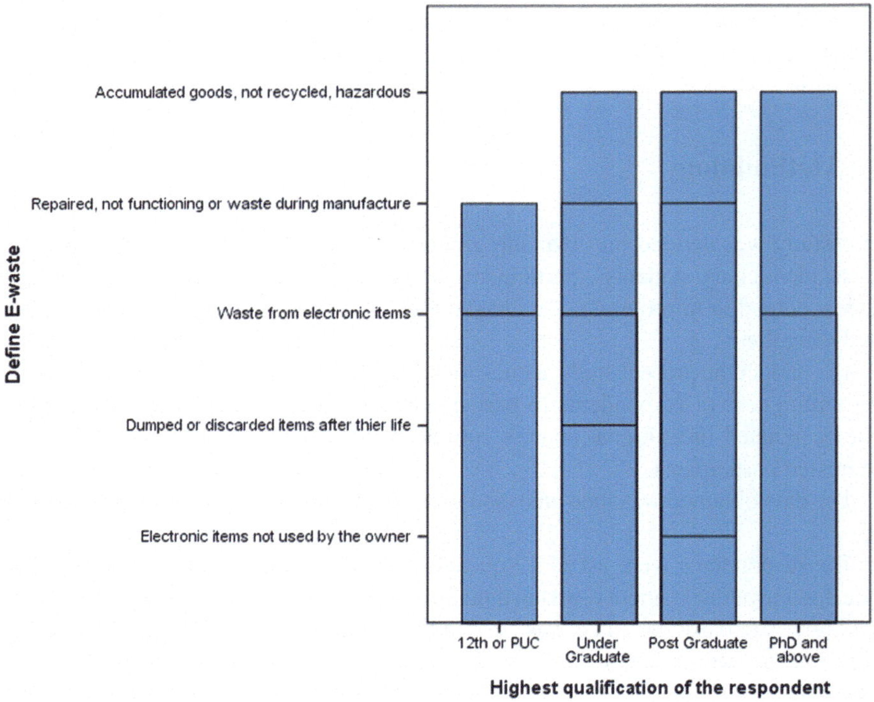

Fig. 1 A pictorial representation of understanding of e-waste by respondents of varying educational qualification

confirmed by all the respondents. However, the fact that it should be discarded at the proper place has not reached many respondents.

Figure 2 provides an understanding of how respondents with different level of qualification respond to environmental-friendly disposal methods of e-waste. Majority of the respondents are interested in this method of disposal of e-waste. Hence, it is a good opportunity to start about the changes required to get the change.

It is important to understand the e-waste disposal attitudes of people with good qualification. Figure 3 provides us with such a co-relation. The figure shows the categories of the methods of disposal of e-waste the respondents wished to choose. The irony is that a little percentage of all the categories of people chose to dispose the e-waste either to the 'Kabadiwalas' or at the regular dustbins. This seems to pose a serious threat on the long run. However, there are many awareness and educational camps, arranged by different volunteers to educate the people on how to practise the safe disposal of e-waste. A considerable number of respondents have also chosen the disposal of e-waste at the recycling centres.

Figure 4 represents the co-relation of the general level of awareness of e-waste of the educated mass. The responses expect the root level answer regarding the knowledge of e-waste. Clearly, majority of the people knew about e-waste, and

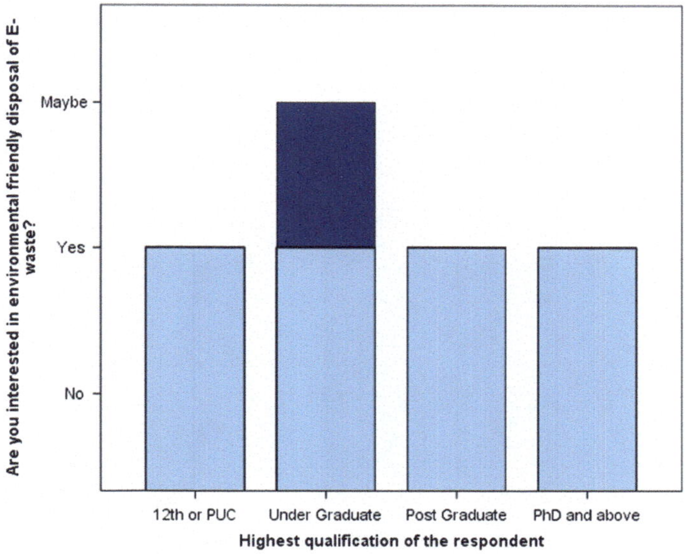

Fig. 2 Pictorial representation of willingness of environmental-friendly e-waste management relating to the respondents' educational qualification

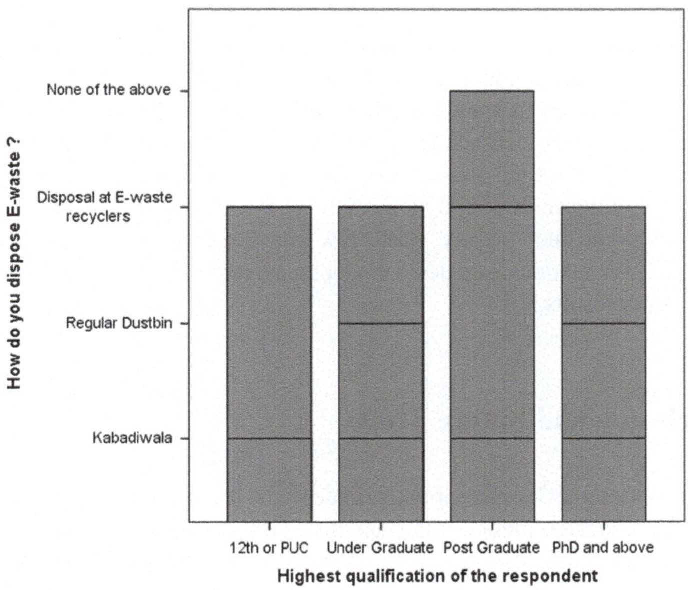

Fig. 3 Current e-waste disposal behaviour of respondents in relation to their qualification

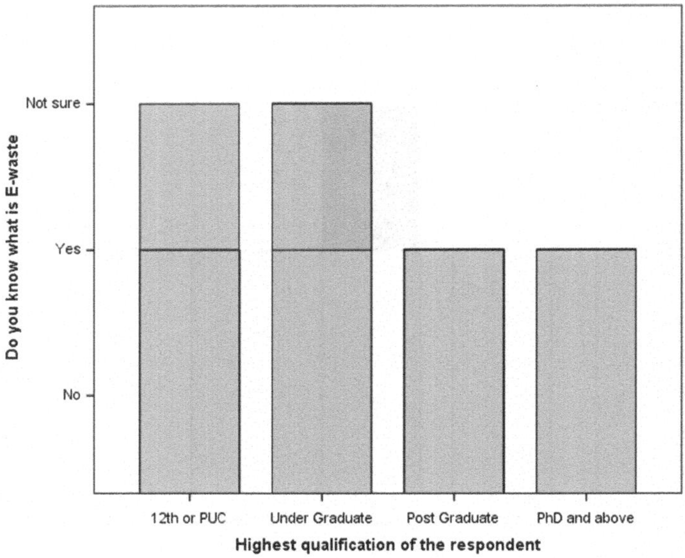

Fig. 4 General understanding of the term e-waste by respondents related to their educational qualification

anyway, a few exceptions state that they were not sure. This was just a basic analysis of how many of the respondents knew about the term 'e-waste'.

This is one of the most wanted comparisons of knowledge of e-waste and their corresponding disposal methods (Fig. 5). Of all the respondents who knew about e-waste, each of 25% of the whole chose each of 'kabadiwalas', 'regular dustbins', 'disposal at recycling centres' and 'none of the above'. This conveys that even among the respondents who knew about e-waste, few were not aware of the ways to adopt for disposing it. This situation again demands for the effective awareness and educational camps for better results.

4 Conclusion and Future Scope

In this paper, an attempt has been made to understand the awareness level and disposal attitude of the educated mass in the Bangalore city. A survey was conducted using structured questionnaire in google form, and the responses were analysed using SPSS software. The findings were very much alarming as it showed the awareness level to be high enough, yet it was painful to find their disposal attitude to be non-complaint with rules. As recorded, some preferred disposal via kabadiwalas too. In the future studies, the participants and sampling size can be increased and also the staffs who serve as house-keepers, laboratory support staff, etc., can also be asked to take the

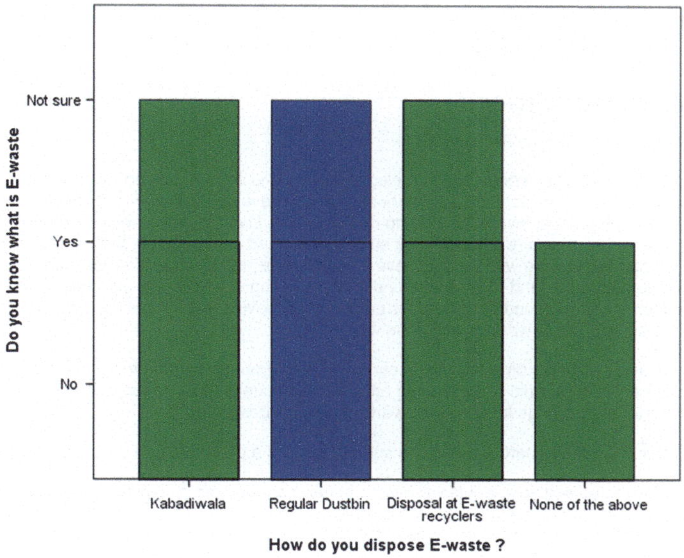

Fig. 5 Pictorial representation of relation between awareness and behaviour towards e-waste management

survey and their responses recorded and compared with other educated mass to come up with different types of awareness programmes for the target groups.

Acknowledgements The authors would like to acknowledge International Society of Waste Management, Air and Water (ISWMAW) for partially funding the project. The authors would also like to acknowledge support from the community of Global Academy of Technology, both in terms of providing ethical consent to move forward with the research as well as participate in the study. Additionally, the help and co-operation from the Centre for Quality Management Systems (CQMS) Jadavpur University, Kolkata; Consortium of Researchers for Environmental Protection, Sustainability and Climate Change (CREPSCC) are gratefully acknowledged.
This may be noted that all the procedures performed in studies involving human participants were in accordance with ethical standards of the institutional research and informed consent was obtained from all individual participants in the study. Authors appreciate the organisation for generous support towards the successful completion of the study.

Appendix

E-Waste Management

Dear fellow Homo Sapien (Wise Man/Woman),

Efficient management of E-waste or electrical & electronic waste is a great concern all over the world. With the rapid change in technology, exponential rate of development, day-in & day-out updates of the smart phones, software, visualizing devices, etc., I wonder how are we going to manage the E-waste which we are generating at such great speed and amounts. With this said, I also believe that together we can work and achieve a sustainable plan for E-waste management. Planning comes before action, and thus understanding the existing status of our understanding of E-waste management in our surroundings becomes crucial for achieving any kind of success in our attempts towards sustainable E-waste management.

So, this survey is an attempt to comprehend the current status of our understanding of E-waste management, which will further aid in developing programs and suggestions for concerned authorities to take up and arrange for creating awareness about E-waste

Also, thank you for choosing to participate in this survey. You are a green warrior!

This survey is anonymous and your inputs will help in better management of E-waste generated in the country. Thus, its a humble request to take the survey with utmost honesty, such that the information obtained from this survey can be used for further studies without any ambiguity.

Thanks in advance for your valuable inputs and time.

* Required

1. **Email address** *

Respondent

2. **Name of the Institution you belong to.** *

3. **Gender** *
Mark only one oval.

- ◯ Female
- ◯ Male
- ◯ Prefer not to say
- ◯ Other:

4. **Your highest qualification** *
Mark only one oval.

- ◯ 12th standard; 2nd PUC
- ◯ Undergraduate (E.g. B.E., B.Com, etc.)
- ◯ Post Graduate (E.g. M.Tech, M.E., MSc, etc.)
- ◯ PhD and above
- ◯ Other:

5. **City/Town/Village you live in** *

6. **Average Annual income** *
Mark only one oval.

- less than INR 50000
- INR 50000- INR 300000
- INR 300000 -INR 500000
- more than INR 500000
- Other:

How much do you know about E-waste

7. **Do you know what is E-waste?** *
Mark only one oval.

- Yes
- No
- Not sure

8. **What is E-waste? Please write here.** *

9. **Do you know about E-waste management rules 2016?** *
Mark only one oval.

- Yes
- No
- Maybe

10. **Are you aware of the health hazards due to E-waste?** *
Mark only one oval.

- Yes
- No
- Maybe

11. **Do you think E-waste is toxic to the environment?** *
Mark only one oval.

- Yes
- No
- Maybe

12. **Are you aware of Extended Producer Responsibility (EPR)? ***
Mark only one oval.

- ◯ Yes
- ◯ No

13. **What are the electronic items you have in your home? ***
Check all that apply.

- ☐ OLD TV SET
- ☐ LCD TV
- ☐ LED TV
- ☐ RADIO
- ☐ CD/DVD PLAYER
- ☐ SMART PHONE
- ☐ FEATURE PHONE
- ☐ FRIDGE
- ☐ AC
- ☐ WATER PURIFIER
- ☐ AIR COOLER
- ☐ LAPTOP
- ☐ PC/DESKTOP
- ☐ SOUND SYSTEM
- ☐ LED BULB
- ☐ CFL BULB
- ☐ PRINTER
- ☐ SCANNER
- ☐ MICROWAVE OVEN
- ☐ INDUCTION OVEN
- ☐ ELECTRIC KETTLE
- ☐ TAPE RECORDER
- ☐ SPEAKER FOR PC/BLUETOOTH SPEAKER
- ☐ HEADPHONE/EARPLUG
- ☐ Other: ______________________

14. **Do you have any E-waste in store currently in your house?**
Mark only one oval.

- ◯ Yes
- ◯ No

15. **Please mention the number of e-waste in store. ***

____

16. **List the E-waste items you currently have in your house.** *

____

____

____

____

____

17. **How do you dispose E-waste ?** *

Mark only one oval.

- ◯ Kabadiwala
- ◯ Regular Dustbin
- ◯ Disposal at E-waste Recycling Centre
- ◯ Other: ______________________________

18. **Do you prefer selling E-waste to kabadiwala than formal recyclers?** *

Mark only one oval.

- ◯ Yes
- ◯ No

19. **Which factors would affect you trading in your old products for a cash reward?**

Mark only one oval.

- ◯ Emotional attachment to the product
- ◯ Initial cost of the product
- ◯ The repairability of the product
- ◯ Logistics of the service (i.e. postage, collection)
- ◯ Other (please specify)

20. **Are you interested in environmental friendly disposal of E-waste?** *

Mark only one oval.

- ◯ Yes
- ◯ No
- ◯ Maybe
- ◯ Other: ______________________________

21. **Do you wish to participate in formal recycling awareness program?** *

Mark only one oval.

- ◯ Yes
- ◯ No

22. **Do you buy refurbished electronics online?** *

Mark only one oval.

- ◯ Yes
- ◯ No

23. **If the answer to the previous question is no, please mention why?**

24. **Do you buy electronic items from e-commerce site?** *
Mark only one oval.

- Yes
- No

25. **Is it legal to dump any kind of 'not in use' electronic devices in the trash or landfills?** *
Mark only one oval.

- yes
- no
- I don't know

What can we do about E-waste?

26. **According to you what the government can do to solve this problem?** *

27. **What is the status of e-waste recycling at your home town/city?** *
Mark only one oval.

- Does not exist
- I am not aware of any kind of E-waste management activity in my surrounding
- Exist, but its working is average
- Excellent E-waste management in my surroundings
- Other:

28. **According to you what you can do to solve this problem?** *

29. **Would you like to participate in such surveys again?** *

Mark only one oval.

- ◯ Yes
- ◯ No
- ◯ Maybe

☐ Send me a copy of my responses.

Powered by
Google Forms

References

Baldé, C. P., Wang, F., Kuehr, R., et al. (2015). *The global e-waste monitor–2014*. United Nation University.

Balde, C. P., Forti, V., Gray, V., Kuehr, R., & Stegmann, P. (2017). *The global e-waste monitor 2017: Quantities, flows and resources*. Bonn, Geneva, Vienna: United Nations University (UNU), International Telecommunication Union (ITU) and International Solid Waste Association (ISWA). ISBN Electronic Version: 978-92-808-9054-9.

Basel Convention. (2014). *Basel convention on the control of transboundary movements of hazardous wastes and their disposal protocol on liability and compensation for damage resulting from transboundary movements of hazardous wastes and their disposal texts and annexes*. UNEP.

Bhatt, G., Khanna, M., Pani, B., & Baweja, R. (2017). Awareness and sensitivity of mobile phone consumers on electronic waste in Delhi-NCR region. In *Sustainable smart cities in India* (pp. 433–442). Springer, Cham.

Borthakur, A., & Govind, M. (2017). Emerging trends in consumers' E-waste disposal behaviour and awareness: A worldwide overview with special focus on India. *Resources, Conservation and Recycling, 117,* 102–113.

Catalyst, C. (India) Pvt. Ltd. (2015). *A brief report on electronics industry in India*. Available online at: http://www.cci.in/pdfs/surveys-reports/Electronics-Industry-in-India.pdf. Accessed January 4, 2016.

Debnath, B., Chowdhury, R., & Ghosh, S. K. (2018). Towards circular economy in E-waste recycling via metal recovery from E-waste (MREW) facilities. In *ISWA 2018 World Congress*. Kuala Lumpur, Malaysia.

Dwivedy, M., & Mittal, R. K. (2013). Willingness of residents to participate in e-waste recycling in India. *Environmental Development, 6,* 48–68.

Mehta, Y., & Rajan, A. J. (2017). Manufacturing sectors in India: Outlook and challenges. *Procedia engineering, 174,* 90–104.

Purification Technologies of Bioreactor Landfill Gas and Its Sustainable Usage: Current Status and Perspectives

Anaya Ghosh, Jyoti Prakas Sarkar and Bimal Das

Abstract Bioreactor landfill (BRLF) has turned into one of the promising treatment technologies of municipal solid waste management (MSWM) globally, mostly in developing countries. A potential amount of landfill gas is produced which is collected, stored and upgraded into biomethane for further utilization of bioreactor landfill gas (BRLFG). This biogas is not only resembled for eco-friendly procedure for instantaneous production of renewable energy but also established a sustainable approach to waste to energy process. Commercialization of BRLFG with high efficiency requires an appropriate upgrading and purification treatment process which includes also the energy value of off-gas. The current practices for biogas upgradation involve high operational expenses, high energy exhaustive and limited business opportunities for biogas. The present work therefore is focused on the available methodologies for purification and upgradation of BRLFG and enrichment particularly concentrated on the promising biomethanation processes. This work contains a complete explanation of different biogas purification and improvement techniques, and their consequences are associated with the biomethanation efficiency, utilization challenges, cost analysis, impacts, sustainability and feasibility of these improvement conceptions of landfill gas. This present study illustrates a general discussion on the future perspectives for conquering the issues and challenges related to the purification technologies which may be beneficial for the policymakers, environmentalists, stockholders for the betterment of an integrated solid waste management and sustainable development.

Keywords Bioreactor Landfill Gas (BRLFG) · Purification Technologies · Upgradation · Biomethane

A. Ghosh (✉) · J. P. Sarkar · B. Das
Department of Chemical Engineering, National Institute of Technology Durgapur, Durgapur, India
e-mail: ghoshanaya@gmail.com

S. K. Ghosh (ed.), *Urban Mining and Sustainable Waste Management*,
https://doi.org/10.1007/978-981-15-0532-4_26

1 Introduction

Municipal solid waste management had become a challenging job due to rapid increase of waste in present days. A sustainable municipal solid waste management involves an integrated technique which includes waste minimization, waste segregation at source, material recovery facility (MRF), resource utilization, suitable eco-friendly technology and treatment processes, recycling and valourization, reduction of open dumping, waste to energy plants and steady administration (Ghosh et al. 2018). The ultimate discarding of solid waste in a suitable planned and designed sanitary landfill (SLF) has become an essential part of an integrated solid waste management system (ISWMS) in present day. Bioreactor landfill (BRLF) is an evolving resource utilization treatment technology where the prime concern is to discard open dumps globally. On the other side where several nations are focusing on zero landfilling waste management arrangement, this BRLF will be a suitable waste to energy process for landfill mining and a way for waste utilization from open dumps (Chiemchaisri et al. 2010; Kumar et al. 2011). BRLF is planned SLF which includes the converting and stabilization processes of biodegradable organic wastes quickly by involving anaerobic digestion system which contains a proper leachate recirculation system and landfill gas (LFG) collection system (Cossu et al. 2016). Municipal solid waste (MSW) from developing countries largely contains organic waste (60–70%) with higher amount of moisture content, especially in tropical states (Ghosh et al. 2019). The United States Environmental Protection Agency assesses that if 50% of the waste are being appropriately landfilled in controlled BRLFs at present, it can offer more than 270 billion cubic feet of methane a year (Lakshmikanthan et al. 2017).

BRLFG or biogas is the main product of such kind of landfill operations which involves anaerobic digestions. It is a mixture of flammable gases which contains methane (CH_4) at a concentration of 40–65%, carbon dioxide (CO_2) in a range of 35–55%, with little bits of hydrogen sulphide (H_2S) (0.1–3%), moisture and other traces of impurities. Comparing the calorific values in between biogas and natural gas, natural gas is having 35.8 MJ/m^3 while biogas is 21.5 MJ/m^3 (Angelidaki et al. 2018). The variance in between two calorific values largely depends on the quantity of the CO_2, the inflammable part of biogas which reduces the heating value and transportation cost as well as prevents the feasibility of biogas for power generation. Correspondingly, containing the gaseous impurities, the raw biogas may damage the downstream machinery because of the corrosive nature of H_2S. The purified and upgraded BRLFG has numerous usages in commercial purpose as compressed methane gas (CMG) along with electricity generation and automobile fuels (Singhal et al. 2017). Decontaminated BRLFG not only leads to less greenhouse gas emissions but also has additional ecological profits by emitting low amount of hydrocarbons, nitrogen oxide and carbon monoxide compared to fossil fuels (Patil and Singh 2017). Figure 1 shows the flow diagram of waste to biogas production and utilization. Presently, developing countries like Sri Lanka, Thailand, Malaysia, China and India

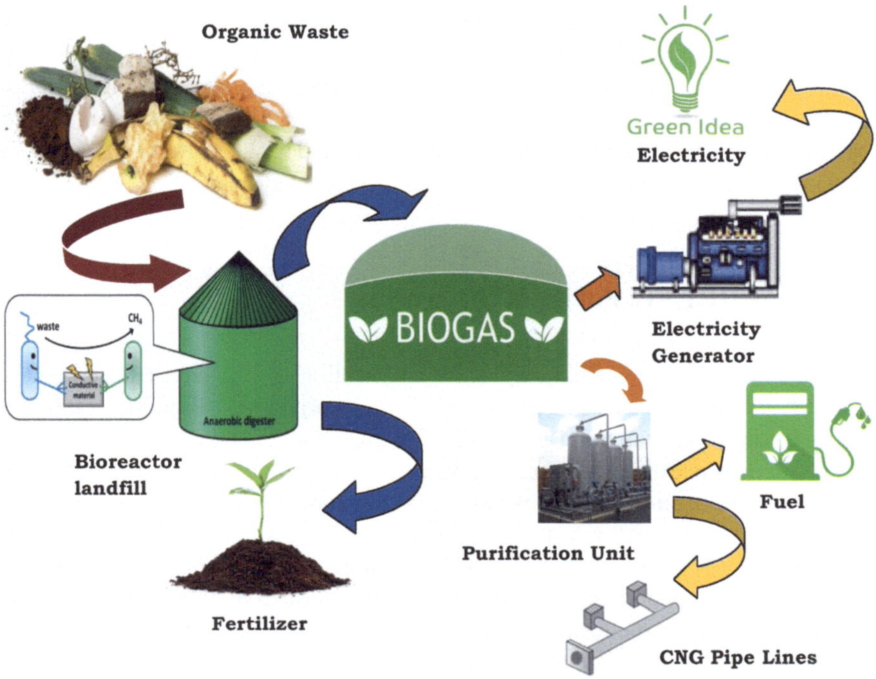

Fig. 1 Flow diagram of waste to biogas production and utilization (Sahota et al. 2018)

have engaged to mitigate the energy demands in respective nations and transformation of open dumps to BRLF.

The objective of the existing assignment is to propose a combined outline of numerous biogas advancement and decontamination technologies and to create further improvement for developing purification and upgradation perceptions. This assessment illuminates more precisely the key principles of commercial and emerging physical, chemical and biological separation practices. Specifically, at this moment, the main efforts have been given to the development of bioenergy production from biomass waste disposal and resource recovery from bioreactor landfill. Finally, a brief review has been presented on environmental and economical aspects which enhance the sustainability standpoint of biomethanation and biogas improvement perceptions.

2 Study Methodology

This study has emphasized the literature evaluations and report assessments. Primarily, numerous journals and reports were carried out to identify the status of biogas generation from MSW, bioreactor landfill management and bioreactor landfill gas

(BRLFG) in developing countries. Different technologies for the purification of biogas generated from anaerobic digester or BRLFs were recognized by networks and Internet search to realize the current status of the waste to energy system. Finally, based on the literature findings and report outcomes, a general discussion of the upgradation and utilization of biogas has been suggested for the betterment of energy crisis mitigation and integrated solid waste management system.

3 Biogas/BRLFG Purification Process

Currently, four separation technologies for upgradation of CH_4 have been commercially carried out which involve the processes of absorption, adsorption, membrane and cryogenic separation described in Fig. 2 (Angelidaki et al. 2018). Additionally, this cryogenic process or chemical hydrogenation process and some other technologies have also been emerging day by day depending upon the investment cost, process efficiency as well as requirements of purification of biogas or BRLFG (Sahota et al. 2018). A brief process description and a compassion of biomethanation efficiency of these biogas upgradation technologies have been described and also represented in Table 1. Primarily, the methane recovery can be accomplished greater than 96% from physicochemical upgradation processes. Additionally, it will be concentrated on the

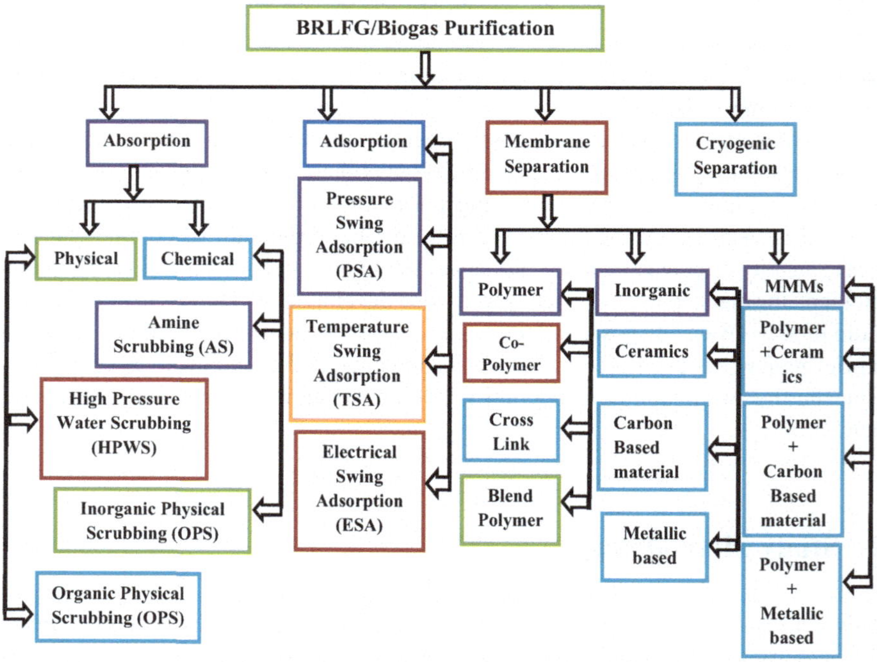

Fig. 2 Available current practices of biogas upgradation (Angelidaki et al. 2018)

Table 1 Detailed process of the different technologies (Angelidaki et al. 2018; Sahota et al. 2018)

Sl no.	Separation technologies	Absorbent/adsorbent	Operational pressure	Temperature (°C)	CH_4 yield	CH_4 lass %	Cost	Energy consumption (kWh/Nm3)	H_2S removal
1.	Water scrubbing	Water	4–12 bars	–	96–98%	<2	Moderate	0.23–0.54	Yes
2.	Pressure swing adsorption	Molecular sieves	3–10 bars	–	>96–98%	<4	Moderate	0.16–0.43	Possible
3.	Chemical absorption	Amines and alkali solutions	1 atm	100 to 180	>96–98%	<0.1	Expensive	0.05–0.18	Contaminant
4.	Physical absorption	Organic solvents	6–8 bars	55 to 80	90%	2–4	Moderate	0.10–0.33	Possible
5.	Membrane separation	Polymeric, ceramic, inorganic, hollow fibber, mixed matrix membrane	5–20 bars	–	96–98%	<0.6	Expensive	0.18–0.35	Possible
6.	Cryogenic separation	Not required	80 bars	−110 to −196	97–98%	2	Expensive	0.18–0.25	Yes

efficiency increment of biomethanation with incorporation of raised temperature, high pressure or addition of chemicals (Vijayanand and Singaravelu 2017).

3.1 Absorption

Absorption is a physical or chemical process where the gas phase constituents get chance to diffuse into liquid when it goes through the interface. Appropriate solvent selection is very imperative stage and that will enhance this absorption process faster. Water scrubbing (WS) is very earliest practice for gas purification (Abdeen et al. 2016). For landfill gas upgradation, different scrubbing processes such as amine scrubbing (AS), high-pressure water scrubbing (HPWS), inorganic physical scrubbing (IOPS) and organic physical scrubbing (OPS) need to perform to separate the impurities from biogas for biomethane. The BRLF gas form contaminations due to containing H_2S and CO_2 which require to be more soluble in absorbers compare to CH_4 during seperation and that will help the absorption process successfully (Achinas et al. 2017).

3.2 Adsorption

Adsorption is another separation practice for biogas upgradation where the gaseous impurities adhere to the surface of a micro-porous solid according to their molecular size. The method is categorized into physisorption and chemisorption subject to the applied force on the process. In case of physical adsorption, existence of frail Van der Waals forces among the adsorbate and the adsorbent helps to detach the impurities by diminishing or accumulation of pressure, whereas in chemical adsorption, strong chemical bonds are recognized as Langmuir adsorption, present within the adsorbate and adsorbent (Ryckebosch et al. 2011). Chemisorption is imperfect for monolayer, while physisorption may be multilayer. Without any problem, the physisorption process is upturned easily but chemical adsorption could not be inverted. Depending on the adsorbent recharging and reinforcement method, different types of biogas upgradation processes are pressure swing adsorption (PSA), purge gas stripping, electrical swing, temperature swing, etc (Bauer et al. 2013). Commercially, PSA is mostly used process because of its less energy prerequisites, process security and higher productivity among the other upgradation processes. The PSA process principle depends on the molecular properties and the adhesion affinity of the adsorbent solid and also on the belongings of the compelled gases which should adhere to the solid surface. Consequently, at high-pressure condition huge quantities of gaseous impurities will be arrested on the solid adsorbent surface. The adsorbents can be carbon molecular sieve, activated carbon, zeolites and other substantial solid, containing high surface area (Sahota et al. 2018).

3.3 Membrane Separation

Membrane separation process engages the principle of selective permeability of membrane module. Concentration gradient of permeate (liquid or gas) allows the gas to go through the membrane. The variation in chemical affinity and particle size of different molecules are the two major constituents of membrane separation process. Because pressure gradient created both sides of the membrane, gases can easily passage through the membrane (Sun et al. 2015). If permeation rate will high, the coefficient of solubility and coefficient of diffusion will be achieved higher. The proportion of permeation membranes varies based on the adsorption constants of the gaseous impurities and on the membrane fabrication substances. Therefore, having smaller molecular size and higher solubility H_2, H_2S and CO_2 penetrate quicker than CH_4 and easily separate into two streams (Awe et al. 2017). Main drawbacks of this membrane separation process are extremely expensive and delicate to handle.

3.4 Cryogenic Separation Process

This practice is a progressing separation process involving some commercial industries, operating on it (Khan et al. 2017). This process includes slow diminution of gas temperature which assists to separate the liquefied CH_4 from both CO_2 and other impure constituents. Primarily, the biogas upgradation process is executed by drying followed by squeezing the raw BRLFG up to 80 bars and then temperature decay up to -110 °C. Therefore, the minor amount of impurities, i.e. vapour, H_2S, siloxanes, halogens, etc., and the major quantity of CO_2 are slowly separated for recovery of cleaned biomethane (<97%) (Angelidaki et al. 2018). In spite of this productivity of the cryogenic purification practice, it is still under improvement condition and very limited amenities are in working state for commercial usages. Higher investment costs, operating and maintenance expenses and clogging problem during the process are the major drawbacks of this cryogenic technology.

3.5 Other Processes

Biological gas separation process is other emerging purification and upgradation technology of BRLFG, and it is categorized into chemoautotrophic and photosynthetic process (Miltner et al. 2017). The most beneficial part of this process is that CO_2 can be transformed by the help of micro-organisms into other higher value-added yields during the operation. During chemoautotrophic process with the involvement of hydrogenotrophic methanogens, H_2 can be converted into CO_2 and then into methane. Hybrid separation process is another developing process where both membrane separation and conventional purification treatment such as amine absorption, water scrubbing, PSA, cryogenic separation are associated simultaneously (Kadam and Panwar 2017). These hybrid technologies not only improve the biomethanation efficiency but are also cost-effective at commercial level.

4 Discussions

A comparative study of the various biogas purification technologies has been described in Table 1. Water scrubbing and PSA are two most common technologies which are used largely in commercial aspects and economically feasible also (Budzianowski 2016). While membrane, cryogenic and chemical absorption separation processes offer high efficiency at the same time huge include high investment cost expenses but operational cost is low in case of membrane separation process. Energy consumption and loss of CH_4 have vigorous impacts while describing the environmental sustainability aspects. Water, organic and amine scrubbing involve higher energy than PSA and membrane separation processes, but the methane loss is low compared to PSA (Angelidaki et al. 2018). Consumption for raw biogas in cryogenic separation process is higher 0.76 kWh/Nm^3, comparing with PSA process where the consumption is 0.23–0.30 kWh/Nm^3. PSA can be recommended as most commercially suitable and eco-friendly process because this process also prerequisites less energy demands and manpower (Xu et al. 2015).

5 Conclusion

Biomethane is an emerging renewable energy possibility and a worthy replacement of compressed natural gas (CNG) in power generation units and automobile applications. Practices of purification and upgradation of biogas are beneficial to dropping global warming potential and endorsing a more cleaner and sustainable atmosphere. Some innovative approaches such as cryogenic separation methods and hybrid processes are in their promising phase for more BRLFG/biogas quality improvement. Due to the lack of expertise in between pilot scale and laboratory scale, the operations need to take care of biogas enrichment in broader aspects. Moreover, the hybrid upgradation processes have an approaching option to solve the issues and challenges of existing upgradation and purification practices.

Acknowledgements The authors would like to acknowledge the help and cooperation from International Society of Waste Management, Air and Water (ISWMAW) and Centre for Quality Management Systems (CQMS), Jadavpur University, Kolkata.

References

Abdeen, F. R., Mel, M., Jami, M. S., Ihsan, S. I., & Ismail, A. F. (2016). A review of chemical absorption of carbon dioxide for biogas upgrading. *Chinese Journal of Chemical Engineering, 24*(6), 693–702.

Achinas, S., Achinas, V., & Euverink, G. J. W. (2017). A technological overview of biogas production from biowaste. *Engineering, 3*(3), 299–307.

Angelidaki, I., Treu, L., Tsapekos, P., Luo, G., Campanaro, S., Wenzel, H., & Kougias, P. G. (2018). Biogas upgrading and utilization: Current status and perspectives. *Biotechnology Advances.*

Awe, O. W., Zhao, Y., Nzihou, A., Minh, D. P., & Lyczko, N. (2017). A review of biogas utilisation, purification and upgrading technologies. *Waste and Biomass Valorization, 8*(2), 267–283.

Bauer, F., Persson, T., Hulteberg, C., & Tamm, D. (2013). Biogas upgrading–technology overview, comparison and perspectives for the future. *Biofuels, Bioproducts and Biorefining, 7*(5), 499–511.

Budzianowski, W. M. (2016). A review of potential innovations for production, conditioning and utilization of biogas with multiple-criteria assessment. *Renewable and Sustainable Energy Reviews, 54,* 1148–1171.

Chiemchaisri, C., Weerasekara, R., Joseph, K., Kumar, S., & Visvanathan, C. (2010). Application of bioreactor landfill technology to municipal solid waste management: Asian perspective.

Cossu, R., Morello, L., Raga, R., & Cerminara, G. (2016). Biogas production enhancement using semi-aerobic pre-aeration in a hybrid bioreactor landfill. *Waste Management, 55,* 83–92.

Ghosh, A., Debnath, B., Ghosh, S. K., Das, B., & Sarkar, J. P. (2018). Sustainability analysis of organic fraction of municipal solid waste conversion techniques for efficient resource recovery in India through case studies. *Journal of Material Cycles and Waste Management*, 1–17.

Ghosh, A., Sarkar, J. P., & Das, B. (2019). Sustainable Energy Recovery from Municipal Solid Waste (MSW) using Bio-reactor Landfills for Smart City Development. In *IEEE International Conference on Sustainable Energy Technologies (ICSET)* (pp. 242–246). IEEE.

Kadam, R., & Panwar, N. L. (2017). Recent advancement in biogas enrichment and its applications. *Renewable and Sustainable Energy Reviews, 73,* 892–903.

Khan, I. U., Othman, M. H. D., Hashim, H., Matsuura, T., Ismail, A. F., Rezaei-DashtArzhandi, M., et al. (2017). Biogas as a renewable energy fuel–A review of biogas upgrading, utilisation and storage. *Energy Conversion and Management, 150,* 277–294.

Kumar, S., Chiemchaisri, C., & Mudhoo, A. (2011). Bioreactor landfill technology in municipal solid waste treatment: An overview. *Critical Reviews in Biotechnology, 31*(1), 77–97.

Lakshmikanthan, P., Santhosh, L. G., & Babu, G. S. (2017). Evaluation of bioreactor landfill as sustainable land disposal method. In *Sustainability Issues in Civil Engineering* (pp. 243–254). Springer, Singapore.

Miltner, M., Makaruk, A., & Harasek, M. (2017). Review on available biogas upgrading technologies and innovations towards advanced solutions. *Journal of Cleaner Production, 161,* 1329–1337.

Patil, B. S., & Singh, D. N. (2017). Simulation of municipal solid waste degradation in aerobic and anaerobic bioreactor landfills. *Waste Management and Research, 35*(3), 301–312.

Ryckebosch, E., Drouillon, M., & Vervaeren, H. (2011). Techniques for transformation of biogas to biomethane. *Biomass and Bioenergy, 35*(5), 1633–1645.

Sahota, S., Shah, G., Ghosh, P., Kapoor, R., Sengupta, S., Singh, P., et al. (2018). Review of trends in biogas upgradation technologies and future perspectives. *Bioresource Technology Reports, 1,* 79–88.

Singhal, S., Agarwal, S., Arora, S., Sharma, P., & Singhal, N. (2017). Upgrading techniques for transformation of biogas to bio-CNG: A review. *International Journal of Energy Research, 41*(12), 1657–1669.

Sun, Q., Li, H., Yan, J., Liu, L., Yu, Z., & Yu, X. (2015). Selection of appropriate biogas upgrading technology-a review of biogas cleaning, upgrading and utilisation. *Renewable and Sustainable Energy Reviews, 51,* 521–532.

United States Environmental Protection Agency, Bioreactor Landfills. Available at https://www.epa.gov/landfills/bioreactor-landfills.

Vijayanand, C., & Singaravelu, M. (2017). Installation of a biogas upgradation system. *Agric International, 4*(1), 12–15.

Xu, Y., Huang, Y., Wu, B., Zhang, X., & Zhang, S. (2015). Biogas upgrading technologies: Energetic analysis and environmental impact assessment. *Chinese Journal of Chemical Engineering, 23*(1), 247–254.

NOx Reduction from Diesel Engine Consuming Diesel-Waste Plastic Oil Blend as Substitute Fuel with Antioxidant and SCR

B. Sachuthananthan, R. L. Krupakaran, A. Sumanth and T. Manikandan

Abstract Emission in Internal Combustion Engines can be achieved through fuel modification, combustion control or by outlet gas treatment. Comparatively, the third approach is easy since there is little or no need of modifying the engine. Similar one technique is selective catalytic reduction (SCR). In this paper, simultaneous attempt is made to control the NOx pollution through fuel modification and exhaust treatment techniques. Plastic pyrolysis oil (PPO) as alternate fuel added with 100, 200, 300, 400 and 400 mg/l of Butylated hydroxyanisole (BHA) antioxidant and injection of constant quantity of aqueous solutions of urea in the tail pipe fitted with Vanadium-based nodule as SCR catalyst was carried out on a 4 stroke, mono barrel CI power train. The results reveal that with urea vaccination conforming to 8–9% of the fuel, the NOx emission dropped by 69% with no significant drop in BTE and rise in other pollutants.

Keywords Plastic pyrolysis oil · Fuel antioxidant · SCR · Emission reduction

Nomenclature

WPO	Waste plastic oil
NOx	Oxides of nitrogen
PPO	Plastic pyrolysis oil
BHA	Butilated hydroxi anisole
CI	Compression ignition
SI	Spark ignition
UBHC	Unburned hydrocarbon
TPO	Tyre pyrolysis oil
USD	Ultra-low sulphur diesel
WPDF	Waste plastic diesel fuel

B. Sachuthananthan (✉) · R. L. Krupakaran · A. Sumanth · T. Manikandan
Mechanical Engineering Department, Sree Vidyanikethan Engineering College, 517102 Tirupati, Andhra Pradesh, India
e-mail: bsachu7@yahoo.co.in

S. K. Ghosh (ed.), *Urban Mining and Sustainable Waste Management*,
https://doi.org/10.1007/978-981-15-0532-4_27

TDF Tyre derived fuel
DEF Diesel exhaust fluid
DICI Direct injection compression ignition

1 Introduction

In an average, the European people shred out 36 kg of plastics per annum. A total of 42% of the waste plastic comes out from the packaging industries, and very negligible quantity undergoes recycling (Vogler 1984). The total waste plastics recovered for recycling were only 8% during the year 2014. Naima et al. analysed to discover alternative fuel for Internal Combustion engines and found that the power train could keep running using neat WPO. NOx is higher by around 17% for WPO than diesel. CO pollutant expanded by 5% for WPO in comparison with diesel fuelled engine. UBHC pollutant is more by around 15%. Engine runs with fuel as like WPO exhibits more BTE till 75% of the full load (Naima et al. 2013). Chindaprasert at el. utilized PPO and TPO in CI power train in the form of neat and blend with diesel and observed that both tested oils can replace the diesel oil. The PPO brings down performance of the engine, the quantity of waste plastic is tremendous, and it should be properly used to lessen environment issues (Chindaprasert and Wongkhorsub 2013). Neelesh Soni at el. primarily focused on finding alternative energy treasures and utilizing those fuels to eradicate the unwanted effect of occurrence to restrict the petroleum usage. It is justified that the created diesel from Plastic pyrolysis oil, waste engine oil and waste tyre oil is in fact technically suitable, monetarily feasible and less responsible to pollute environment (Soni and Sharma 2013). Pawar Harshal et al. decided like PPO represents a better option energy source for CI engine and along these lines must be thought about later on for transport reason. Likewise inferred that engine was continued operating with neat WPO. The CI engine operated with WPO as fuel reveals higher BTE up to 75% of rated power. Likewise BSFC is higher when mixed with diesel (Pawar Harshal and Shailendra 2013). Anup et al. analysed the WPO, WPO of gasoline and USD grade and their mix with petrol and diesel individually was presented as an elective energy basis. He revealed SI power train could keep running using neat WPO. Power train powered with waste plastic oil exhibits more BTE (Anup and Watwe 2014). Mohd. Herzwan Hamzah et al. analysed the burning properties of two fuels from waste known as PDF and TDF have been compared with diesel. Finally, the power train can work with neat PDF and TDF. From overall result, demonstrated PDF has similar combustion characteristics as like diesel than TDF (Hamzah et al. 2015). Sunbong Lee et al. utilized the oil which is derived through the pyrolysis technique. The pure Plastic pyrolysis oil was mixed along with ULSD fuel in 20 and 40% by volume composition. PPO can be utilized as part of diesel engine with 20% mix without any restriction and 40% mix at a contented engine rpm (2450 rpm) (Lee et al. 2015). Vijay Kumar et al. observed that nearly 10 crore tons of plastics are manufactured worldwide in a year, and used things have converted into a poisonous compound and landfills. Likewise the different sort of pollution because of waste plastic can be minimized (Vijay Kumar and Patil 2015). To overlook the above-specified drawbacks of PPO, few spices are mixed in energy sources for minimizing the pollution in specific issues, and few approaches employed are nano fuel additives to CI fuel, which was consistently conveyed for reasonable BTE and minimizing the outlet pollutions from CI engines.

2 Methodology

2.1 *Plastic Pyrolysis Oil Preparation*

Catalytic cracking of plastics can be performed in the range from 400 to 800 °C, and the yield is maximum between 300 and 350 °C. This research attempt uses waste plastic of various categories which were heated in a closed container as shown in Fig. 1. With dimensions of 30 cm height × 25 cm diameter at 300–350 °C for 90 minutes the reactor starts melting the raw material. The waste plastic is heated, and the output is cooled to yield low-sulphur products. To avoid forming of toxic gases, the catalyst has been used. Since the waste plastics are treated without oxygen in the range of 300–350 °C, all hazardous gases are combusted.

Refining process set up used low to high boiling temperature energy source. Gasoline-type energy source was collected from 1st distillation stage between 104 and 141 °C. Diesel-type energy source has been collected from 2nd distillation stage, and it was between 250 and 285 °C.

Fig. 1 Plastic pyrolysis reactor, pyrolysed oil, distilled pyrolysed oil

2.2 SCR Unit and DEF Solution

Selective catalytic reduction (SCR) is a method of changing oxides of nitrogen, with the support of an inducer into N_2 and H_2O by more than 70–95%. DEF (Diesel exhaust fluid) is a semi-liquid with 32.5% urea, 67.5% de-ionized aquatic (Eq. 1).

The conversion output of the SCR is shown by

$$\text{RE}(\%) = (\text{Upstream pollution} - \text{Downstream pollution}) \times 100/\text{Upstream pollution} \quad (1)$$

The outline of the entire layout is illustrated in Fig. 2. The entire reformation of DEF solution occurs in three stages. Water in DEF is primarily vaporized at exhaust temperature liberating melted urea (Eq. 2).

$$NH_2{-}CO{-}NH_2(\text{aqueous}) \rightarrow NH_2{-}CO{-}NH_2(\text{molten}) + H_2O(\text{gas}) \quad (2)$$

Urea further disintegrates to give NH_3 and isocyanic acid (HNCO). The further work is to deal with HNCO which decomposes to yield next NH_3 atom and CO_2 molecule. The other NH_3 atoms are found for diminution. Almost 20% of urea is converted to HNCO and NH_3 in the vapour phase at 330 °C, and the remaining is converted at 400 °C (Eq. 3).

$$NH_2{-}CO{-}NH_2(\text{molten}) \rightarrow NH_3(\text{gas}) + HNCO(\text{gas}) \quad (3)$$

A pneumatic pump with diaphragm styled is utilized to absorb the NH_3 solution and sends it to the injectors. The maximum capacity of the pump to force the liquid

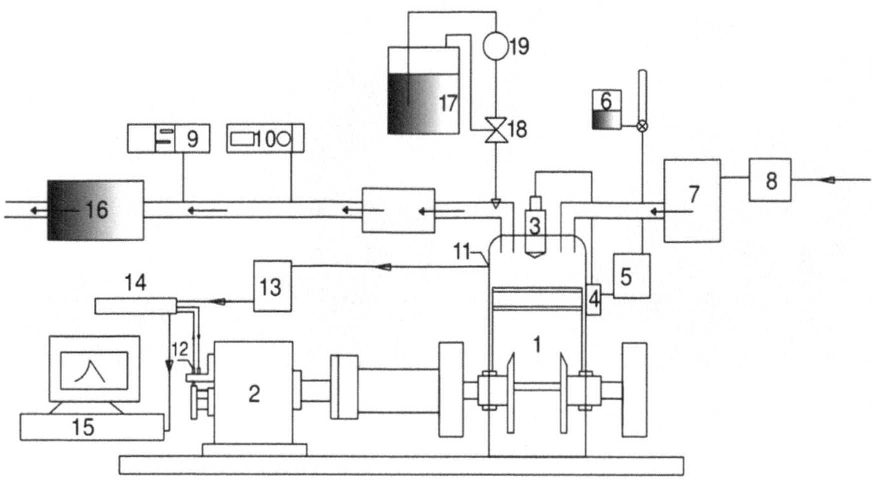

Fig. 2 Layout of the experimental setup

ammonia is 0.5 lit/min. The ejector with 12 V electric-operated solenoid valve is used to eject the DEF solution. The duty-cycle is around 1.5–6 ms.

3 Experimental Setup

Experimental runs have been performed on a Kirloskar five HP single cylinder naturally breathing aqua cooled DICI engine attached with an electrical generator resisted by a varying loading device. The outline of the engine setup is shown in Fig. 2. The specification of power source is given in Table 1. In-cylinder pressure has been quantified by AVL-GH12D pressure sensor. A calibrated five gas emission analyzer, AVL Digas 444 assesses the constituents of exhaust out pollution. AVL makes smoke measuring device was employed to gauge the smoke intensity. The in-cylinder information of the combustion gases can be obtained from P-θ diagram. The performance of the power train was arrived using BSFC, BTE, the combustion parameters such as P-θ, HRR and the pollution features such UBHC, CO, smoke were evaluated using the appropriate measuring units. The experimental runs were repeatedly performed for different loads in steps of 25%. The fuel properties are presented in Table 2.

Table 1 Specification of the test engine

Brand and type	Kirloskar and TV1
Power	5.2 kW @ 1500 rpm
Number of barrel	Single
Chamber type	Hemispherical
Piston shape	Shallow bowl
Ratio of Compression	17:5:1
Speed (rated)	1500 rpm
Diameter of bore	87.5 mm
Length of stroke	110 mm
Pressure of injection	220 bar
Timing of injection	23 deg CA BTDC
Fuel injection type	Direct
Holes in nozzle	3
Hole diameter	0.25 mm
Spray cone angle	120°
Cubic capacity	661.45 cc
Loading type	Electrical load
Cooling type	Water cooling
Ignition type	Compression ignition

Table 2 Characters of diesel, PPO and PPO with BHA

	Properties	Diesel	PPO	PPO + 100 BHA	PPO + 200 BHA	PPO + 300 BHA	PPO + 400 BHA
1	Calorific value (kJ/kg)	44,500	42,000	42,245	42,300	42,340	42,400
2	Fire point °C	65	72	73	75	72	73
3	Viscosity (Cst)	2.12	2.52	2.71	2.85	2.95	3.10
4	Specific gravity	0.850	0.835	0.846	0.843	0.855	0.844
5	Flash point °C	52	62	63	65	65	65
6	Density (kg/m^3)	835	793	796	800	803	805
7	Specific heat (kJ/kg K)	2.2	55	–	–	–	–
8	Cetane number	55	44	44	44	45	45
9	Aromatic content	20	55	–	–	–	–
10	Carbon residue	0.20%	0.01%	–	–	–	–
11	Sulphur content	<0.05	<0.02	–	–	–	–
12	Gum content (gm/m^2)	–	36	–	–	–	–
13	Colour	Orange	Pale black	Pale black	Pale black	Pale black	Pale black

4 Results and Discussion

The recital, burning and emission patterns of the CI power train using ULSD, neat plastic oil, plastic oil with dissimilar quantity of BHA are ascertained. The presentation attributes considered in this investigation are BTE, BSFC and pollution pattern like as NOx, HC, CO and smoke intensity are created with respect to BP.

4.1 Performance Characteristics

4.1.1 Brake Thermal Efficiency

Figure 3 illustrates deviations of thermal efficiency against BP for fuels used like Diesel, PPO, PPO + 100 mg/l of BHA, PPO + 200 mg/l of BHA, PPO + 300 mg/l of BHA, PPO + 400 mg/l of BHA, PPO + 400 mg/l of BHA + SCR fuels at non constant load situation. The BTE is found to increase with increase in load for all tested fuels. The BTE for neat ULSD was 28.25%, for PPO the thermal efficiency was observed as 26%, for PPO along 100 BHA the BTE was noted as 25.6%, for PPO along 200 BHA the BTE was noted as 25.3%, for PPO along 300 BHA the BTE was 25%, for PPO with 400 BHA the BTE was 24.8%, and for PPO with 400 BHA + SCR the BTE was noted to be 24.5%. As the BHA quantum raises, the thermal efficiency decreases compared to the diesel and raw PPO. This is because the antioxidant has higher reduction potential which creates larger quantity of fuel not to respond with O_2 which leads to lower the engine output. This may be because of improper burning attributes of antioxidant (Ramesha et al. 2015; Rajappan et al. 2015).

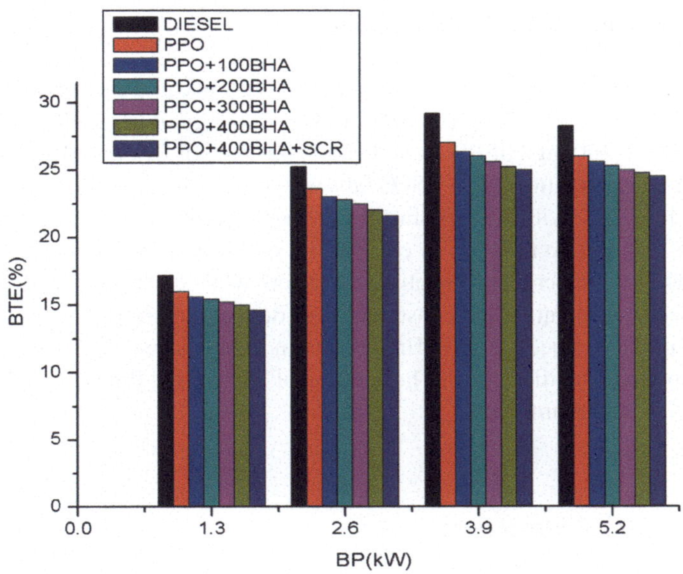

Fig. 3 Variation of brake thermal efficiency with brake power

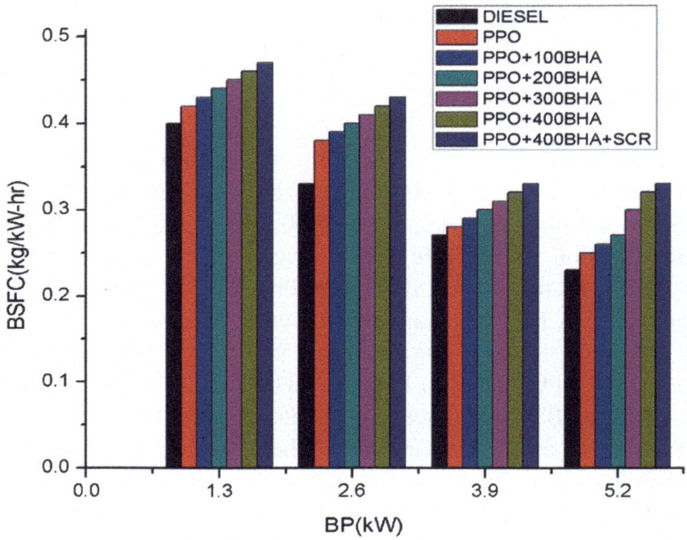

Fig. 4 Variation of BSFC with brake power

4.1.2 Brake-Specific Fuel Consumption

Figure 4 illustrates change of fuel consumption along brake power of tested fuels such as ULSD, PPO, PPO + 100 mg/l of BHA, PPO + 200 mg/l of BHA, PPO + 300 mg/l of BHA, PPO + 400 mg/l of BHA and PPO + 400 mg/l of BHA + SCR tested trials at various burden situations. The BSFC falls with raise in load for all the tested fuels. It is noted that BSFC with the blending of BHA antioxidant to the PPO drops equated to the neat PPO due to the drop in combustion delay duration (Senthilkumar et al. 2015). PPO has a lesser heat liberating characters while comparing with the diesel fuel and hence to maintain the constant power delivery, more fuel was utilized for pure PPO usage (Sanjeev 2015). This results in an enhanced PPO fuel utilization while comparing with diesel fuel. The addition of BHA to the PPO resulted in 3.56% rise in SFC at maximum load.

4.2 Emission Characteristics

4.2.1 NOx Emissions

Figure 5 illustrates deviations of NOx pollution in contradiction of load for ULSD, PPO, PPO + 100 mg/l of BHA, PPO + 200 mg/l of BHA, PPO + 300 mg/l of BHA, PPO + 400 mg/l of BHA and PPO + 400 mg/l of BHA + SCR. The oxides of nitrogen emission decrease due to blending of BHA antioxidant to all samples when comparing with the raw diesel fuel. The NOx pollutions are getting decreased

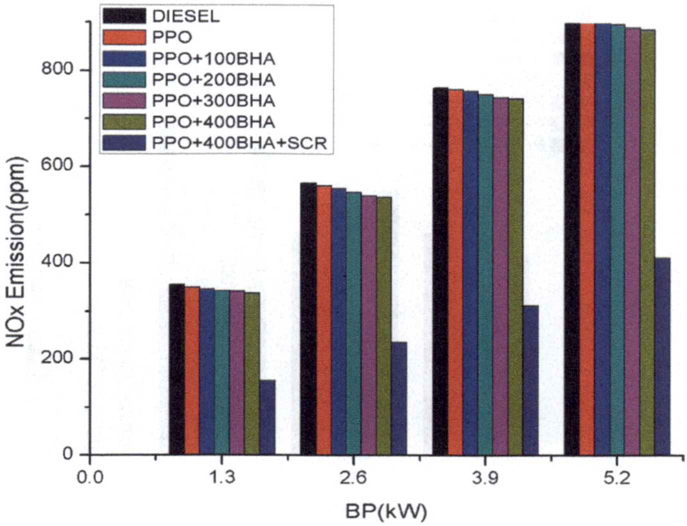

Fig. 5 Variation of NOx emissions with brake power

with decrease in the maximum cylinder heat. Maximum in-cylinder heat is getting increased with raise in load due to more fuel combustion temperature. NOx varies as air–fuel ratio varies. Longer burning period and more oxygen level in the combustion chamber were the main factors for NOx formation (Shaafi and Velraj 2015; Sahoo and Das 2009). The addition of antioxidant leads to incomplete combustion because the BHA behaves like oxygen-reducing agent. NOx pollution decreased for BHA because of high release rate and maximum in-cylinder compression (Vishnu Prakash and Hariram 2015). It is noted that the oxides of nitrogen pollution for diesel were noted to be 826 ppm, for neat PPO it was noted as 818 ppm, the NOx pollution as 815, 810, 807 and 802 ppm for various dosage of BHA, respectively.

4.2.2 CO Emissions

Figure 6 illustrates change of CO effluence in contradiction of the load for ULSD, PPO, PPO + 100 mg/l of BHA, PPO + 200 mg/l of BHA, PPO + 300 mg/l of BHA, PPO + 400 mg/l of BHA and PPO + 400 mg/l of BHA + SCR at various load situations. The CO pollution falls with increase in burden at all fuels tested. Less temperature of flames, less dwelling time, too less or too high equivalence ratio, effluent fuel–air ratio, unfinished combustion due to insufficient oxygen or flame slaking is prime factor for CO pollutions from CI engines (Rolvin et al. 2016). It is noticed that CO pollutions raise with the blending of BHA antioxidant (Selvan et al. 2014). Further addition of BHA increases the CO emissions because normally antioxidant performs reduction process which increases CO during combustion. The antioxidant blends, reduces the delay duration and rises the mixture of A:F and

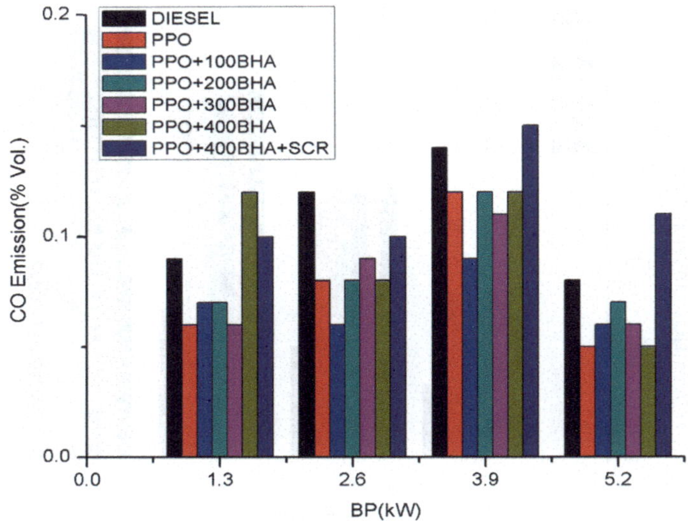

Fig. 6 Variation of CO emissions with brake power

activation of carbon burning resulting high quality burning (Selvaganapthy et al. 2013). It is observed for diesel the CO emission was found to be 0.08% by vol., for PPO the CO pollution was observed to be 0.05% by vol., the CO pollution was noted to be 0.06% by vol., for PPO + 100 BHA, for PPO with 200 BHA CO pollution was 0.06% by vol., the CO pollution was noted to be 0.07% by vol. for Plastic pyrolysis oil with 300 BHA, for Plastic pyrolysis oil with 400 BHA the CO pollution was 0.06% by vol., and it was 0.12% by vol., for 400 BHA + SCR.

Figure 7 demonstrates the change in smoke density in contradiction of brake power for pure HSD, pyrolysis oil, PPO + 100 mg/l of BHA, PPO + 200 mg/l of BHA, PPO + 300 mg/l of BHA, PPO + 400 mg/l of BHA and PPO + 400 mg/l of BHA + SCR at various load situation. The main factor for the generation of smoke pollution is the improper burning in the diffusive burning zone (Krupakaran et al. 2016). On mixing of antioxidant to PPO, the pollution of smoke received trails the trend as revealed by (Datta and Mandal 2016). The considerable rise in smoke may be because of the antioxygenated BHA exists in the fuel (Moffat 1988). Smoke pollution for ULSD was 2.1 BSU, for neat Pyrolysis oil it was noted to be 2.5 BSU, for PPO with 100 BHA it was 3.4 BSU, for PPO with 200 BHA it was 3.1 BSU, for PPO with 300 BHA it was 3.7 BSU, for PPO with 400 BHA it was 3.7 BSU, and for PPO with 400 BHA + SCR it was 4.5 BSU.

4.2.3 HC Pollution

Figure 8 illustrates the change of HC pollution in contradiction to load for ULSD, PPO, PPO + 100 mg/l of BHA, PPO + 200 mg/l of BHA, PPO + 300 mg/l of

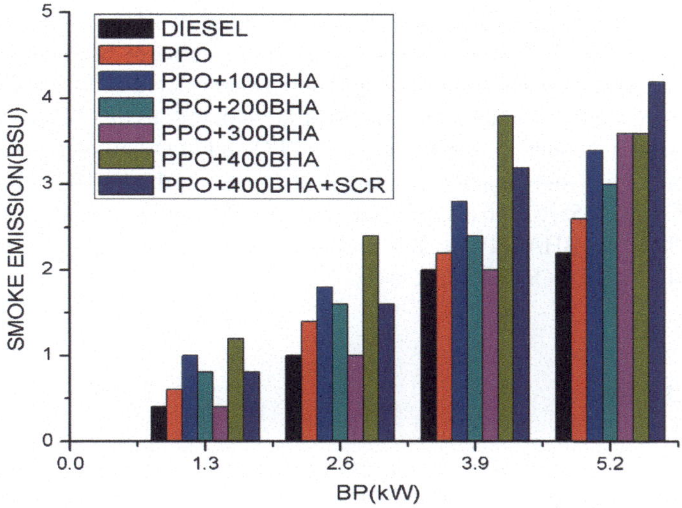

Fig. 7 Change of smoke pollution against BP

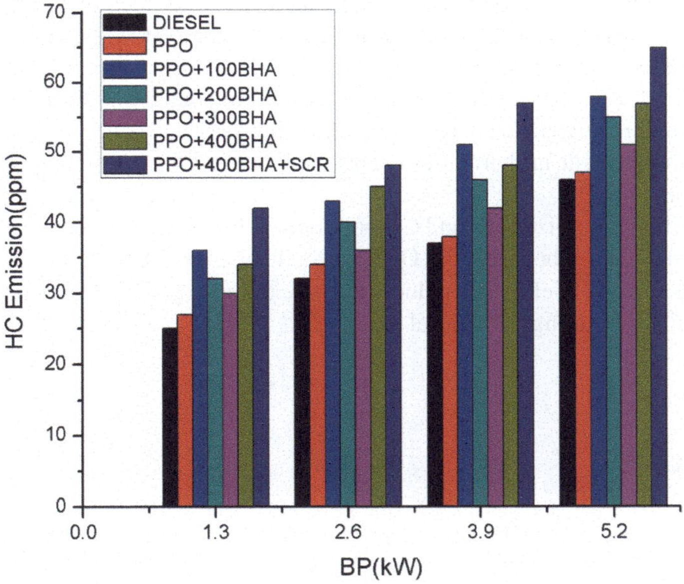

Fig. 8 Change of UBHC pollution against BP

BHA, PPO + 400 mg/l of BHA and PPO + 400 mg/l of BHA + SCR at diverse load situations. The HC pollution raises because of mixing of BHA into the PPO. The BHA antioxidant behaves as an oxygen suppressing agent and stops oxygen for combustion (Sachuthananthan et al. 2018). The attempt reveals that BHA antioxidant shall be castoff as reduction agent with HSD, PPO and in DP oil blends to improve emission of the fuel significantly. At peak load situations, the UBHC pollution was noted as 45, 47, 57, 55, 52, 56 and 63 ppm for Diesel, PPO, PPO + 100 mg/l of BHA, PPO + 200 mg/l of BHA, PPO + 300 mg/l of BHA, PPO + 400 mg/l of BHA, PPO + 400 mg/l of BHA + SCR, respectively.

5 Conclusion

- The experimental runs for characters of PPO reveal that most of the properties are similar to diesel fuel.
- The engine function effortlessly using neat PPO at zero to peak load condition with no variation in operating characteristics.
- Plastic pyrolysis oil has more fuel usage, due to less heat liberating property. Due to mixing of BHA to PPO, a mild rise in fuel usage up to 200 mg/l BHA compared to neat PPO operation.
- BHA blending to PPO revealed rise in the burning stage and drop in the flash temperature relative to base fuel and plastic oil process.
- A small decrement in thermal efficiency was noticed due to mixing of BHA to plastic oil.
- BHA getting minimized HC and CO pollution up to 200 mg/l associated with neat plastic oil, since it behaves like O_2 storage element and donates O_2 for the HC, CO oxidation. NOx effluence reduces because of the use of BHA antioxidant with PPO while comparing with diesel fuel.

References

Anup, T. J., & Watwe, V. (2014). Waste plastic pyrolysis oil as alternative for SI and CI engines. *International Journal of Innovative Research in Science, Engineering and Technology, 3*(7), 14680–14687.

Chindaprasert, N., & Wongkhorsub, C. (2013). A comparison of the use of pyrolysis oils in diesel engine. *Energy and Power Engineering, 5*(4), 350–355.

Datta, A., & Mandal, B. K. (2016). Numerical investigation of the performance and emission parameters of a diesel engine fuelled with diesel—biodiesel—methanol blends. *Journal of Mechanical Science and Technology, 30*(4), 1923–1929.

Hamzah, M. H., Abdullah, A. A., Sudrajat, A., Ramlan, N. A., Jaharudin, N. F. (2015). Analysis of combustion characteristics of waste plastic disposal fuel (WPDF) and tire derived fuel (TDF). In *Applied Mechanics and Materials* (Vol. 773, pp. 600–604).

Krupakaran, R. L., Hariprasad, T., & Gopalakrishna, A. (2016). The performance and exhaust emissions investigation of a diesel engine using γ-Al_2O_3 nanoparticle additives to biodiesel. *Journal of Carbon Management, 7,* 1–9.

Lee, S., Yoshida, K., & Yoshikawa, K. (2015). Application of waste plastic pyrolysis oil in a direct injection diesel engine: For a small scale non-grid electrification. *Energy and Environment Research, 5*(1), 18–32.

Moffat, R. J. (1988). Describing the uncertainties in experimental results. *Experiment Thermal Fluid Science, 1,* 3–17.

Naima, K., Liazid, A., & Mnaouer, B. P. (2013). Waste oils as alternative fuel for diesel engine: A review. *Journal of Petroleum Technology and Alternative Fuels, 4,* 30–43.

Pawar Harshal, R., & Shailendra, M. L. (2013). Waste plastic pyrolysis oil alternative fuel for CI engine—A review. *Research Journal of Engineering Sciences, 2*(2), 26–30.

Rajappan, R., Suresh, V., Udhayakumar, K., & Anbuselvan, D. (2015). Performance and emission characteristic of a variable compression ratio direct injection diesel engine using plastics oil. *Journal of Chemical and Pharmaceutical Sciences, 7*(7), 40–43.

Ramesha, D. K., Kumara, G. P., Mohammed, A. L., Mohammad, H. A., & Kasma, M. A. (2015). An experimental study on usage of plastic oil and B20 algae biodiesel blend as substitute fuel to diesel engine. *Environmental Science and Pollution Research, 23*(10), 9432–9439.

Rolvin, D., Vinoothan, K., Binu, K. G., Thirumaleshwara, B., & Raju, K. (2016). Effect of titanium dioxide and calcium carbonate nanoadditives on the performance and emission characteristics of C.I. engine. *Journal of Mechanical Engineering and Automation, 6*(5A), 28–31.

Sachuthananthan, B., Balaji, G., & Krupakaran, R. L. (2018). Experimental exploration on NO*x* diminution by the combined effect of antioxidant additives with SCR in a diesel engine powered by neem biodiesel. *International Journal of Ambient Energy, 45,* 1–14.

Sahoo, P. K., & Das, L. M. (2009). Combustion analysis of Jatropha, Karanja and Polanga based biodiesel as fuel in a diesel engine. *Fuel, 88,* 994–999.

Sanjeev, K. (2015). Experimental investigation on performance of direct injection diesel engine fuelled with jatropha methyl ester. *Waste Plastic Oil and Diesel Oil, 2*(6), 72–75.

Selvaganapthy, A., Sundar, A., Kumaragurubaran, B., & Gopal, P. (2013). An experimental investigation to study the effects of various nanoparticles with diesel on DI diesel engine. *ARPN Journal of Science and Technology, 3*(1), 34–42.

Selvan, V. A. M., Anand, R. B., & Udayakumar, M. (2014). Effect of cerium oxide nanoparticles and carbon nanotubes as fuel-borne additives in diesterol blends on the performance, combustion and emission characteristics of a variable compression ratio engine. *Fuel, 130,* 160–167.

Senthilkumar, S., Sivakumar, G., & Manoharan, S. (2015). Investigation of palm methyl-ester biodiesel with additive on performance and emission characteristics of a diesel engine under 8-mode testing cycle. *Alexandria Engineering Journal, 54*(3), 423–428.

Shaafi, T., & Velraj, R. (2015). Influence of alumina nanoparticles, ethanol and isopropanol blend as additive with diesel–soybean biodiesel blend fuel: Combustion, engine performance and emissions. *Renewable Energy, 80,* 655–663.

Soni, N., Sharma, M. C. (2013). Production of alternative diesel fuel from waste oils and comparison with fresh diesel—A review. *4*, 54–58.

Vijay Kumar, B., & Patil, B. M. (2015). Fuel from plastic waste. *International Journal on Emerging Technologies, 6*(2), 121–128.

Vishnu Prakash, R., & Hariram, V. (2015). Characterization of waste plastic oil derived from pyrolytic batch reactor and analysis of performance and emission parameters in a direct injection compression ignition engine. *Journal of Chemical and Pharmaceutical Research, 7*(1), 488–498.

Vogler, (1984). Recent advances in bio energy research, conversion of plastic wastes into liquid fuels—A review. *National Institute of Renewable Energy, India, 14*(1), 10–16.

A Study of Solid Waste Management in Indore City: With Special Reference to Biomedical Waste

Mohini Jadon

Abstract Solid waste management is the process of collection, treatment and disposal of solid material that is discarded because it is no longer useful. Solid waste management is important for the protection of public health, safety and environment quality. Indore city is holding it's first position in Swacch Bharat Abhyian for cleanliness and is managing its waste sustainably. The waste generated in the area, i.e. dry, wet and biomedical waste is managed by the IMC. If the waste generated is not managed properly it may lead to serious health hazard, infectious diseases, environmental issues, climatic changes. The Indore city has started collection of waste door to door with segregation by which they can easily manage it. There are various methods used to dispose the waste. Study area is trying to manage it's waste sustainably for which Indore will host the eighth 3R (Reduce, Reuse and Recycle) a strategic platform for sharing best practices, including new and emerging issues of concern in waste management area.

Keywords Solid waste management · Biomedical waste · Environmental issues · Disposal · Health hazard · 3R · IMC

1 Introduction

In India, most of the urban local bodies (ULBs) are facing the major problem of solid waste management (SWM), where per person solid waste generation is increasing due to urbanization, industrialization and economic growth. Nowadays it is major challenge for effective solid waste management in densely populated cities. Waste management is a good initiative for achieving sustainable development in a country which focuses on optimum utilization of resources. India is at tenth position for waste generation in the world due to growing urbanization and high consumption. Urbanization and modernization are the key factors responsible for generation of solid waste in the country. The cities like Mumbai, Nagpur, Chandigarh, Delhi, Surat, Ahmedabad and Indore are best at managing their municipal waste. In India,

M. Jadon (✉)
24, Kumharkhadi, Marimata Square, Indore, India
e-mail: Mohinijadon16@gmail.com

S. K. Ghosh (ed.), *Urban Mining and Sustainable Waste Management*,
https://doi.org/10.1007/978-981-15-0532-4_28

Madhya Pradesh is the first state which has 100% started door to door collection of waste per day. Indore is ranked as first and second Bhopal in India for cleanliness in the city. In the year 2001, the population of Indore city was 10,86,673 and the waste generated in the area was 600 mt per day; in year 2011, the population of Indore city was 19,94,398 and waste generated was about 885 mt per day and in the year 2018, the expected population of Indore city is 34 lakhs approximately and waste generated in the city is 1100 mt per day which is segregated as dry waste-500 mt per day, wet waste-600 mt per day. With the rapid increase in population, the waste is also generating at a higher rate which can be hazardous to the environment if not managed sustainably. With increasing population, the health issues are also increasing and so there are many communicable diseases which are spread in the area. For the treatment of these diseases, the medical facilities are increasing and thus, the biomedical waste generated is also increasing which is a big hazard to society. To dispose off this waste safely, there are many organizations working which collect the waste from hospitals, laboratories, clinics and other places. In India, about 550 t per day medical waste is generated and in the western part of Madhya Pradesh, the medical waste generated is 5200 kg per day, from which 2400 kg waste is only from the Indore city. Improper management of the solid waste and biomedical waste can lead to environmental issues and serious health issues. Rubbish waste can cause air and water pollution.

Objective The objective of the study is to analyse the:

I. The dry and wet waste generated in the study area and to prefer the 3R (Reduce, Reuse, Recycle) model.
II. Defining the biomedical waste generated in the area and the process used to dispose it.

Methodology The study is based on secondary as well as primary survey. The survey is done on the trenching ground, material recovery centre and Hoswin incinerator Pvt. Ltd. Indore (for biomedical waste). Other information is collected from Indore municipal corporation and other organizations.

Study area Geographically, Indore is located on the Malwa plateau where the latitudinal extent is between 22° 37′ 29.66″ N 75° 46′ 86″ E and 22° 48′ 34″ N 75° 56′ 32″ E at an average altitude of 553 metres. The area of Indore District is 3898 KM^2 and the area of Indore city is 13,717 ha. Indore city is a commercial capital of Madhya Pradesh which is most populated city of the State.

Discussion The waste generated in Indore city is collected door to door from household, industrial and commercial areas. The waste is segregated as dry waste, wet waste and biomedical waste which includes (testing, diagnosis, immunization or treatment of human being or animals). The solid waste generated is about 1100 mt per day which after segregation is about, wet waste-600 mt per day and dry waste-500 mt per day. The method used to dispose the wet waste is composting method. Waste after collection is brought to transfer station, after compression process it is filled in the capsules (big trucks). Thus, the waste is dumbed on the site.

Wet waste The method used to decompose wet waste is the Win drop composting method, which is 35 days process in which the dumbed waste is let opened for seven days, then the turning process is done in which the waste is turned to supply it organic materials—water, oxygen and the other micro-organisms which breaks the material in fine particles. The material is further moved to filtration process in which the compost is obtained and the remaining material is moved to the landfill sites. The compost is used in the gardens as a manure. Thus, the wet waste generated in the city is fully decomposable. At some places in Indore, where the wet waste is generated in bulk such as temples, schools, hotels and marriage gardens produces about 40–50 kg waste per day such organizations have installed an organic waste compost machine in which the waste is compost and it breakdowns into humus after a period of 30 days, the waste is composted daily which is good for the environment. To motivate the organizations IMC has rebate them on tax about 5–6%.

Dry waste The waste after compression is dumbed on the material recovery centre. On the site, the rag-pickers collect the plastic and other reusable material such as metals, bottles, cans etc which is send to the industries for its recycling. 3–4 years ago this rag-pickers move from one place to another in search of this recyclable material and after collection, they used to sell it to the Kabbadis for which they even don not get fair prices. But now, the IMC has brought these people on the MRC where they collect the reusable items and sell it there for which they get fair wages and this people are uplifted on their social as well as economic level. The remaining valueless inorganic non-recyclable waste is referred to as residuals. The polythene as a residual is a big hazard to environment. IMC has banned the use of polythene in the city but still polythenes are being used for specific purposes so to recycle this polythene there is a process:

(1) Phatka machine—in this machine, dust, moisture, stone and other materials are removed from the polythene so that it could be proceeded further.
(2) Aglo machine—now the plastic is processed into the aglo machines in which granules are formed by the polythene.
(3) Gatta machine—the granules which are being formed are now further proceeded in this machine by which gattas are formed and thus, these gattas are sent to industries for the formation of pipe.

(pp polythene is processed in grandier machine and the crushed plastic is further send to M.P.R.D.A. for use in road construction).

Lastly, some e-waste and residual waste are prepped and sent in the combustor for clean disposal. The waste which was collected from last decades was dumbed on the ground openly but from the year 2017, this waste is disposed by bio-mining process.

Indore had host the eighth 3R (Reduce, Reuse And Recycle) Regional Forum that introduced a strategic platform to implement best practices with innovative ideas in waste management. The summit was organized by the Ministry of Housing and Urban Affairs Government of India, the Ministry of the Environment, Government of Japan (MOEJ) and the United Nations Centre for Regional Development (UNCRD). The theme of the summit was "Achieving Clean Water, Clean Land and Clean Air through 3R and Resource Efficiency". The concept of 3R—"Reduce, Reuse, Recycle" is to be implemented in the city so that waste can be treated easily. Unfortunately, rapid urbanisation and attendant challenges have gradually pushed this concept to the background of urban consciousness. The launch of the Swachh Bharat Mission by Government of India in 2014, with its objective of 100% scientific management of Municipal Solid Waste synergises well with "Mission Zero Waste", and so this is the next step which is taken by the IMC for sustainable solid waste. The IMC has banned the use of poythenes, use of disposals on the tea stall, etc., also the next target is to reduce the waste generated in the city about 100 mt in a day.

Biomedical waste Biomedical waste contains infectious material which is generated during the diagnosis, treatment and immunization of human beings or animals or in research activities pertaining thereto. The main source of biomedical waste is hospitals, laboratories, clinics and pathologies. In the Western part of Madhya Pradesh, there is a private organization which deals with biomedical waste named Hoswin Incinerator Pvt. Ltd, Indore. The company is granted authorization to operate health care facility for the disposal of biomedical waste till 2019 and then they will get renewed their license.

The organization has divided the districts in division namely—**Indore, Khandwa, Barwani, Burhanpur and Khargon**.

Working The waste is collected daily from hospitals, clinics, laboratories, etc. The total number of vehicles engaged in this work is 20 which collect the medical waste from the five divisions and within the city there are six vehicles which are engaged in collecting the waste. The waste collected from Indore city within a day is about 2400–2600 kg of medical waste which is to be disposed within 24 h and if there is any technical problem in the plant then the waste has to be disposed off within 48 h. The waste is disposed quickly otherwise it would be hazardous to health and environment. The waste is categorized in four categories.

2 Types of Medical Waste and the Methods Used to Dispose It

1. **Yellow clinical waste bag**—This bag are used for the storage of soft clinical waste contaminated with infectious or potential infectious blood or bodily fluids. It consists of human anatomical waste, animal waste, chemical liquid waste and other clinical waste. The method used to dispose this waste is the incineration process in which the waste is burnt in a controlled process. Incineration is an old technology and was widely used in the past for all sorts of waste. After the process of incineration the ashes are sent for landfill process at ramky plant at pithampur.

2. **Red clinical waste bag**—The bag contains the contaminated waste (recyclable) such as bottles, intravenous tubes, urine bags, etc. The process used to manage the red bag waste is autoclaving. Autoclaving method is being used for sterilizing medical equipments for reuse in which the equipments are closed under big chambers that apply heat, pressure and steam at 120 °C in which after the sterilization process the waste is crushed or cut into smaller parts so that it could be sent to industries for recycling. The recycled waste is sold to authorized industries for resell from which polythenes are made for packing the medical products.
3. **Blue clinical waste bag**—The blue bag contains of the discarded and contaminated glass including medicine vials and ampoules. Even the process used to discard the blue clinical waste bag is the autoclaving process in which the glass is sterilized at 120 °C temperature and then the glass is crushed and sold to industries for recycling process.
4. **White clinical waste bag**—The white bag contains the metals like syringes, needles, burners, scalpels, blades or any other contaminated sharp object which is discarded. Then this bag in the plant is disposed off by landfill in which the ground is digged for about 50 feet and then this material is dumbed in the ground.

All the four categories of medical waste are disposed off by different methods, which is essential for a healthy and safe environment.

3 Conclusion

This study explores the importance of solid waste management for sustainable development. There were two objectives to be achieved in the study. The first objective was to study the dry and wet waste generated in the area with preference to 3R model. The researcher investigated the amount of waste generated and the methods used to decompose it also the IMC would prefer the model so that the waste could be reduced and managed easily. The second objective was defining the biomedical waste generated in the area and the process used to dispose it. The researcher investigated the division covered by the organization to treat the waste, categories of the waste and methods used to dispose it. Further, this study incorporates the organization to explore the level of environmental friendly methods.

4 Suggestions

1. An integrated municipal solid waste management system may prioritize its waste management by minimization of waste, recycling/recovery of more products. Reduction in waste is most important for any municipal corporation to deal with it sustainably.

2. Hospitals, clinics and laboratories should segregate the medical waste as described by the Biomedical waste rules 2016. So that it could be disposed off fast.
3. There should be awareness program by the government and NGO'S for the segregation of and reduction of waste.

References

Agarwal, R. Waste management initiatives in India for human well being.
Nag, A. (2006). Environment education and solid waste management. (ISBN-9788122416909).
Newspapers—The Hindu, Times of India.
Smart City Project Report. (2017).

Hydrothermal Carbonization—A Sustainable Approach to Deal with the Challenges in Sewage Sludge Management

Vicky Shettigondahalli Ekanthalu, Gert Morscheck, Satyanarayana Narra and Michael Nelles

Abstract Hydrothermal carbonization (HTC) is emerging as a most promising technology to effectively manage the extensively produced sewage sludge by converting it into high energy density bio-coal and bio-products. The main objective of the current review is to briefly compare the existing sewage sludge management solutions with the HTC process. Also, this review clearly explains the effectiveness of HTC in terms of environmental sustainability, considering the existing legislation (European and national level (Germany)). Further, an attempt is made to explain the eco-innovative strategies of HTC to fulfill the principle concept of the circular economy "from waste to resource" as the most suitable waste management approach. The importance of sewage sludge as a valuable resource of matter and energy has been highly appreciated. Besides updating the knowledge on the effectiveness of HTC as a technology to manage sewage sludge, this review briefly summarizes economically the feasibility and specifies some of the most appropriate future research prospects for the technical development of HTC in sewage sludge management.

Keywords Hydrothermal carbonization · Sewage sludge · Sewage sludge management · Environment sustainability

V. Shettigondahalli Ekanthalu (✉) · G. Morscheck · S. Narra · M. Nelles
Faculty of Agriculture and Environmental Science, Universität Rostock, Rostock, Germany
e-mail: vicky.ekanthalu@uni-rostock.de

G. Morscheck
e-mail: gert.morscheck@uni-rostock.de

S. Narra
e-mail: satyanarayana.narra@uni-rostock.de

M. Nelles
e-mail: michael.nelles@uni-rostock.de

S. Narra · M. Nelles
Deutsches Biomasseforschungszentrum (DBFZ), Leipzig, Germany

S. K. Ghosh (ed.), *Urban Mining and Sustainable Waste Management*,
https://doi.org/10.1007/978-981-15-0532-4_29

1 Introduction

The German sewage sludge ordinance prohibits the direct use of sewage sludge to agriculture and landscaping from 2029 (AbfKlärV 2017). In this concern, the current number of incineration plants in various states of Germany will be enough to handle the produced sewage sludge in future, and this further can trigger a new problem to store and process the produced sewage sludge. Additionally, the option for disposing the sewage sludge is becoming progressively limited as the results of advanced environmental policy and laws made in Germany over the last decade. In this concern, increase in number of incineration plants might seem to be the possible solution, but this alternative solution would not only be costly but also creates political contention. In the year 2016 in Germany, >25% of sewage sludge was used for agricultural purpose, and ~65% of produced sewage sludge was incinerated ((Destatis) 2018). The best possible way to tackle this problem of sewage sludge management is by introducing the technology where sewage sludge can be transformed as a value-added product, not only by recovering the nutrients (viz. phosphate) but also by producing different products which have the market value. By implementing such a strategy, not only the problem will be solved but also it is possible to make additional profit by treating the sewage sludge. Provided the potential of the produced value-added products is technically proved and with the practical market value.

Dewatering is the primary challenge in the management of sewage sludge, and most of the current nutrient recovery technologies mandatorily require dewatering; this makes the recovery process expensive and problematic. Further, if sewage sludge is incinerated, then the operating costs will be very high due to the presence of high moisture content. Further, incineration can potentially emit toxic air pollutants such as NOx, SO_2 and dioxins which can make the situation worse. In this concern, hydrothermal carbonization is a technology that demonstrates a high potential to treat the moist sewage sludge without having it to be dewatered. HTC uses the moisture present in sludge as the reaction medium to process the sewage sludge without pre-drying, and the increase in the temperature and pressure will aid the moisture to serve as a solvent, reactant and even as catalyst for the treating sewage sludge. The product produced (Biochar) from this process is hygiene, essentially free of pharmaceuticals and easily dewaterable (Crocker 2010). The HTC produced biochar at controlled environment is reported to be well qualified and likely to be of the coal quality (Saetea and Tippayawong 2013). However, the liquid phase is rich in dissolved organic compounds must be treated as wastewater or discarded.

2 Laws and Regulation in European Union Regarding Sewage Sludge Disposal

The EU Sludge Directive regulates the use of sewage sludge (especially in agriculture) EU-wide (EU Commission 1986). The directive commissioned the limit values for heavy metals in sludges and soils, and for the limit in the quantity of heavy metals that must be applied to the soil per year. Nevertheless, the use of sewage sludge in farming is completely prohibited if the concentration of any of the heavy metals exceeds the specified limits. The directive also includes application restrictions depending upon the intended use. For example, no sewage sludge may be applied for fruit and vegetable crops during their growing season, and also the application is restricted on grasslands or forage crops.

The German Waste Management Law (KrWG—Closed Substance Cycle Act) is the legal basis for the disposal of waste, and thus, also of sewage sludge (KrWG 2012). The KrWG should promote the circular economy to ensure the environmentally sound management of wastes. The sewage sludge ordinance is based on Section 11 of the Closed Substance Cycle Act. The German sewage sludge ordinance (AbfKlärV) regulates the disposal of sewage sludge in agriculture, horticulture and landscaping (AbfKlärV 2017). Further, the quality of the sewage sludge must also satisfy the requirements of the German Fertilizer Act (epidemic hygiene, phytohygiene, pollutant limit values) for its soil-related use. Since 2017, the new sewage sludge regulation is in force in Germany and makes it mandatory to recover P from sewage sludge or sewage sludge incineration ash.

According to this regulation, at least 50% of the phosphorus in the dry mass must be recovered from the sewage sludge, the phosphorus content should be lowered to less than 20 g/kg DM, and at least 80% of the phosphorus must be recovered from sewage sludge ashes. Further, this regulation amends that, wastewater treatment plants with more than 100,000 populations equivalent (PE) must only recover phosphorus from the effect of 1 January 2029 (soil-related disposal of sewage sludge is no longer permitted), and for the wastewater treatment plants with more than 50,000 PE must only recover phosphorus by 1 January 2032. Only sewage sludge from smaller wastewater treatment plants ($\leq$50,000 PE) may be used for soil amendment in the near future.

In this concern, technology like HTC can be highly appropriate. HTC has a high potential to reach the requirement of KrWG and AbfKlärV in managing the sewage sludge. HTC can produce value-added products from sewage sludge which has commercial value, rather than using sewage sludge directly for agriculture purposes.

3 Sewage Sludge Processing and Conventional Sewage Sludge Management Technique

In Europe, it is estimated that approximately 50% of secondary sewage treatment plants operating cost is associated with the sewage sludge treatment and disposal. In addition to the quality of the raw sewage sludge, several processes and process parameters within the sewage sludge treatment at wastewater treatment plants (WWTP's) influence on final character sewage sludge and the cost associated with the sludge management (Kacprzak et al. 2017).

The typical process is summarized below (Burton et al. 2013):

- Preliminary treatment by mechanical screening
- Primary thickening (flotation, gravity, drainage, centrifuges, belt)
- Liquid sludge stabilization (aerobic and anaerobic digestion, lime treatment)
- Secondary thickening (flotation, gravity, drainage, centrifuges, belt)
- Conditioning (thermal and chemical)
- Dewatering (belt press, plate press, drying bed, centrifuge)
- Final treatment (wet oxidation, composting, drying, lime addition, incineration, pyrolysis, disinfection)
- Storage (liquid sludge, compost, dry sludge and ash)
- Transportation (road, pipeline, sea)
- Destination (agriculture, forest, land reclamation, landfill and other use).

3.1 Conventional Techniques in Sewage Sludge Management

German municipal sewage treatment plants generate about two million tons of dry sewage sludge annually, with the quantity of thermally treated sewage sludge increasing from 31.5% in 2004 to more than 65% in 2016. The conventional sewage sludge management route in Germany has been explained in Table 1.

Table 1 Conventional strategies in managing sewage sludge (Kacprzak et al. 2017)

Principle strategies of sewage sludge management	
Organic recycling Management of sludge by converting it into fertilizer and potential of sewage sludge to be used in amending the soil	**Recycling the energy and material** Associated with using the fuel property of the sludge and recovery of nutrients and mineral
Agriculture • Direct use of treated sewage sludge as soil fertilizer Reclamation • Using the sludge in various forms to restore or give new utility values for the devastated or degraded land Composting • Using the sewage sludge for the production of compost for soil fertilization that fulfils the criteria of the organic fertilizers Mechanical biological treatment • To prepare the sludge to recover the nutrients (phosphorus)	Incineration • Managing the sludge directly by thermally treating it in the plant designed mainly to incinerate sewage sludge Alternate thermal methods • Using the sewage sludge in the solid fuel production process, such as • Pyrolysis • Quasi-pyrolysis • Gasification • Thermochemical of sewage sludge ashes to recover the nutrients Co-incineration • In cement plants, power plants, etc. • Fuel property of the sewage sludge can be used by co-incinerating the sludge with other fuels

4 Hydrothermal Carbonization of Sewage Sludge

HTC can be referred to as a wet pyrolysis process for the conversion of organic feedstock with high moisture content into a solid product denoted as biochar or hydrochar. HTC is performed in an autogenous pressure and at an elevated temperature between 160 and 250 °C (Sirén Ehrnström 2016, Libra et al. 2011). In hydrothermal carbonization, chemical dehydration of the biomass takes place, which eliminates the hydroxyl groups from the biomass leading to a decreasing hydrophilic functional group. The process results in significantly enhanced mechanical dewatering properties of the produced HTC product compared to sewage sludge or initial biomass (AVA-CO_2, sustainability consult 2016).

HTC is a "wet pyrolysis process", meaning sewage sludge can be used without prior, costly drying process. HTC has been efficiently demonstrated to save 65% of electric energy and 60% thermal energy on laboratory scale compared to the conventional sewage sludge treatment process. Additionally, the high carbon efficiency of the process aids in minimizing emissions of greenhouse gases (vom Eyser et al. 2015). The HTC produced biochar at a controlled environment is reported to be well qualified and likely to be of the coal quality (Saetea and Tippayawong 2013). The observation made on the energy content of the biochar shows that HTC process increases the carbon content of the sludge, providing the higher calorific value by enhancing the energy content of the produced biochar from 17 to 19 MJ/kg. Further, the fuel ratio of raw sludge has been reported to increase from 0.02 to 0.18 in the HTC conversion process (He et al. 2013).

Figure 1a, b depicts the SEM images of hydrochar and sewage sludge which explain the surface morphology, illustrating the rupture of the hydrochar surface due to the effect of HTC. The hydrochar seemed to be brownish-black in colour, signifying that hydrochar is not completely carbonized. Hydrochars have the potential to be used as an adsorbent for wastewater treatment. Despite their scant porosity, the tenable surface chemistry can enhance adsorption characteristics. Moreover, the combined chemical and physical optimization can improve their adsorption performance as a result of pore development and surface activation. In general, equilibrium adsorption data of hydrochar fit well with Langmuir and Freundlich models, and the values of adsorption capacity of hydrochar are in the same order of magnitude than those obtained using commercially activated carbons.

Iodine number is typically used as an indication of the adsorption capability of a material. The experiments carried out by (Saetea and Tippayawong 2013) clearly explain that HTC significantly increases the iodine number of raw sewage sludge from 93 $\pm$ 4 mg/g to 222 $\pm$ 12 mg/g at 1 h reaction time. Nevertheless, with the increase in HTC reaction time, iodine absorption capacity dramatically decreases, as the small pore structure gets destroyed which paves the way for larger pores with decreased ability to adsorb iodine (see Fig. 1c) (Saetea und Tippayawong 2013).

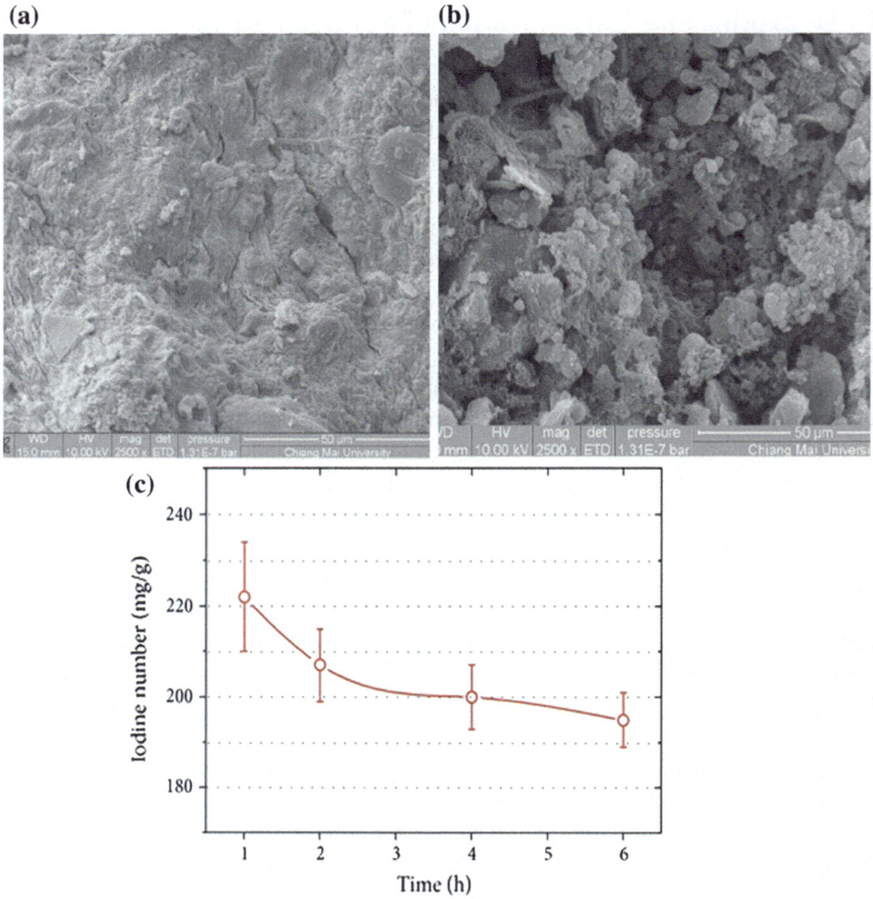

Fig. 1 SEM images of **a** dried sewage sludge and **b** resulting hydrochars and **c** Effect of HTC time on iodine adsorption ability of hydrochars (Saetea and Tippayawong 2013)

4.1 Advantages of HTC Against the Incineration of Sewage Sludge

HTC has several advantages comparing with the conventionally available technology (incineration). The primary advantage of HTC against incineration is cost effectiveness. The management/disposal of sewage sludge with HTC is highly economical than incineration, and therefore, the price/ton disposed using HTC will be significantly lower. Based on the test results and preliminary estimation done by AVA-CO_2, HTC is priced at 50 €/ton disposal costs while incineration costs are between 80 and 110 €.

The HTC method has a high potential to dewatering wet biomass efficiently than any other technology. The research was performed by the Zurich University for

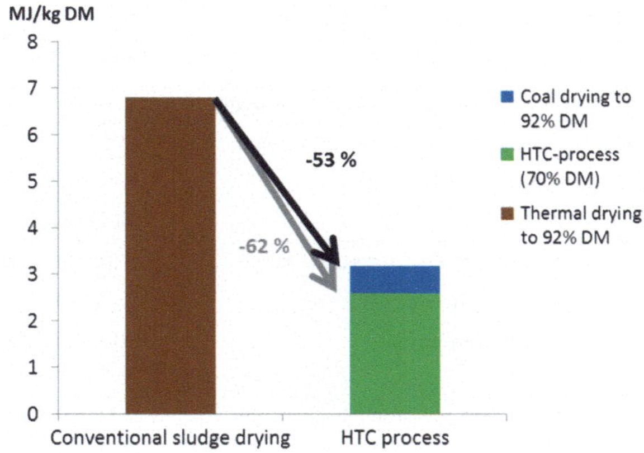

Fig. 2 Results comparing energy required to reduce the DM of sewage sludge from 21.3 to 92%, using conventional sludge drying process and HTC process (Stucki et al. 2015)

Applied Sciences to compare energy required to reduce the DM of sewage sludge to 92%, using conventional sludge drying process and HTC process. The results obtained prove that the HTC process saves up to 53% thermal energy and up to 69% electrical energy compared to conventional sludge drying methods (see Fig. 2) (Stucki et al. 2015).

AVA-CO_2, a Swiss-based company with the headquarters in Zug, Switzerland, and subsidiaries in Germany and Switzerland have introduced the world's first industrial size HTC plant in Karlsruhe, Germany. In 2012, the company has set another minestrone by operating the industrial plant for the hydrothermal carbonization of biomass produced worldwide. This HTC plant has as overall capacity of 14.400 L and the capacity to produce 8.400 DM ton/year (AVA-CO_2, Business Wire, Inc. 2012b). Recently in 2017, AVA-CO_2 has officially launched the commercial hydrothermal carbonization plant in Innovationspark Vorpommen in Relzow, Germany. The plant consists of two different phase, phase I with two HTC reactors and phase II with six HTC reactors. The phase I reactor has the capacity to process 8000 ton of biomass (sewage sludge) per year and targets to produce the 2.664 ton bio-coal every year (HTCycle 2017; AVA-CO_2 2012a).

4.2 *Cost Effectiveness in Using HTC in WWTP's to Handle Sewage Sludge*

In municipal wastewater treatment, chemical phosphate recovery from the sludge phase normally needs more capital than that from the liquid phase. Despite the fact of having several conventional technologies for recovering the phosphate, it can still be concluded that phosphate recovered in WWTPs are costlier compared to the available commercial phosphate in the market (Ye et al. 2017). Hence, making the WWTP's energy neutral still is a challenge. The analysis made by (Sun et al. 2013) using the Japanese tripping fee for treating sewage sludge in a WWTP of 30 ton/day

Table 2 Cost-effectiveness of hydrothermal process for sewage (Sun et al. 2013)

Sewage sludge treated	**30 ton/day**
Total investment for a full treatment plant	2 M US$
The moisture content of sewage sludge	80%
Annual operation period	330 days
Daily operation	24 h/day
Daily solid fuel production (dry base)	6 ton/day (30 ton/day × 0.2)
Daily liquid fertilizer production	30 ton/day
Boiler fuel consumption	100% of produced solid fuel will be utilized as a boiler fuel
Maintenance cost per annum (5% of the capital cost)	0.1 M US$/year
Labour and utility costs per annum	0.2 M US$/year
Capital cost (5 years depreciation) per annum	0.4 M US$/year
Total expenditure per annum	0.7 M US$/year
Tipping fee income per annum	1.0 M US$/year (100 US$/t × 30 ton/day × 330 days/year)
Fertilizer and value-added product sales income per annum	1.0 M US$/year (100 US$/t × 30 ton/day × 330 days/year)
Total income	**2.0 M US$/year**

capacity is shown in Table 2. This theoretical analysis proves that using the HTC method, the WWTP can not only be made energy neutral but also revenue can be generated. The detail assumptions and calculations are shown in Table 2.

5 Concluding Remark and Future Research Prospect

HTC can potentially fit into the eco-innovative trend to treat the sewage sludge by fulfilling the main concept of the European Commission "Reduce, reuse and recycle". Further, HTC also adapts some concept of the circular economy by transferring the waste to resource. During sewage sludge treatment, the elimination of prior drying makes HTC a most promising technology to directly recover energy and resources from sewage sludge by providing significant environmental advantages compared to conventional thermal sewage sludge methods. Many researchers have identified the potential to mitigate greenhouse gas emissions by substituting natural gas in auxiliary sewage sludge incineration with HTC coal. In such plants, downstream phosphorus recovery is also possible from the combustion residues. HTC is the most appropriate technology for treating the sewage sludge in the country like German because (1) the excessively produced sewage sludge is creating storage and handling problem and (2) German Sewage Sludge Ordinance has planned to prohibit the direct use of sewage sludge for agricultural purpose in the near future. Using HTC, excessively produced sewage sludge can be transformed as a source of income by not only recovering nutrients but also by the production of various value-added products (Fig. 3).

Nevertheless, there are many research lines that are required to be understood in detail, and the process must be optimized to make HTC as highly efficient technology to treat and manage the sewage sludge. In this concern, the University of Rostock with two HTC reactors of capacity 1 and 100 L (see Fig. 3) is starting the research to detailly understand the diversified application of the produced biochar, especially to understand the adsorption characteristic of biochar and to improve their performance by optimizing the physical and chemical characteristics. Furthermore, it is obvious that sewage sludge contains many heavy metals and nutrients, which includes nitrogen, phosphorus and potassium, etc., and the detailed investigation will also be made to understand the possibility to recover these valuable elements effectively. The outcome of this research will certainly find new innovative reasons to make HTC most robust and reliable process to treat sewage sludge over the conventional treatment methods.

Fig. 3 HTC reactor of capacity 1 L (Right) and 100 L (left) at Universität Rostock

References

AbfKlärV. (2017). Verordnung über die Verwertung von Klärschlamm, Klärschlammgemisch und Klärschlammkompost. Report from German sewage sludge ordinance.

AVA-CO_2. (2012a). *AVA-CO_2 Schweiz AG*. December, 2012. Accessed August 28, 2018. http://duene-greifswald.de/doc/rrr2013/talks/HTC.pdf.

AVA-CO_2. (2012b). Business Wire, Inc. *Business wire, a Berkshire Hathaway company*. October, 2012. Accessed August 27, 2018. https://www.businesswire.com/news/home/20121122005340/en/HTC-1-Industrial-Plant-hydrothermal-carbonization-Worldwide-AVA-CO2.

AVA-CO_2. (2016). *Sustainability consult*. July 06, 2016. Accessed May 10, 2018. https://sustainabilityconsult.com/news/159-press-release-valuable-phosphorus-from-sewage-sludge-ava-cleanphos-pilot-plant-comes-online.

Burton, F. L., Tchobanoglous, G., Tsuchihashi, R., Stensel, H. D., Metcalf, & Eddy. (2013). *Wastewater engineering: Treatment and resource recovery*. McGraw-Hill's.

Crocker, M. (2010). Thermo chemical conversion of biomass to liquid fuels and chemicals.

(Destatis), Statistisches Bundesamt. (2018). *Abwasserbehandlung – Klärschlamm*. Tabellenband 2015/2016, Destatis. https://www.destatis.de/DE/Publikationen.

EU Commission, European. (1986). On the protection of the environment, and in particular of the soil, when sewage sludge is used in agriculture (86/278/EEC). *Council Directive, Official Journal of the European Communities*. Accessed 08 29, 2018. https://eur-lex.europa.eu/LexUriServ/LexUriServ.do?uri=OJ:L:1986:181:0006:0012:EN:PDF.

He, C., Giannis, G., & Wang, J. Y. (2013). Conversion of sewage sludge to clean solid fuel using hydrothermal carbonization: Hydrochar fuel characteristics and combustion behavior. *Applied Energy 111*, 257–266.

HTCycle. (2017, November 20). Accessed August 28, 2018. https://htcycle.ag/en/article/official-launch-of-the-htc-plant-in-relzow_10.

Kacprzak, M., Neczaj, E., Fijałkowski, K., Grobelak, A., Grosser, A., Worwag, M., et al. (2017). Sewage sludge disposal strategies for sustainable development. *Environmental Research, 156*, 39–46.

KrWG. (2012). The Waste Management Act (KrWG - Closed Substance Cycle Act) - Gesetz zur Förderung der Kreislaufwirtschaft und Sicherung der umweltverträglichen Bewirtschaftung von Abfällen (Kreislaufwirtschaftsgesetz - KrWG)." The Waste Management Act. https://www.saarland.de/dokumente/thema_abfall/KrWG.pdf.

Libra, J. A., Ro, K. S., Kammann, C., Funke, A., Berge, N. D., Neubauer, Y., Titirici, M. M., et al. (2011). Hydrothermal carbonization of biomass residuals: A comparative review of the chemistry, processes and applications of wet and dry pyrolysis. *Biofuels, 2*(1). https://doi.org/10.4155/bfs.10.81.

Saetea, P., & Tippayawong, N. (2013). Recovery of value-added products from hydrothermalc arbonization of sewage sludge. *Hindawi Publishing Corporation*. https://doi.org/10.1155/2013/268947.

Sirén Ehrnström, M. (2016). *Recovery of phosphorus from HTC converted municipal sewage sludge*. Masters Thesis, Department of Civil, Environmental and Natural Resources Engineering, Luleå University of Technology.

Stucki, M., Eymann, L., Gerner, G., Hartmann, F., Wanner, R., & Krebs, R. (2015). Hydrothermal carbonization of sewage sludge on industrial scale: energy efficiency, environmental effects and combustion. *Journal of Energy Challenges and Mechanics, 2*.

Sun, X. H., Sumida, H., Yoshikawa, K. (2013). Effects of hydrothermal process on the nutrient release of sewage sludge. *International Journal of Water Resource, 3*(2). https://doi.org/10.4172/2252-5211.1000124.

vom Eyser, C., Palmu, K., Schmidt, T. C., & Tuerk, J. (2015). Pharmaceutical load in sewage sludge and biochar produced by hydrothermal carbonization. *Science of Total Environment*, 180–186.

Ye, Y., Ngo, H. H., Guo, W., Liu, Y., Li, J., Liu, Y., Zhang, X., & Jia, H. (2017). Insight into chemical phosphate recovery from municipal wastewater. *Science of the Total Environment*, 159–171.

Sustainable Bio Medical Waste Management—Case Study in India

Vandana, V. Venkateswara Rao and Sadhan Kumar Ghosh

Abstract The Biomedical waste (BMW) is an area of concern which has a direct impact on the health and human safety. Management of BMW may be an ethical, social and obligation of all supporting and funding health-care activities. The scavengers sort out open, unprotected health-care wastes for recycling, and reuse of syringe, bottles and other medical aides compounded for BMW problems without using personal protective equipment (PPE)s, namely, gloves, masks, shoes etc. It's been ascertained that 10–25% of BMW is dangerous, whereas remaining 75–95% is non-hazardous. The dangerous part of the waste involves the physical, chemical, and/or microbiological risk to the health-care staff that is associated in handling, treatment, and disposal of waste. BWM rules in India were introduced first in the year 1998 and with the recent revision on the Bio Medical Waste management rules 2016 there after the 2018 amendments to boost the segregation, transportation, and disposal strategies, to decrease environmental impact therefore on amendment the dynamic of BMW disposal and treatment in India. The objectives of this article are, (a) assessing the present status of handling and treatment system of hospital bio-medical solid wastes, (b) analysing the requirements in BMW Management Rules 2016, (c) identifying issue and challenges in the practical problems for its effective implementation, the major drawback of conventional technological interventions, the latest technologies for BMW disposal, and potential solutions, (d) comparing the SWM system in India with a few developed countries like USA, UK, Japan and S. Korea and (e) to present a case study on the BMW treatment facility at Mangalgiri, AP BMWM facility. This paper will be helpful for making strategies in different states in India and other developing countries for the management of biomedical wastes.

Keywords Biomedical waste · Treatment · Autoclave · Incineration · Business model

Vandana · V. V. Rao
Safenviron Bio Medical Waste Treatment Facility, Mangalgiri, AP, India

S. K. Ghosh (✉)
Department of Mechanical Engineering, Jadavpur University, Kolkata, India
e-mail: sadhankghosh@gmail.com

S. K. Ghosh (ed.), *Urban Mining and Sustainable Waste Management*,
https://doi.org/10.1007/978-981-15-0532-4_30

1 Introduction

The Health care institution or hospitals which are responsible for generation of BMW are emitting 517 tonnes/day (Central Pollution Control Board 2018) from wards, operation theatre and outpatient areas. Management of hospital waste is crucial to take care of hygiene, aesthetics, cleanliness and management of environmental pollution. Communicable diseases which are caused through water, sweat, blood, body fluids and contaminated organs are vital and need to be prevented. Hospitals generate 10–15% Hazardous waste (which is infectious, injurious, Cytotoxic and Chemical) and 75–85% Non-Hazardous Waste (General Municipal Waste) (Fig. 1). Nonhazardous waste while mixed with hazardous waste entirely becomes hazardous. Biomedical waste *means "any waste, which is generated during the diagnosis, treatment or immunisation of human beings or animals or research activities pertaining thereto or in the production or testing of biological or in health camps.*"

Data from Government of India site (CPCB-as on 13.06.2018) BMW generated in the country in 2016 is 517 TPD (tonnes per day) from 1,87,160 HCFs, when compared to 501 tonnes/day from 1,88,098 HCFs in 2015 (Table 1). Unfortunately, only 501 TPD is treated and 16 TPD is left untreated in 199 CBMWTF and 23 under construction. The number of HCFs using CBMWTFs are 91,061, and approximately 15,281 HCFs have their own treatment facilities on-site. India will generate nearly 775.5 tons of medical waste/day by 2022 growing at a compound annual growth rate (CAGR) of about 7% (ASSOCHAM 2018). Figure 1 shows composition of Bio Medical Waste. The comparison between the Annual Report Information for the year 2015 and 2016 (Table 1) reveals the following:

- 938 (0.49%) numbers of Healthcare Facilities (HCFs) are inoperative in the year 2016 as compared to the numbers of HCFs reported for the year 2015.
- No. of HCFs/CBWTFs violating the provisions of BMW Rules in 2016 has increased to 12,034 in the year 2016 from 6074 in the year 2015.

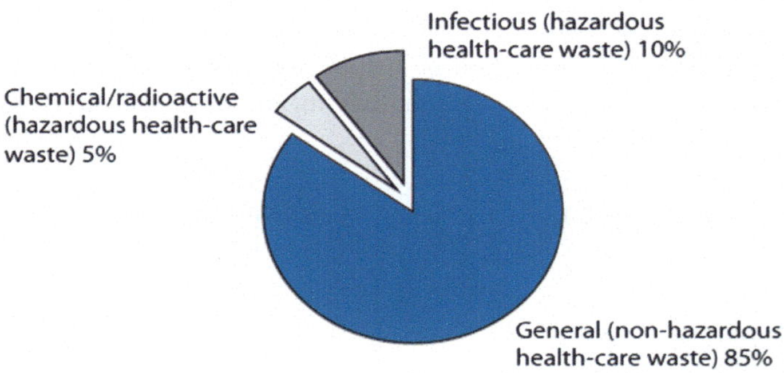

Fig. 1 Composition of bio medical waste

Table 1 Comparison between the annual report for the year 2015 and 2016

Particulars	As per annual report information for the year 2015	As per annual report information for the year 2016
No. of healthcare facilities	1,88,098	1,87,160
No. of beds	17,61,316	18,99,269
No. of common bio-medical Waste treatment facilities (CBWTFs)	203 + 32	199 + 23
No. of healthcare facilities applied for authorization	46,697	95,723
No. of healthcare facilities granted authorization	99,945	91,061
Quantity of bio-medical waste generated in Tonnes/day	501	517
Quantity of bio-medical waste treated in Tonnes/day	486	501
No. of HCFs/CBWTFs violated BMW rules	6074	12,034
No. of show-cause notices/directions issued to defaulter HCFs/CBWTFs	5103	11,272

2 Literature Review

In the twenty first century with accrued use of disposable material and therefore the presence of infectious disease like Hep-B and AIDS, it's utmost vital to require care for infected and dangerous waste to save lots of the group from disaster. Rag pickers within the hospital, looking for the garbage are at a risk of obtaining tetanus and HIV infections. The scattered Bio Medical Waste in and around the hospitals invites infectious and dangerous flies, insects, rodents, cats and dogs that are responsible for the spread of different disease like plague and rabies. The recycling of disposable syringes, needles, IV sets and alternative article like glass bottles without proper sterilization are liable for infectious disease, and alternative diseases. It becomes primary responsibility of Health directors to manage hospital waste in most safe and eco-friendly manner (Sengodan 2014, Mishra et al. 2016).

A study shows that 16 billion injections are administered every year. The indiscipline, unsafe and indiscriminate ways of disposal of needles and syringes within the bio medical wastes creates the opportunity for reuse, possible risk of injury and infection. The efforts have been made to reduce the possibilities of injection with contaminated needles and syringes in developing countries that have reduced infections. Due to use of contaminated syringe, in 2010, 33,800 new HIV infections, 1.7 million hepatitis B infections and 315,000 hepatitis C infections were reported

(WHO 2014). Risk of a person being infected with HBV, HCV and HIV due to use of an infected needle stick is 30, 1.8, and 0.3% respectively.

Segregation is the key for proper biomedical waste management at the source, namely, all the inpatient and out Patient Department (IPD and OPD) in different care activity areas, operation theatres, labour rooms, diagnostic services areas, treatment rooms etc. The biomedical waste generators, e.g., doctors, nurses, technicians and the patient (medical and paramedical personnel) are responsible for the segregation. Biomedical waste is segregated and collected by the cleaning personnel/ayahs in color coded, labelled bags. These bags are filled 3/4th and tied at the top, to prevent accidental spillage. Covered containers/trolleys with wheels are labeled and used to transport the waste to common collection point. Waste is stored in a separately at room temperature. Waste is collected within 48 h from the generation site and transported thru designated vehicles to the final treatment facility.

The impact of mismanaging the bio medical waste enormously affect the environment and human health. Followings are some of the potential hazards out of the mismanaged bio medical waste. Injuries inflicted by Sharps may cause danger to the human and animal health. Exposure to toxic substances, e.g., mercury, dioxins, pharmaceutical products, antibiotics and cytotoxic drugs released into the surrounding environment may cause damage to the health of the bio medical waste handlers for storing, transportation, incineration etc. without PPEs. Health hazards may occur from emission during bio medical waste incineration, chemical burns from disinfection, sterilization and other waste treatment activities, thermal injuries with open burning and radiation burns etc.

3 Rules in India

BMW is a real problem for the community and environment. Management of BMW came into limelight only after 1990s. Some landmark decisions have been made to streamline hospital waste management in the recent past (Sutha Irin 2018, Datta et al. 2018, Shreedevi 2019). These are:

1. Supreme Court judgment for safe disposal of hospital waste: Ordered dated 01.03.1996—all hospitals with 50 beds and above have to install either their own incinerator or an equally effective alternative method before 30.11.1996.
2. The then Ministry of Environment and Forest, Govt. of India: Issued notification for Biomedical Waste (Management and Handling) Rules 1998.

The BMW Rules 1998 (Fig. 2) were modified in the years 2000, 2003, 2011, 2016 and 2018. The draft of BMW rules 2011 remained as draft and did not get notified. Ministry of Environment, Forest and Climate change (MoEF and CC) in March 2016 and subsequently in 2018 has amended the BMWM rules. These rules have simplified the categories (Tables 2, 3 and 4) and defined the roles of the operators of Common Bio Medical waste treatment Facility (CBMWTF) and the occupiers

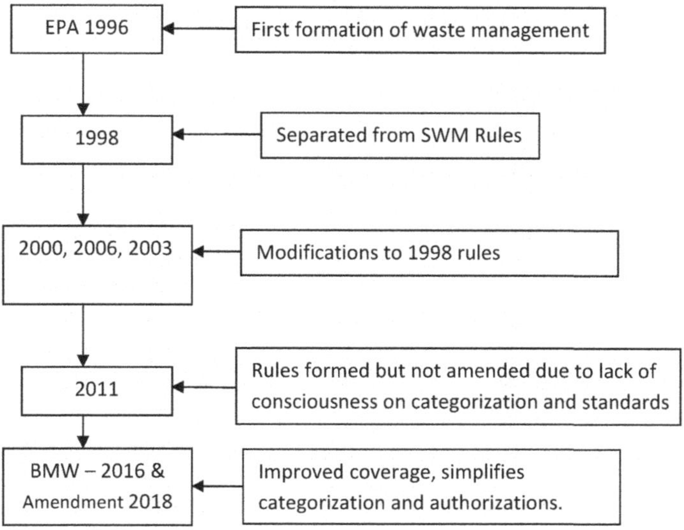

Fig. 2 Chronology of bio-medical waste management rules in India

Table 2 Forms with Biomedical Waste Management 2016

Forms	Use of the form
F1	Accident reporting
F2	Application for authorization or renewal of authorization (submitted by occupier of HCWs of CBMWTFs)
F3	Authorization (for operating, facility) for generation, collection, reception, treatment, storage, transport, disposal
F4	Annual report
F5	Application for filling "appeal" against order pass by the prescribed authority

HCWs healthcare workers, *CBMWTFs* common bio-medical waste treatment and disposal facilities

(Hospitals/Generators) at large. These new rules have improved the segregation, transportation and disposal methods.

These rules have been modified to include bring more clarity in the application and also included the word handling. In addition to improvement in categories, schedules have been reduced and forms for annual report, accident report with biomedical waste have also been changed (Table 2). Usage of non-chlorinated plastic bags and the thickness of the bag (50 μm) have been made mandatory. Efforts have been made for improving the effectiveness of collection, segregation, transport, and disposal of waste. Role of incinerator in increasing environmental air pollution has been controlled and checked by introducing stringent parameters with online monitoring systems. The new monitoring condition increases the retention time in secondary chamber. Figure 2 shows the evolution of BMW Management Rules in India.

Table 3 Categories as per Biomedical Waste Management Rules—1998

Option	Waste category	Treatments disposal
Category No. 1	Human anatomical waste	Incineration/deep burial
Category No. 2	Animal waste	Incineration/deep burial
Category No. 3	Microbiology and biotechnology waste	Local autoclaving/microwaving/incineration
Category No. 4	Waste sharps	Disinfection by chemical treatment/autoclaving/microwaving and mutilation/shredding
Category No. 5	Discarded medicines and cytoxic drugs	Incineration/destruction and drugs disposal in secured landfills
Category No. 6	Solid waste	Incineration/autoclaving/microwaving
Category No. 7	Solid waste	Disinfection by chemical treatment/autoclaving/microwaving and mutilation/shredding
Category No. 8	Liquid waste	Disinfection by chemical treatment and discharge into drains
Category No. 9	Incineration ash	Disposal in municipal landfill
Category No. 10	Chemical waste	Chemical treatment and discharge into drains for liquids and secured land for solids

These rules described the duties of the Occupier or Operator of a Common BMW Treatment Facility as well as the identified authorities. Every occupier or operator handling BMW, irrespective of the quantity, must obtain authorisation from the prescribed authority, i.e. State Pollution Control Board and Pollution Control Committee, as the case may be. These rules consist of four schedules and five forms.

4 Waste Handling and Treatment Systems

The objective of biomedical waste management is to reduce generation of waste. A secure and reliable technique for handling of BMW is crucial to save lots of human beings from the "adverse effects of health care waste". Effective management of BMW is not legal necessity however additionally a "social responsibility". Segregation is the key to proper management of biomedical waste. To achieve an effective Bio-medical waste management four steps are to be followed accurately and routinely. The steps are as follows. Trollies and vehicle used for outside transport are designed to facilitate minimum damage to structural integrity of waste. Waste should be transported only through the pre-designated route. The waste handlers should use personnel protection Equipment. Proper documentation of the type and quantity of waste being transported should be logged as per the prescribed format. To avoid illegal reuse of needles and other kind of waste barcoding systems have been introduced for waste collection.

Table 4 Categories as per Biomedical Waste Management Rules—2016

Colour code	Category	Treatment
Yellow	Anatomical waste, Animal waste, Soiled waste, Solid waste, Discarded linen, expired and discarded medicine, chemical waste	Incineration, plasma pyrolysis or deep burial
	Microbiology, biotechnology and other clinical waste	Pre-treat to sterilize with non-chlorinated chemicals on-site thereafter for Incineration
	Chemical liquid waste	Separate collection system should lead to effluent treatment system. On completion of the resource recovery, the chemical liquid waste shall be pre-treated before mixing with other wastewater
Red	Contaminated waste recyclables	Autoclaving or micro-waving/hydroclaving followed by shredding or mutilation or combination of sterilization and shredding. Treated waste has to be sent to registered or authorized recyclers
White	Waste sharps including metals	Autoclaving or micro-waving/hydroclaving followed by shredding or mutilation or combination of sterilization and shredding. Treated waste has to be sent to registered or authorized recyclers
Blue	Glassware	Disinfection (by soaking the washed glass waste after cleaning with detergent and Sodium Hypochlorite treatment) or through autoclaving or microwaving or hydroclaving and then sent for recycling

5 Issues and Challenges

Improper segregation—Even after a long time of implementation of these rules there are still some hospitals and some healthcare institutions which do not segregate the waste in desired standards of segregation. This may be due to insufficient funds, confusion due to large number of categories and lack of awareness in healthcare facilities. Improper segregation of BMW causes Corrosion and also accidents while incinerating.

Usage of Chlorinated Plastic Bags—Due to low cost of chlorinated bags, hospitals are still using them. Which increase the particulate matter and the flu gases while incineration. These bags are also not labelled as per the Schedule-IV.

Storage and Packing—The hospital waste, namely, body parts, organs, tissues, blood and body fluids along with soiled linen, bandage, cotton, plaster casts from infected and contaminated areas must be properly collected, segregated, stored, transported, treated and disposed of in safe manner to prevent nosocomial or hospital acquired infection. No proper designated storage is provided at the hospitals. Bags are not being tied at the top, which will attract flies and rodents.

Internet connection—As per the new rules, Air pollution Equipments are to be connected to the CPCB/SPCB websites. This is major problem, since most of the providers do not provide wired connection to remote areas. As per the new BMW rules, a CBMWTF is to be located 2 km away from habitation. Since there is no wired connection, a dongle has to be used. Dongle being a wireless connection and the location of CBMWTF, network is on-off most of the times. Also, every state has its own rules. Now in AP, CBMWTF are to connect webcams and GPS Tracking systems of Vehicles also to SPCB site.

6 Comparing the SWM System in India with a Few Developed Countries like USA, UK, Japan and S. Korea

In 2012, WHO conducted a survey on the BMWM status of twenty four West Pacific countries on five main areas, namely, management, training, policy and regulatory framework, technologies implemented, and financial resources. In the field of management, training, and policies regarding BMWM, except Micronesia, Nauru, and Kiribati all other West Pacific countries were satisfactory. It has been observed that the best available technologies for BMW logistics and treatment were used only in Japan and Republic of Korea, which were well maintained and regularly tested. Most of the countries had very less financial resources for BMWM. In 2015, a joint WHO/UNICEF assessment just over half (58%) of sampled facilities from those 24 countries of west pacific area. It was found that adequate systems have been followed for the safe disposal of health care waste.

The pharmaceutical and healthcare companies are major medical waste generators. When it comes to the handling of medical waste, European and American regulatory bodies are stringent. By 2018, the Medical Waste Management market is anticipated to reach a value of $10.3 billion, at a CAGR of 4.9%. Pharmaceutical waste management with an expected market size of $5.8 billion by 2018 will lead waste management (Reddiar Janagi et al. 2015). As seen from the Table 5, the

Table 5 Waste generated in different countries as per 2015

Country	Quantity (kg/bed/day)
UK	2.5
USA	4.5
France	2.5
Spain	3.0
India	1.5

Medical Waste Management market is highest for US which are the followed by Europe.

With the growing awareness in health care waste management, there has been an gradual increase in the no of CBMWTFs setup till 2014. But in 2015, authorizations have been issued to all/most of the applicants without proper evaluation of need, because of which there has been a steep growth in no of CBMWTFs setup. When there is a competition, the quality of the service drops. Which led to closure of facilities since they couldn't with stand the competition and the daily needs. The last century Andhra Pradesh (AP) witnessed the rapid mushrooming of hospital in the public and private sector for serving the growing demands of expanding population. The advent and acceptance of "disposable" has made the generation of hospital waste a significant factor in current scenario. Figure 3 and Tables 6 and 7 demonstrate the growth and status of BWM systems in the ate of AP, India.

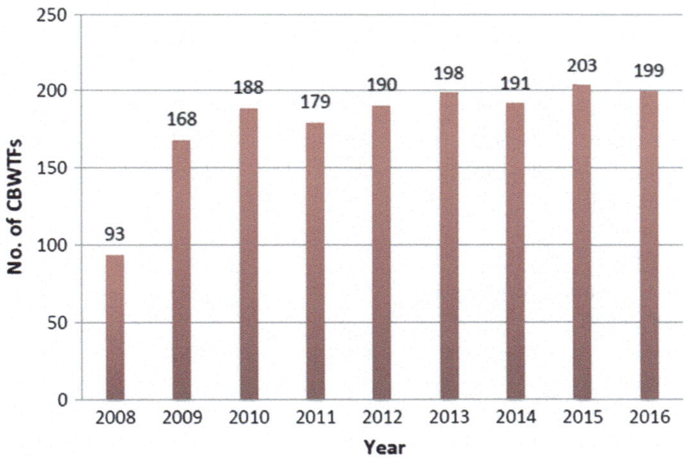

Fig. 3 No. of CBWTFs in Andhra Pradesh from FY 2008 to 2016

Table 6 Growth in no. of hospitals in Andhra Pradesh

S. no	Year	No. of HCE's	NO. OF BEDS	QTY waste kg/day
1	2013	2861	61,266	5491
2	2014	3096	67,484	6494
3	2015	5016	96,300	8360
4	2016	5721	102,464	9841.7

Table 7 Common bio-medical waste treatment and disposal facilities in AP (for the year 2016)

Sl. no	Name of CBMWTF at health care facilities (HCFs)	District	Total no of HCFs covered	Total no of beds covered	Quantity of BMW collected (kg/day)
1	Rainbow Industries	Srikakulam and Vizianagaram	370	6732	438.2
2.	Maridi Eco Industries Pvt.	Visakhapatnam	704	13,917	2000
3.	EVB Technologies Pvt Ltd.	East Godavari	525	11,589	679.5
4.	Safenviron and Associates	West Godavari	524	6068	510
5.	Safenviron Unit II	Krishna	752	12,848	1468
6.	Safenviron	Guntur	700	13,683	896
7.	S S Bio Care	Nellore	601	7921	1011
8.	Ongole Medical Waste Treatment Facility	Prakasam	357	4638	462
9.	Sriven Environ Technologies	Ananthapur	354	9593	712
10.	AWM Consulting Ltd	Chittoor	449	9552	975
11.	Medical Waste Solutions	Kurnool	385	5923	690

7 Case Study on the BMW Treatment Facility at Mangalgiri, AP

M/s Safenviron, a full-fledged Bio Medical Waste (BMW) Treatment facility was established in the year 2001 at Chinakakani Village in Guntur District, about 15 km away from Vijayawada, M/s Safenviron has been functioning fully complying with the norms of A.P. Pollution Control Board from time to time. M/s Safenviron is an operator dealing with collection, transportation and disposal of BMW with trained manpower, fleet of vehicles and well established plants, for more than 1399 Health Care establishments in Krishna and Guntur Districts. With a fleet of 15 vehicles registered with the concerned SPCB/PCC BMW is collected from enrolled members within 48 h of generation. Since 2014, with its sister concern M/s Safe Medi Aids, bags and bins used for BMW disposal are supplied to the enrolled hospitals.

Fig. 4 The 8th IconSWM delegates from India, Nepal, Kenya, Sri Lanka, Japan and members of ISWMAW and APPCB visited the facility on Nov 27, 2018

BMW collection bags at the M/s Safenviron are labelled and manufactured as per Schedule-III and IV of the Bio-medical waste (Management and Handling) Rules 2016. Vehicles are labelled with the bio-medical waste and cytotoxic symbols (as per the Schedule-III of the Rules) and the name, address and telephone number of the CBWTF (Fig. 4).

Coloured bags handed over by the healthcare units are collected in similar coloured containers with proper cover. Waste storage cabins in the vehicles have provisions for the sufficient openings in the rear and/or sides for easy loading and unloading. Vehicles used for collection are partitioned according to norms with FRP coating, to withstand any possible damage that may occur during loading, transportation or unloading. The base of the waste cabin is made leak proof to avoid pilferage of liquid during transportation. A manifest is issued to the HCE during collection; manifest contains details like name of the HCF, date, no of bags and Weight disposed. All vehicles are having GPS tracking systems. Vehicles used for collection of BMW are cleaned using jet pumps and disinfected daily. Waste water generated from vehicle wash and during the treatment process is cleaned in ETP.

Treatment and disposal of segregated BMW is done in conformity with A.P. Pollution Control Board norms/rules and regulations with quality and commitment. All the vehicles unload the waste collected from the HCE's in their respective storage areas at the treatment facility. In and Out timing and the total weight unloaded at the facility are recorded. Yellow bags from the storage room are incinerated immediately. Incineration, a high temperature thermal process, employs combustion of the waste under controlled condition for converting them into inert material and gases. Incinerators have primary and secondary combustion chambers with refractory lining to ensure optimal combustion. The primary chamber has pyrolytic conditions with a temperature of 800 °C. The secondary chamber operates under excess air conditions at about 1050 (±)50 °C. Flue gasses from incinerator are passed via Air Pollution

Control Equipment for removal of particulate matter. Quality of incineration is determined by Online monitoring systems attached to Air Pollution Control Equipment. The chimney is 30 m above ground level.

Red, White and Blue waste from the storage room is treated in an autoclave. Autoclaving is a sterilization method using high-pressure steam. The working principles of autoclaving based on the concept of boiling point of water (or steam) increases under higher pressure. It is a thermal process where steam comes in direct contact with waste in a controlled environment for significant time duration to disinfect the wastes as stipulated under the Bio-medical Waste Management Rules. For ease and safety in operation, a horizontal type autoclave is used. Pre-vacuum based system is gives optimum results than the gravity type system. A PLC is attached to the autoclave and incinerator separately with efficient display and recording devices for recording critical parameters such as time, temperature, pressure, date and batch number etc. as required under the BMWM Rules. Figure 5 shows the visitors in the facilities from different countries.

Red bags after autoclave are shredded and then sold to authorized recyclers. Waste are de-shaped or cut into smaller pieces by shredding which make the wastes unrecognizable. It helps in prevention of reuse of bio-medical waste and also acts as identifier that the wastes have been disinfected and are safe to dispose. M/s Safenviron has established waste treatment facility with essential machinery and trained manpower to endure day-to-day operation of collection, transportation and disposal of BMW in accordance with A.P. Pollution Control Board norms/rules and regulations. The

Fig. 5 Prof. Sadhan K Ghosh, Mentor of M/s Safenviron and the Chairman, IconSWM (4th from right), Mr. Venkateswara Rao, Owner of the plant, (5th from right) and other personnel in the plant

Fig. 6 Blooming flowers in the tank using treated water from ETP in M/s Safenviron

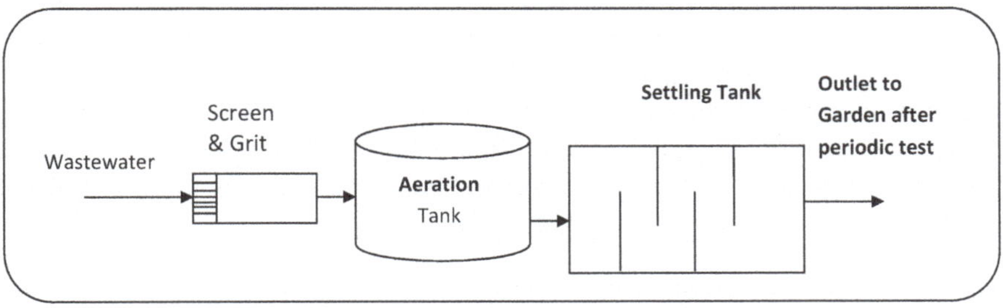

Fig. 7 Schematic diagram of Effluent treatment plant at M/s Safenviron

operation is being continued for more than 14 years which has modernized the facility recently for better quality service to meet ever-increasing loads. After incineration, ash generated is stored in secure landfill on the premises. Secured landfill is a concrete construction lined with HDPE sheets to prevent perforation of leachate into the ground water. But as per the new BMW rules, 2016, Incineration ash is to be disposed of at TSDF. Figure 6 shows the effluent Treatment Plant (Fig. 7).

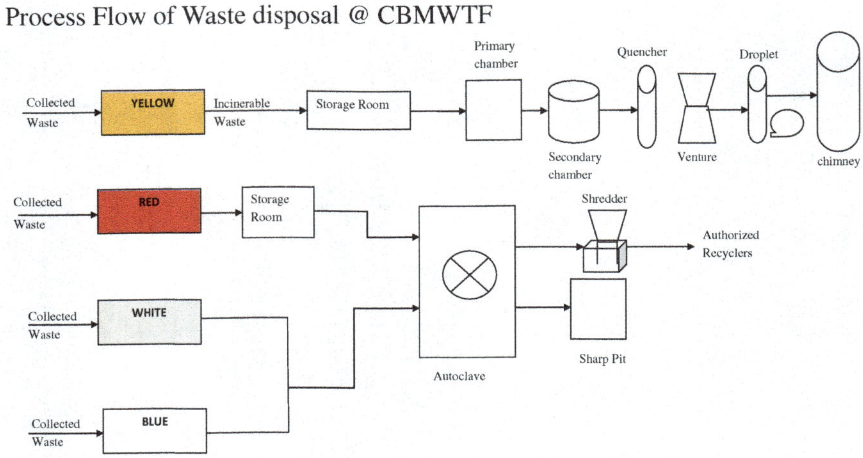

8 Conclusions

Medical Waste handling is a hazardous waste management process which requires a high standard of preparation. Biomedical waste management is the need of the hour, scientific and cost effective methods have to be developed. The waste should be segregated at the source rather than end of pipe approach. Specific training should be given depending on the nature of the work, responsibilities in the hospital and worker experience of individual workers. Training for health-care staff is essential in the efforts to minimize the transmission of secondary infections. Staff training helps to achieve higher standards of infection control measures.

Acknowledgements All procedures performed in this study involving human participants were in accordance with ethical standards of the organisational committee and informed consent was obtained from all individual participants in the study. Authors appreciate the organisation as well as the International Society of Waste Management, Air and Water (ISWMAW) for generous support towards the successful completion of the study.

References

ASSOCHAM. (2018, April). *ASSOCHAM Bulletin, 43*(04). https://timesofindia.indiatimes.com/business/india-business/indias-medical-waste-growing-at-7-annually-assocham/articleshow/63415511.cms.

Central Pollution Control Board. (2018). *Status on bio-medical waste management scenario and recommendations for ensuring compliance to the bio- medical waste management rules, 2016.* Central Pollution Control Board (Waste Management Division-Bio-Medical Waste), Delhi—110 032 as on 13.06.2018.

Datta, P., Mohi, G. K., Chander, J. (2018). Biomedical waste management in India: Critical appraisal. *Journal of Laboratory Physicians*. http://www.jlponline.org/temp/JLabPhysicians1016-327972_090637.pdf.

Government of India Ministry of Environment, Forest and Climate Change. (2016). *Bio-medical waste management rules*. Extraordinary, Part II, Section 3, Sub-Section (i). Published in the Gazette of India. Notification; New Delhi, the 28 March, 2016.

http://cpcb.nic.in/rules-3/.

Janagi, R., Shah, J., Maheshwari, D. (2015). Scenario of management of medical waste in US and UK: a review. *Journal of Local Trends in Pharmaceutical Sciences, JGTPS, 6*(1), 2328–2339. ISSN: 2230-7346; https://www.jgtps.com/admin/uploads/ERmDV8.pdf.

Ministry of Environment and Forests. (1998). Bio-medical waste (management and handling,) rules, 1998 (pp. 276–84). New Delhi: Government of India Publications. Ministry of Environment and Forests Notification.

Mishra, K., Sharma, A., Sarita, Ayub, S. (2016). A study: biomedical waste management in India. *IOSR Journal of Environmental Science, Toxicology and Food Technology (IOSR-JESTFT), 10*(5), ver. II, 64–67. e-ISSN: 2319-2402, ISSN: 2319-2399. www.iosrjournals.org.

Sengodan, V. C. (2014). Segregation of biomedical waste in a South Indian tertiary care hospital. *Journal of Natural Science, Biology and Medicine, 5*(2). http://www.jnsbm.org on Monday, July 8, 2019, IP: 122.15.82.86.

Shreedevi, D. (2019). Hazardous waste management at healthcare facilities, http://www.indus.org/healthcare/Secientific%20Sessions/Dr.%20D.%20Shree%20Devi%20-%20Hazardous%20Waste%20Management%20at%20Healthcare%20Facilities.pdf. Accessed on 07.07.2019.

Sutha Irin, A., (2018). An analytical study on medical waste management in selected hospitals located in Chennai city. *Environmental and Waste Management and Recycling, 1*(1).

World Health Organization. (2014). Blue book-2014, Safe management of wastes from health-care activities.

A Study on Plastic Waste Management by Stakeholders Using Reverse Logistics

R. Madhura

Abstract Plastic has become a benefit for the mankind because of its functionality and it is a commonly used material in their day to day activities. The indispensable habit of using the plastic by the present society for various purposes like carry bags, storage bags, packing materials, and so on is leading to the large-scale production of such products. After its usage or the effective life, it is regarded as waste. The sustainable disposal of these waste plastic is a critical issue and the very essence of the present-day waste management. The role of stakeholders (waste generators, the municipal corporations, and manufacturers of plastic) in disposal, on responsible terms is very crucial. The annual waste generated in India is about 62 million tons of which the plastic waste is 5.6 million tons. If the plastic waste is not collected, this leads to haphazard dumping and clogging of the waste. If it is collected and dumped to landfill sites without any treatment, it causes underground water pollution and other consequences. This paper aims to study the role of reverse logistics in plastic waste management and to explore sustainable solution to the problem of plastic waste management by applying reverse logistics. The present study is mainly conceptual in nature. The study suggests the embracing of corporate social responsibility (CSR) for creating awareness and to educate the waste generators. The study also recommends that the municipal corporations can tag on the public–private partnership (PPP) for collecting and segregating such plastic waste. Extended Producer Responsibility (EPR) may be recommended for the manufacturers as an eventual tool for the efficient waste management. Thus, the proactive collection and source segregation of the locally generated plastic waste and the implication of the initiatives by the municipal corporations can act as a primary solution to this problem.

Keyword Stakeholders · Reverse logistics · Sustainability

R. Madhura (✉)
CMR University, Bengaluru, India
e-mail: sastry.maddy@gmail.com

S. K. Ghosh (ed.), *Urban Mining and Sustainable Waste Management*,
https://doi.org/10.1007/978-981-15-0532-4_31

1 Introduction

The traits of plastic such as ease of manufacture, adaptability, and immunity to water, has led to the multiple uses of plastics such as carry bags, storage bags, packing materials, liquid storage cans, food containers, toys and playthings, show items, and so on. The greatest advantage of plastic is its durability. India has seen considerable growth both in the production and consumption of plastic. As a result of improved living standards, plastic products are inseparable fraction of the present society. The increased production of plastic products has definitely contributed to economic development of the country, employment opportunities, and quality of life.

Plastic has almost replaced metals in components of many electronic goods and also it has become vital in packaging since it reduces cost and provides better protection for the items. Plastic usage is more or less indispensible for the households. However, after its usage or the effective life, it is regarded as waste. Improper disposal of plastic creates problems to plant life, wild life, and even the human population (source: www.conserve-energy-future.com). Utmost care has to be taken for the proper disposal of plastic; otherwise, this leads to problems in collection, segregation, and recycling. Rapid urbanization coupled with increased production and consumption of plastic materials has alarmed the need for a sustainable solution for the problem of plastic waste management.

The main objective of the study is to explore sustainable solution to the present-day problem of plastic waste management. The study revolves around the plastic waste generated by the household that is the municipal plastic waste. This paper is a descriptive—qualitative research. The source of information is secondary sources and observation of best practices.

2 Different Types of Plastics

The term plastic is used in general for all types of products. However, these are made of different polymers (source: https://www.plasticsmakeitpossible.com). Hence, for the purpose of the study, it is essential to understand the classification of the plastics based on the usage of the resins for manufacturing it. The Resin Identification Codes are developed by the Society of the Plastics Industry Inc., in the year 1988 which helps in the identification of the plastic resin used in manufacturing the plastic product (Table 1).

3 Problems Relating to Plastic Waste

Due to the wide range of usage of plastic products in India, the generation of plastic waste and the consequences of improper management of this waste are aggravating.

Table 1 Table showing the different types of plastics, their household uses, and recyclability

Code and plastic type	Household uses	Recyclability
1. Polyethylene terephthalate—PET or PETE	Beverage bottles, medicine jars, rope, clothing and carpet fiber, shampoo and soap bottles and so on	Yes
2. High-density polyethylene—HDPE	Milk containers, snack and food boxes, toys, buckets, crates, and so on	Yes
3. Polyvinyl chloride—PVC or V	Pipes, fittings, tiles, window and door frames, wire and cable outer layer credit cards, and so on	Yes
4. Low-density polyethylene—LDPE	Carry bags, bin liners, flexible bottles, bubble wraps, irrigation/flexible pipes, toys, and so on	Yes
5. Polypropylene—PP	Plastic bottle caps, lunch boxes, biscuit wrappers, drinking straws, and so on	Yes
6. Polystyrene—PS	Disposable coffee cups, plastic food boxes, cutlery sets, styrofoam, and so on	Difficult to recycle
7. Other	Layered or multi-material mixed polymers, headlight senses	Difficult to recycle

Source Different types of plastics and their classification; https://www.ryedale.gov.uk/attachments/article/690/Different_plastic_polymer_types.pdf

The average per capita consumption of plastic in India is 11 kg and the per capita consumption in the world is 28 kg. India generates approximately 40,000 tons of plastic waste per day as compared to 15,342 tons in 2013 that is approximately there is an increase of 160% (source: www.greensutra.in/news/plasticwaste/). An article dated tenth December 2017 in BBC News Website states a survey which highlights that 192 coastal countries are contributing to the ocean plastic. The article also mentions that the beverage bottles are the most common type of plastic waste.

The catastrophic effect of poor waste management lies not only on environment and people but also on the economy of a country. The improper handling of plastic waste has a long-lasting effect such as water contamination, air contamination, and ocean life damage, effect on flora and fauna, severe weather due to climate contamination and also damage on human.

It is highly essential to know the time taken by the plastic waste to decompose in order to understand the seriousness of employing sustainable solution to the problem of plastic waste. Plastic waste takes a long time to decompose varying from ten years to 1000 years. Hence, it is extremely essential for finding solution to the problem of plastic waste.

4 Stakeholders of Plastic Waste Generation

According to Section 3(y) of Plastic Waste Management Rules (2016) "Waste generator means and includes every person or group of persons or institution, residential and commercial establishments including Indian Railways, Airport, Port and Harbor and Defense establishments which generate plastic waste." The generators of plastic waste may be individual or bulk generators. However, for the purpose of present study, the generators are defined as the household plastic waste generators.

The stakeholders of Plastic Waste Management (PWM) include the generators, local body, and manufacturers of plastic and their responsibilities play a key factor in effective PWM. The major responsibility is on the local body that is the municipal corporations for developing and setting up an instrumental infrastructure for collection, segregation, storage, transportation, processing, and disposal of plastic waste. The Rules mention that the infrastructure for the same can be established either on its own or with the help of agencies or producers. Being in line with this provision of the Rules, the study suggests that reverse logistics is one such tool which helps in creating a loop among these stakeholders.

5 Responsibility of Plastic Waste Generators

The Plastic Waste Management Rules (2016) clearly lays down certain responsibilities of waste generator where in it states that he shall take steps to reduce the generation of plastic waste, not to litter the plastic waste and most importantly to segregate the waste at source (at generation point) and hand it over to the Local Bodies or any authorized person or waste collection agency or registered recyclers. The Rules also states that the awareness creation responsibility lies with the local body accountable for effective waste management.

5.1 Recommendation

In order to increase the responsibility of the generators, the creation of awareness is essential. Awareness about improper dumping of the plastic waste, the serious problems of non-segregation, and various stages involved in the waste management, the list of authorized agencies for waste collection, the list of registered recyclers, and so on. It is essential for the ULBs to clearly design the awareness programmes in order to mitigate the aggravating problems of plastic wastes. For this, the municipal corporations have to clearly define the problem and assign the responsibility of creating awareness to those organizations which are socially responsible. The municipal corporations can enter into an understanding with those companies who are by law and will be responsible for adopting the corporate social responsibility (CSR).

The Companies advocate the concept of the triple bottom line which emphasizes on people, planet, and profit and project their accountability toward the social and ecological considerations in doing business. It is the ethical duty of the corporate citizens because they have built their business on the foundation of various opportunities provided by the country. Hence, they are likely to contribute toward the development and wellbeing of the society in which they operate. Corporate social responsibility is the initiative of the corporations in order to balance the various facets such as the environment, community, workplace, and marketplace. The Companies Act, 2013 has brought the idea of corporate social responsibility to the forefront. The companies are taking up a lot of socially responsible initiatives for the sake of protecting the environment and also to meet the CSR conditions lay down by the Companies Act.

These companies may form an expert panel within the organization and/or appoint a non-government organization for the purpose of creating and spreading awareness among the households about the various aspects related to plastic waste management right from reduce till the recycling.

6 Responsibility of Local Bodies

The Plastic Waste Management Rules (2016) clearly lays down certain responsibilities of the local bodies. The major responsibility of the local body is developing and setting up an operational infrastructure for collection, segregation, storage, transportation, processing, and disposal of plastic waste. They are also responsible for ensuring that no damage is caused to the environment, that recyclable plastic waste reaches the recyclers, that opening burning of plastic does not happen and so on. However, the Rules mention that the infrastructure for the management of the plastic waste can be established either on its own or with the help of agencies or producers.

6.1 Recommendation

In order to provide a better solution to the problem of plastic waste, the local bodies may enter into a partnership with the private business players who are equally interested in venturing into eco-friendly projects. The local bodies may enter into a public–private partnership (PPP) for the various activities of waste management such as collection, segregation, storage, transportation, processing, and disposal of plastic waste.

Public–private partnership (PPP) is financial support model for a public infrastructure project such as communication, transportation, rail and road networks, and so on. This model facilitates providing infrastructure and other services by a public sector (mostly Government) in association with private business sector. In such agreements, the private players take up certain amount of outlay, operational and maintenance responsibility, technical collaboration, risk bearing, and so on and the remuneration is performance based.

The local bodies may identify such private players who are suitable for outsourcing the various activities of the plastic waste management and enter into a memorandum of understanding with such companies to venture into the PWM. Such an arrangement will prevent open dumping and enhances soil productivity and helps in clean urban landscape. An article in the World Bank Website on Legal Issues on municipal solid waste (MSW) PPPs mention that the private players can play a significant role in improving environmental and hygiene issues around solid waste collection and disposal. However, these agreements are not only positive from the environment and local body point of view but are also beneficial to the companies. The projects help in improving the livelihood of all those who are involved in the various activities like collection, segregation, sorting, and transportation. These PPP models also benefit the companies involved in the recycling activities. A well designed and implementable PPP model may help in meeting the triple bottom line approach which states that sustainability lies in balancing the 3Ps that is profit, people and planet. However, for the success of these partnerships, it requires strong governance and policy making from both ends and also necessitates support from waste generators and manufacturers.

7 Responsibility of Producers, Brand Owners, and Importers (Manufacturers)

The Plastic Waste Management Rules (2016) states that the producers, brand owners, and importers of plastic items are responsible for the collection of the plastic waste generated from their products either through their own distribution channel or approach the local body. For this purpose, the Rules mentions the Extended Producer Responsibility may be used as a medium for collection of plastic waste generated.

The application of Extended Producer Responsibility is an intelligent step in involving the corporate citizens toward a clean and good environment. The companies may adopt the EPR as a policy matter and accept the financial as well as operational responsibility for the disposal and treatment of the plastic waste. The concept of EPR makes the producers responsible for what they produce, what they sell; design of the product, packaging materials used, and so on.

7.1 Recommendation

In line with the Extended Producer Responsibility suggested by the Rules, the study makes suggestion to correlate the two major recommendations mentioned in the above paragraphs that is the corporate social responsibility and public–private partnership with the Extended Producer Responsibility (Fig. 1). Those companies which

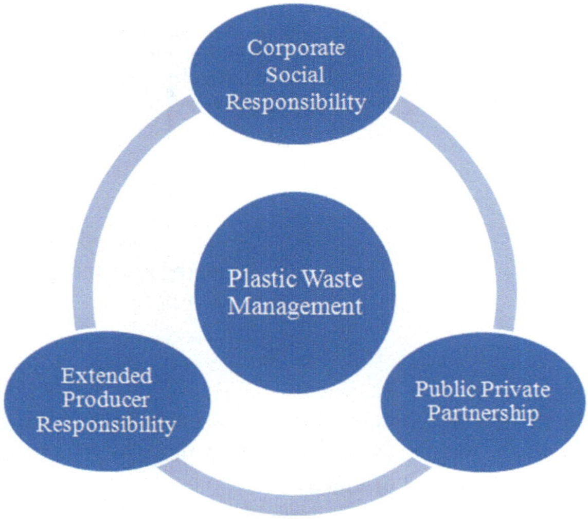

Fig. 1 Figure showing the relationship of CSR, PPP, and EPR toward plastic waste management

are duty-bound to follow the Extended Producer Responsibility may also presume the corporate social responsibility and public–private partnership, which facilitates a proper flow of plastic waste back to the producer from the consumer, thus, facilitating an effective reverse logistics in the plastic waste management.

8 Reverse Logistics

Logistics refers to the forward movement of goods from manufacturers to the ultimate consumers. In general, the term reverse logistics refers to the value addition process that is performed in the return process. Reverse logistics in waste management refers to the process of moving the waste from its generators for the purpose of recapturing value or proper disposal. Reverse logistics is a process whereby a business can become environmental friendly through the waste hierarchy which includes reuse, reduce and recycle. A more exhaustive view of reverse logistics necessitates the reduction of materials in forward logistics so that fewer materials flow back. In a nut-shell, reverse logistics refers to the movement of waste from its generators to the manufacturers and includes value addition to such end-of-life or end-of-use materials by the stakeholders of plastic waste generation. The whole process of reverse logistics handles the movement of goods from manufacturers to consumers, who become the waste generators and reverse flow of the waste products from these waste generators to the manufacturers who act as recyclers as explained in Fig. 2.

In emerging economies, the application of reverse logistics is still in preliminary stage may be because of reasons such as volume of generation, lack of awareness of the concept, lack of awareness on segregation, and so on. There is a need to study reverse logistics in developing countries like India. The stakeholders' involvement in

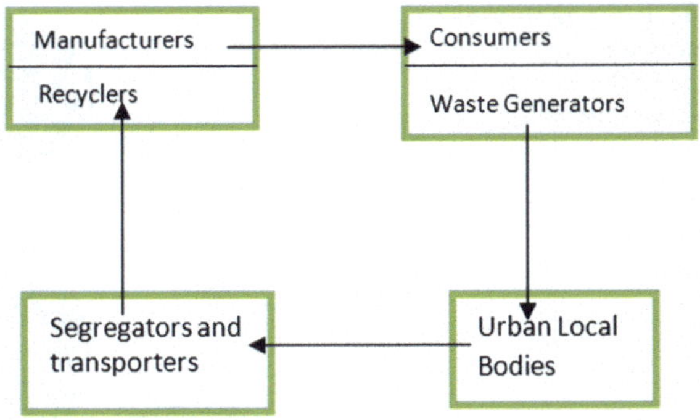

Fig. 2 Figure showing the flow of materials in the reverse logistics

the waste management process is essential to the wellbeing of the country. Stakeholders are the players in the reverse logistics which includes waste generators, municipal corporations, and producers.

In the developing countries, the manufacturers should seriously consider the application of reverse logistics and this would make a remarkable change in the whole scenario of the PWM. The concept of Extended Producer Responsibility (EPR).

9 Suggestions

- The application of reverse logistics in the management of plastic waste may serves the purpose of clean and plastic-free environment.
- Corporate social responsibility objective of the plastic manufacturers may be used as a tool by the municipal corporations to instigate the producers in creating awareness among the primary player of the reverse logistics that is the waste generators.
- The municipal corporations and the producers may enter into an understanding through the public–private partnership to lift the plastic waste from the generators, do the secondary segregation or do the primary segregation if not segregated at source, and then transport it to the waste management service providers or the producers themselves for further processing or final disposal.
- Extended Producer Responsibility is as such mentioned in the Plastic Waste Management Rules (2016) as an initiative by the producers to mitigate the plastic waste. However, the study suggests that if the producers correlate all three concepts, it would act as a sustainable solution (Fig. 3).
- The suggestion of the producer taking up all the activities under different capacities (CSR, PPP, and EPR), it would help them in achieving the responsibility laid down by the Rules which summarizes that producer is responsible for what he produces and sells.

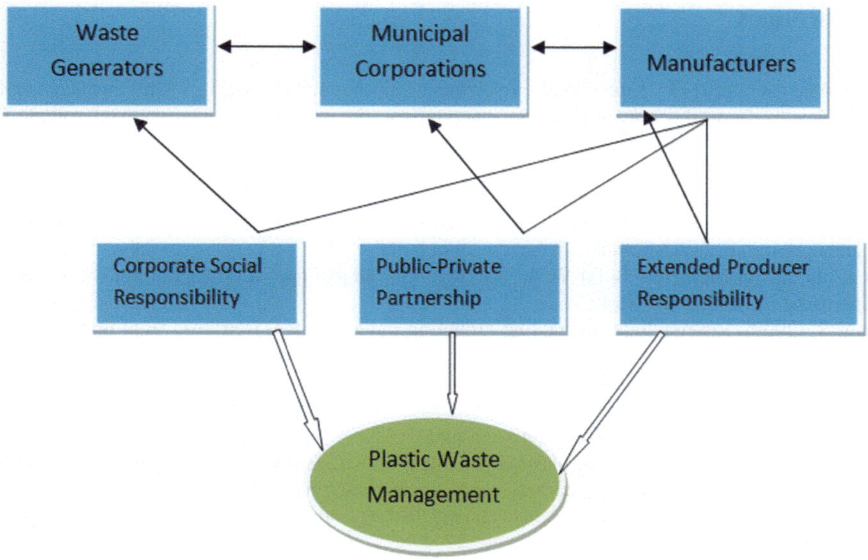

Fig. 3 Figure showing the responsibility of manufacturer under different capacities in reverse logistics

10 Conclusion

In developing countries, generation of plastic waste and its managements is the major highlight for Government and all the municipal corporations. Reverse logistics as suggested above, may include the application of Corporate Social Responsibility, public–private partnership and Extended Producer Responsibility concepts which provides a sustainable solution for the effective plastic waste management. Hence, there is need to study reverse logistics in developing countries. The limitation of the present study is the lack of primary data collection for better understanding of the applicability of the recommendations. However, the methodology as mentioned earlier is only the secondary sources for information.

The study suggests the embracing of corporate social responsibility for creating awareness and to educate the waste generators. The study also recommends that the municipal corporations can tag on the public–private partnership for collecting and segregating such plastic waste. Extended Producer Responsibility in correlation with the CSR and PPP by the manufacturers serves as an eventual tool for the efficient waste management. Thus, the proactive collection and source segregation of the locally generated plastic waste and the implication of the initiatives by the municipal corporations can act as a primary solution to this problem.

References

Abeliotis, K. (2011). Life cycle assessment in municipal solid waste management. Retrieved from https://www.intechopen.com/books/integrated-waste-management-volume-i/life-cycle-assessment-in-municipal-solid-waste-management.

Banerjee, T., Srivastava, R. K., Hung, Y. T. (2014). Plastic waste management in India: An integrated solid waste management approach. https://www.researchgate.net/publication.

Bhattacharya, R. R. N. S., Chandrashekar, K., Deepthi, M. V., Roy, P., Khan, A. (2018). Challenges and opportunities: Plastic waste management in india.

Legal Issues on Municipal Solid Waste (MSW) PPPs; https://ppp.worldbank.org/public-private-partnership/sector/solid-waste.

Nair, S. (2009). Reverse logistics gaining ground in Indian market. Retrieved from https://www.livemint.com.

Plastic Waste Management Rules. (2016). www.envfor.nic.in.

Ruj, B., Pandey, V., Jash, P., Srivastava, V. K. (2015). Sorting of plastic waste for effective recycling. https://www.researchgate.net/publication/305503715.

Singh, P., Sharma, V. P. Integrated plastic waste management: Environmental and improved health approaches. https://www.sciencedirect.com/science/article/pii/S1878029616301578.

Science and Environment—Seven Charts that explain the plastic pollution problem. (2017). https://www.bbc.com/news/science-environment-42264788.

Valle, P. O. D., Menezes, J., Reis, E., Rebelo, E. (2009). Reverse logistics for recycling: The customer service determinants. *International Journal of Business Science and Applied Management, 4*(1).

Wilhelm, R., Resin identification codes. https://www.astm.org/newsroom/astm-plastics-committee-releases-major-revisions-resin-identification-code-ric-standard.

A Study on Pyrolysis of Plastic Wastes for Product Recovery and Analysis

Ankita Mukherjee, Biswajit Ruj, Parthapratim Gupta and A. K. Sadhukhan

Abstract In this twenty-first century, industrialization, urbanization and modernization have led to rapid generation of waste plastics causing a global concern for safe disposal. Owing to the current trend in plastic waste generation, there is a need to utilize the huge amount of high calorific plastic trash into a valuable source of energy as a solution to the waste disposal cum energy crisis issues. In this context, pyrolysis has been acknowledged as an efficient waste to energy technology degrading waste plastics under anaerobic conditions into oil, gas and char products. The current initiative deals with the study of thermal pyrolysis of waste plastic polyethylene in a batch pyrolyser at an inert environment to recover value-added products followed by their characterization. The plastic waste pyro-feedstock of 100 g shredded low-density polyethylene (LDPE) bags, a significant part of plastic waste stream, was used per batch experiment. Thermo-gravimetric analysis (TGA) was done at different heating rates of 10, 20 and 40 °C/min to evaluate the waste degradation profile based on temperature. Pyrolysis experiments were carried out in the temperature range of 450–600 °C at a heating rate 20 °C/min producing 74–84% oil, 11–20% gas and 2–15% char as by-products. The highest oil yield was observed at 550 °C with maximized gas and solid char formation at 600 and 450 °C, respectively. The FTIR data infers that the derived pyro-oil comprises mainly alkanes, alkenes and aromatic groups indicating fuel quality. The gas product was analysed using gas chromatography which highlights the presence of H_2, CO and C_1–C_6 hydrocarbons without any trace of CO_2 after the process. The residual char was also characterized using

A. Mukherjee (✉) · P. Gupta · A. K. Sadhukhan
Department of Chemical Engineering, National Institute of Technology (NIT), Durgapur 713209, India
e-mail: ankita.mukherjee98@gmail.com

P. Gupta
e-mail: parthagupta2000@yahoo.com

A. K. Sadhukhan
e-mail: t_sadhu@yahoo.com

B. Ruj
Environmental Engineering Group, CSIR-Central Mechanical Engineering Research Institute (CMERI), Durgapur 713209, India
e-mail: biswajitruj@yahoo.co.in

S. K. Ghosh (ed.), *Urban Mining and Sustainable Waste Management*,
https://doi.org/10.1007/978-981-15-0532-4_32

BET and FESEM. Thus, LDPE plastic wastes offer a potential energy value after pyrolysis through the sustainable recovery of value-added products. Further research on process scale-up and product applications is anticipated.

Keywords Waste disposal · LDPE plastic wastes · Waste to product · Pyrolysis · Product recovery · Characterization

1 Introduction

The vast accumulation of plastic trash is causing a worldwide alarming concern for their safe disposal attributed to reckless use of plastics in this modern era on account of their wide range of applications, lightweight, durability, flexibility, resistivity to corrosion and chemicals, versatility of types, low moisture content, reusability and cheaper price. The dumping of plastic wastes poses big threat to biosphere since plastic wastes, being non-biodegradable and persistent, enters the food chain and also leads to harsh environmental impacts like choking of drains, leaching of plastic toxic additives like lead and cadmium pigments, land infertility, anaesthetics, hazardous emissions of dioxins, furans, acetaldehyde, benzene, styrene and greenhouse gases like carbon dioxide due to burning of waste plastics. The year 2015 has witnessed a global plastic waste generation of about 6300 million tons, approximately (Geyer et al. 2017). India generates about 5.6 million tons per annum plastic wastes, approximately, which equals to 15,342 tons per day (CPCB 2012). The present scenario of increasing burden of plastic wastes on earth has led to the exploration of feasible and effective ways of sustainable energy recovery from high calorific plastic wastes as an effective solution of issues related to energy crisis and waste disposal. The suitable treatment of plastic wastes in order to utilize their energy content is one of the key factors of waste management and is important from environmental, economical, political and energy aspects (Delattre et al. 2001). A widely accepted plastic waste disposal technique is incineration; however, their poor design allows the release of toxic compounds like dioxins and furans. Pyrolysis, one of the 'waste to energy' technology routes, has been investigated as an emerging and potential alternative for plastic waste disposal, conserving landfill space in addition to energy and resource recovery (Chen et al. 2014). It is advantageous in terms of safe environment-friendly plastic waste disposal, cutting down toxic gas emissions while recovering value-added products in the form of oil, gas and char after thermal degradation of high calorific waste plastics within oxygen-free inert ambience (Lam et al. 2016).

Among the various waste plastic types, low-density polyethylene (LDPE) forms a significant portion of the waste stream. LDPE (Density: 0.92–0.93 g/cm^3) is ethylene polymerization product catalysed by organometallics under high pressure and is less crystalline with higher degree of branching. The global LDPE production in 2016 has been displayed as 20.9 million tons, with Europe accounting for approximately 29% of the share followed by Asia (Plastics Europe 2017; Statista 2018). The annual demand of LDPE has been figured out as 17.5% among the various plastic types around the globe and 23% in India, which reflects the high LDPE waste output in need of suitable disposal means (TCT Magazine 2016). In this context, pyrolysis of waste LDPE plastics in the absence of oxygen can be a waste disposal management solution coupled with valuable product recovery. Researchers like Cit et al. (2010) demonstrated tar, gas and char extraction from LDPE pyrolysis in a fixed bed reactor under inert nitrogen atmosphere in the temperature range of 400–700 °C at 10 °C/min heating rate. They evaluated lower gas production rates at temperatures resulting in high tar formation. Higher proportion of pyrolytic char and gas recovery from waste LDPE plastic waste was observed beyond 425 °C by Onwudili et al. (2009) during their pyrolysis experiments over 300–500 °C temperature range.

The current work is an attempt to evaluate the thermal pyrolysis of waste low-density polyethylene (LDPE) plastics at varied temperatures in a novel batch pyrolyser set-up to recover oil, gas and char as by-products. The effect of temperature on the yield of products has been studied along with a prime focus on product characterization.

2 Materials and Methods

2.1 Pyrolysis Feedstock

The pyrolysis feedstock used was low-density polyethylene (LDPE) wastes collected from campus. The feed properties have been compiled in Table 1. The LDPE bag wastes were shredded into 3–5 cm pieces and fed into the batch pyrolyser unit before the experimental operation.

Table 1 LDPE plastic facts (Wang et al. 2012; Panda et al. 2010; TCT Magazine 2016)

<table>
<tr><th></th><th>Structure</th><th>Applications</th><th colspan="2">Global demand</th><th>Pyrolysis features</th><th>Products</th></tr>
<tr><td>LDPE</td><td>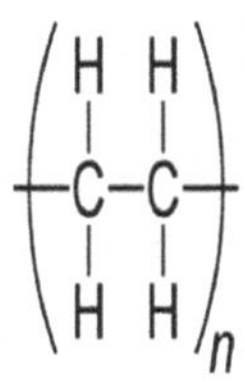</td><td>Films, shopping and garbage carry bags, wrapping stuff, containers, dispensing bottles</td><td colspan="2">17.5%</td><td>• Thermal decomposition through random chain rupture</td><td>Oil, gas, solid char</td></tr>
<tr><td></td><td colspan="2">Proximate analysis (wt%)</td><td colspan="3">Ultimate analysis (wt%)</td><td></td></tr>
<tr><td></td><td>Volatile matter</td><td>Ash</td><td>C</td><td>H</td><td>O</td><td>HHV_{dry} MJ/kg</td></tr>
<tr><td></td><td>99.25</td><td>0.75</td><td>83.72</td><td>13.16</td><td>2.37</td><td>45.97</td></tr>
</table>

2.2 Experimental Set-up and Methodology

The overall plastic waste pyrolysis technology involves a step-wise process methodology as a whole, as depicted in Fig. 1. The process onset is with thermo-gravimetric analysis (TGA) of waste LDPE pyrolysis to evaluate the variation of waste mass loss with temperature at different heating rates of 10, 20 and 40 °C/min. About 10–12 mg of small LDPE pieces were pyrolysed applying nitrogen flow rate of 20 ml/min in thermo-gravimetric analyser system (Make: SHIMADZU; model: DTG-60/60H).

The pyrolysis apparatus works in anaerobic conditions to treat the waste plastic feedstock in the heart of the system, i.e. the stainless steel fixed bed reactor of 500 g capacity which was purged with inert nitrogen at a rate of 50 ml/min for 15 min. The reactor is comprised of electrical heating coils and connected to a condenser unit, through which the condensed oil fraction passes into the oil chamber and the non-condensable pyrogas product was collected in the gas storage component, as represented in Fig. 2. The wall and core temperatures of the reactor were measured using K-type thermocouple. The pyrolysis experiments were performed with reaction time of 90 min under the following operational conditions:

- Feedstock: 100 g shredded waste LDPE plastic bags
- Pyrolysis temperature: 450, 500, 550, 600 °C
- Inert N_2 atmosphere

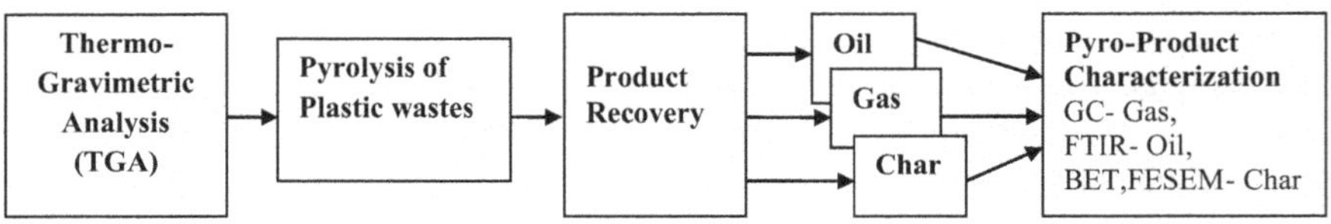

Fig. 1 Schematic view of methodology

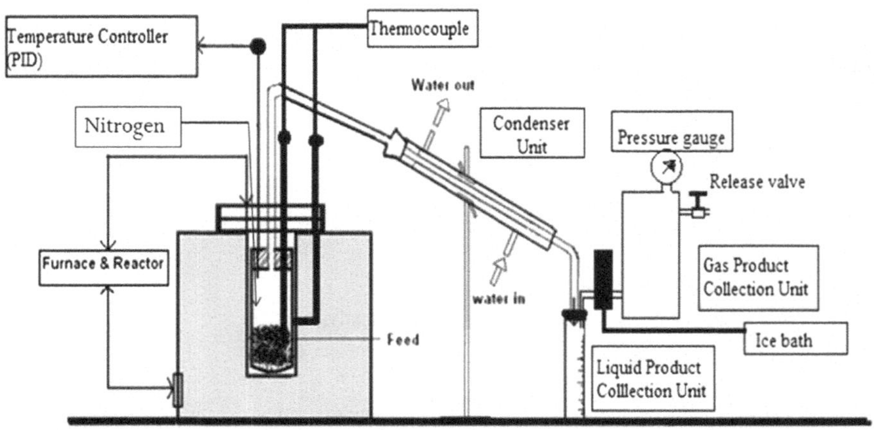

Fig. 2 Pyrolysis apparatus

- Heating rate: 20 °C/min
- Products: Oil, gas, char.

The gas was sampled for analysis, and the residual char at the reactor bottom was collected and weighed. The amount of gas was obtained by the difference after measuring the oil and char weight.

2.3 Analytical Techniques

- Initially, the thermo-gravimetric analysis (TGA) was performed to study temperature-wise waste LDPE pyrolysis at different heating rates which previewed the ideal temperature range and heating rate for LDPE plastic waste pyrolysis.
- The non-condensable pyrogas produced was analysed through sample injection (auto-sampler) into gas chromatograph from Thermo-fisher Scientific (Model: Trace GC 1110) equipped with thermal conductivity detector (TCD) for hydrogen, carbon monoxide, carbon dioxide estimation and flame ionization detector (FID) for detection of methane and other hydrocarbons.
- The pyrolysis char residue collected from the reactor after the experiment was assayed by BET for specific surface area and FESEM for the study of surface morphological features.

3 Results and Discussion

3.1 Thermo-gravimetric Analysis (TGA)

Thermo-gravimetric Analysis (TGA) was performed to identify the optimum pyrolysis temperature range and a suitable heating rate for the waste LDPE experiments which would enable the early start of waste LDPE degradation uniformly with a sufficient residence time. The plot in Fig. 3 evaluated the thermal degradation behaviour of waste LDPE and the mass loss with temperature over the course of pyrolysis reaction utilizing different heating rates of 10, 20 and 40 °C/min. The key findings are as follows:

- It was observed that there was an early start of waste degradation with uniform residence time at a heating rate of 20 °C/min while higher heating rates delayed the breakdown process.
- The degradation occurred over a narrow temperature range at higher heating rates.
- The optimum degradation range for waste LDPE pyrolysis tends to be from 400 to 550 °C with degradation getting stabled at 600 °C, at 20 °C/min heating rate, as revealed from graph (Fig. 3). This was preferred for the pyrolysis set-up experiments.

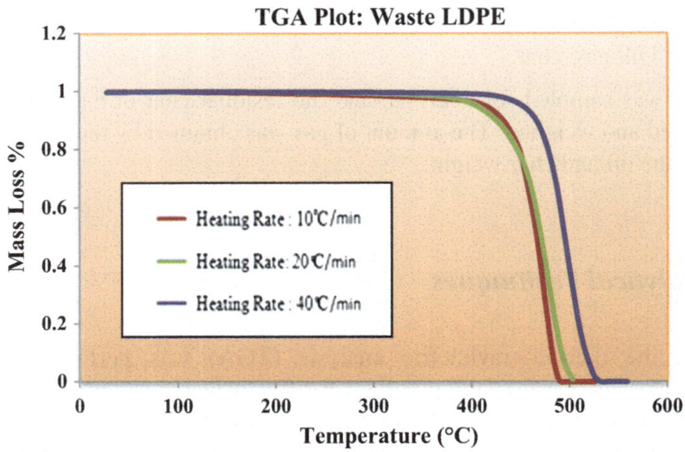

Fig. 3 TGA graph of mass loss (%) versus temperature (°C) for waste LDPE at heating rates of 10, 20 and 40 °C/min

3.2 *Effect of Temperature on Pyrolysis Product Yield*

The pyrolysis experiments for waste polyethylene (LDPE) were carried out at operating temperatures, 450, 500, 550, 600 °C fixing 20 °C/min as heating rate (TGA curve, Fig. 3), and the temperature-dependent variation in yield of products, oil, gas and char after the process, was studied (Fig. 4).

- The plastic waste was thermally decomposed into oil, whose production elevated with temperature rise but suddenly dropped from 84% at 550 °C to 76% at 600 °C with an increase in char content. The oil yield pattern observed in this case hints that plastic conversion to oil is favoured at higher temperatures.
- The pyrogas yield for waste LDPE was analysed to reach from 11% at 450 °C to the figure of 20% at 600 °C, following a linear increasing trend due to good cracking effect on waste LDPE for a high residence time in the pyrolyser favouring

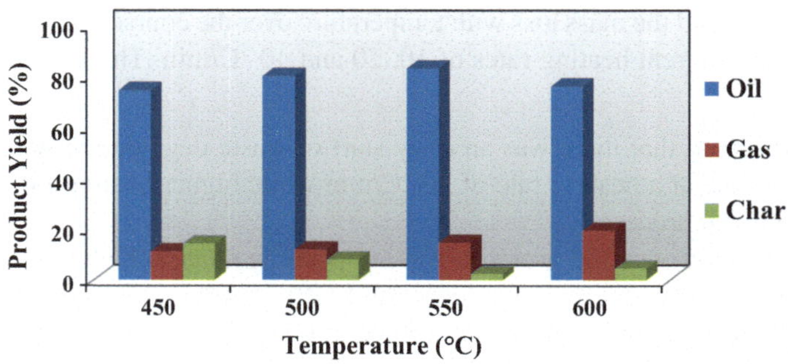

Fig. 4 Product yield (%) from pyrolysis of waste LDPE at different operating temperatures (°C)

secondary reactions to form high amount of gas species at higher temperatures, as revealed from Fig. 4.
- The solid residual char formation followed a declining trend from 15 to 2 g by weight from 450 to 550 °C, respectively, though it was observed that the quantity slightly increased to 4.5 g with further 50 °C rise in temperature which might be assumed due to destabilization and fusion of free radicals formed after polymer degradation at high temperature. On the contrary, reported data reveals reduced char formation at 600 °C (Cit et al. 2010).

It can be reported that the highest oil recovery was at 550 °C with the lowest char residue. However, Cit et al. (2010) confirmed the maximum tar yield from LDPE waste and low char residue to be at 600 °C. Also, Onwudili et al. (2009) revealed a decreasing trend in oil yield from 72.4% at 450 °C to 37.5% at 500 °C, which is unlike our case. The FTIR analysis was executed for oil produced at 550 °C which assessed the presence of aliphatic hydrocarbons like linear alkanes and alkenes with low aromatic content.

The gas production is assumed to be optimized at a process temperature of 600 °C with the highest formation of lighter gases and hydrocarbons ($<C_{10}$), as discussed in the following section related to gas chromatography of gas product stream.

3.3 *Gas Chromatographic Analysis*

Gas chromatography was done for the gas sampled in the last phase of experiment at 600 °C for waste plastic polyethylene using thermal conductivity detector (TCD) and flame ionization detector (FID) through Chemito Chrom-Card software for detection of gas components, provided by the peaks in the chromatogram. TCD confirmed the presence of hydrogen, methane and carbon monoxide, without traces of carbon dioxide while hydrocarbons in C1–C6 range were reported through FID chromatogram of the combustible gas product from plastic waste pyrolysis. However, Scheirs and Kaminsky (2006) reported the presence of only C1–C4 lower hydrocarbons from polyolefinic waste plastics like LDPE. Cit et al. (2010) analysed and concluded the detection of C4 and C6 + C7 hydrocarbons as principal components in the product gas stream from waste LDPE pyrolysis, but C1–C3 was detected as important gas constituents in our research. The gas characterization data, TCD and FID chromatograms for the product at 600 °C as well as the instrument are presented in Table 2, Figs. 5, 6 and 7, respectively.

Table 2 Gas chromatography data

TCD chromatogram results			
Sl. no.	Component name	Retention time (mins)	Amount (%)
1.	Hydrogen	2.128	10.25
2.	Carbon monoxide	2.493	4.3
3.	Carbon dioxide	–	Nil
FID chromatogram results			
1.	Methane	2.74	12.03
2.	Ethane	3.35	12.08
3.	Propane	5.72	10.83
4.	i-Butane	10.03	0.26
5.	*n*-Butane	10.51	4.92
6.	i-Pentane	15.18	0.10
7.	*n*-Pentane	15.65	2.52
8.	Hexane	20.20	0.51

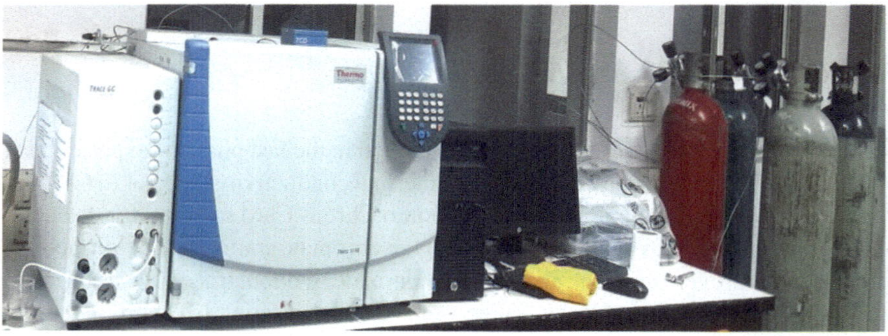

Fig. 5 Gas chromatogram

3.4 Characterization of Pyrolysis Char from LDPE Plastic Wastes

The generation of char residue and structure after pyrolysis is acted upon by key factors like heating rate, reactor temperature, residence time (Paethonom and Yoshikawa 2012). The black coloured fine powdery char residue left at reactor bottom after pyrolysis of LDPE plastic wastes was characterized for its surface area applying Brunauer-Emmett-Teller (BET) analysis after degassing the sample at 250 °C for 3 h. The BET surface area of the waste LDPE pyro-char product obtained at 550 °C was reported to be 8.76 m^2/g.

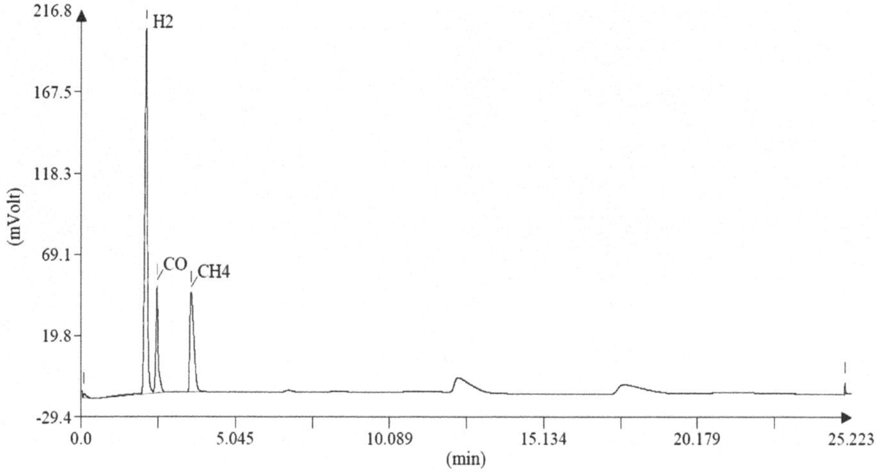

Fig. 6 TCD chromatogram of pyrogas from LDPE waste

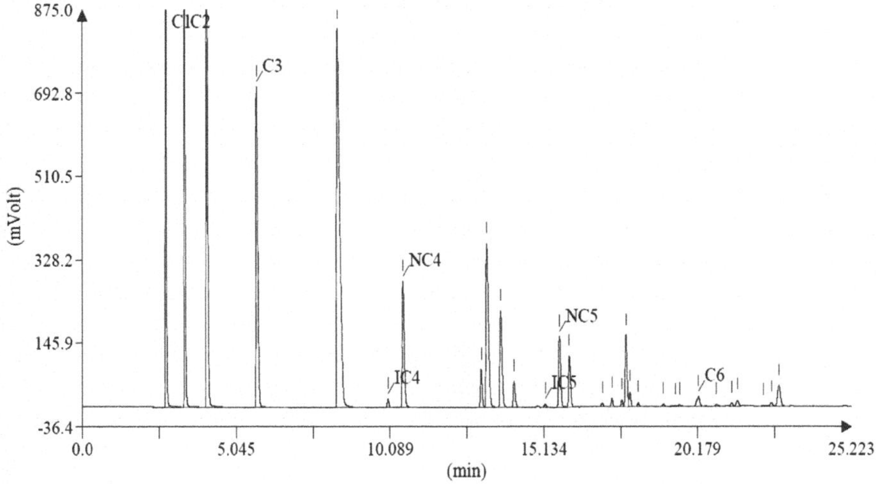

Fig. 7 FID chromatogram of pyrogas from waste LDPE waste

The pyrolytic char was also morphologically examined through field emission scanning electron microscopy (FESEM), as picturized in Fig. 8 which reveals a rough surface of agglomerates and different microstructures with non-uniform pore distribution. The decreased surface area may be likely due to pore surface blockage by LDPE derived hydrocarbon residuals after pyrolysis on account of short residence time in the reactor, thereby forming shallow pores, visualized as bound white spots in the image.

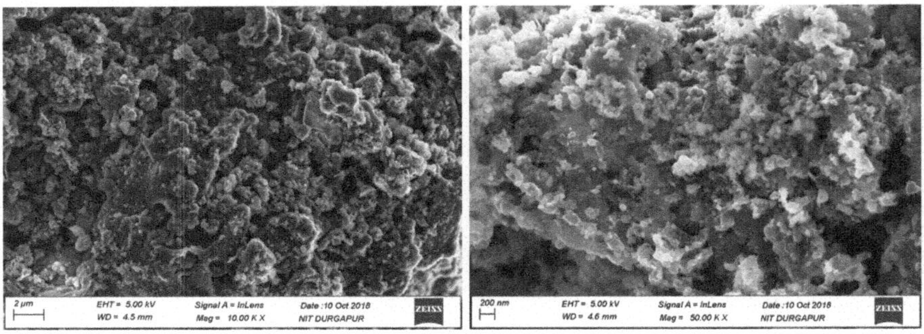

Fig. 8 FESEM images of pyrolysis char from LDPE plastic wastes

4 Conclusions

The thermal pyrolysis technology, in the current research work, was applied to low-density polyethylene (LDPE) plastic wastes to recover oil, gas and char as by-products followed by product characterization and analysis. TGA study along with a detailed characterization of pyrolytic products from LDPE wastes has been presented herein. The product yield varies with temperature and can be optimized as per product choice. Highest oil recovery of 84% has been done at 550 °C while gas and char productions are significant at 600 and 450 °C, respectively, unlike that figured out by other researchers.

The plastic oil and gas may be used as fuel while the char may be activated to increase its surface area for further applications as carbon black. The high value products of pyrolysis, if further processed, can offset the initial waste facility set-up expenses leading to an economical process. There is a need to explore the technological potential and role of 'plastic waste to energy' in the circular economy so as to place back waste plastics as resource in their life cycle for sustainable plastic waste management. Process scale-up and reactor modifications are certain future tools for commercialization that may take this 'energy from plastic waste' sector to great heights.

Acknowledgements The authors are grateful to Director, CSIR-CMERI, Durgapur, and Director, NIT Durgapur, for supporting the research work. The University Grants Commission (UGC), New Delhi, is being sincerely thanked for providing NET Fellowship for pursuing the Ph.D. research work.

References

Central Pollution Control Board, CPCB. (2012). Material on waste plastic management, pp. 1–23, India.

Chen, D., Yin, L., Wang, H., & He, P. (2014). Pyrolysis technologies for municipal solid waste: A review. *Waste Management*. https://doi.org/10.1016/j.wasman.2014.08.004.

Cit, I., Sınag, A., Yumak, T., Ucar, S., Mısırlıoglu, Z., & Canel, M. (2010). Comparative pyrolysis of polyolefins (PP and LDPE) and PET. *Polymer Bulletin, 64,* 817–834.

Delattre, C., Forissiera, M., & Pitault, I. (2001). Improvement of the microactivity test for kinetic and deactivation studies involved in catalytic cracking. *Chemical Engineering Science, 56*(4), 1337–1345.

Geyer, R., Jambeck, J. R., & Law, K. L. (2017). Production, use and fate of all plastics ever made. *Science Advances, 3*(7), e1700782.

Lam, S. S., Liew, R. K., Jusoh, A., Chong, C. T., Ani, F. N., & Chase, H. A. (2016). Progress in waste oil to sustainable energy, with emphasis on pyrolysis techniques. *Renewable and Sustainable Energy Reviews*, 53741–53753.

Onwudili, J. A., Insura, N., & Williams, P. T. (2009). Composition of products from the pyrolysis of polyethylene and polystyrene in a closed batch reactor: Effects of temperature and residence time. *Journal of Analytical and Applied Pyrolysis, 86,* 293–303.

Paethonom, A., & Yoshikawa, K. (2012). Influence of pyrolysis temperature on rice husk char characteristics and its tar adsorption capability. *Energies, 5,* 4941–4951.

Panda Achyut, K., Singh, R. K., & Mishra, D. K. (2010). Thermolysis of waste plastics to liquid fuel: A suitable method for plastic waste management and manufacture of value added products—A world prospective. *Renewable and Sustainable Energy Reviews, 14,* 233–248.

Plastics Europe. (2017). An analysis of European plastics production, demand and waste data. Plastics.

Scheirs, J., & Kaminsky, W. (2006). Thermal and catalytic conversion of polyolefins. Wiley.

Statista. (2018). https://www.statista.com/statistics/858556/global-low-density-polyethylene-production-distribution-by-region/.

TCT Magazine, (2016, December) https://www.tctmagazine.com/3d-printing-news/recycling-plastic-high-performance-3D-printing-filament/.

Wang, J., Liu, Y., Yan, R., Dong, Z., & Tay, J. H. (2012). Pyrolysis characteristics and kinetics of MSW in Singapore. In: 3rd International Conference on Industrial and Hazardous waste management, CRETE 2012.

The manufacturer's authorised representative in the EU is Springer Nature Customer Service Centre GmbH, Europaplatz 3, 69115 Heidelberg, Germany. If you have any concerns regarding our products, please contact ProductSafety@springernature.com

Printed and bound by CPI Group (UK) Ltd, Croydon, CR0 4YY
30/09/2026
02225556-0007